Inorganic Chemistry Concepts
Volume 1

Renata Reisfeld
Christian K. Jørgensen

Lasers and Excited States of Rare Earths

With 9 Figures
and 26 Tables

Springer-Verlag
Berlin Heidelberg New York 1977

Renata Reisfeld

Department of Inorganic and Analytical Chemistry
Hebrew University, Jerusalem, Israel

Christian Klixbüll Jørgensen

Département de Chimie minérale, analytique et appliquée
Université de Genève, Switzerland

ISBN 3-540-08324-3 Springer-Verlag Berlin Heidelberg New York
ISBN 0-387-08324-3 Springer-Verlag New York Heidelberg Berlin

Library of Congress Cataloging in Publication Data. Reisfeld, Renata. Lasers and excited states of rare earths. (Inorganic chemistry concepts; v. 1) Includes bibliographies. 1. Rare earth lasers. 2. Earths, Rare — Spectra. I. Jørgensen, Christian Klixbüll, joint author. II. Title. III. Series. QC688.R44 535.5'8 77-24052

Typesetting: Elsner & Behrens, Oftersheim.
Printing and bookbinding: Zechnersche Buchdruckerei, Speyer.
2152/3140-543210

Preface

The possibility of stimulated light emission was discussed by Einstein in 1917, eight years before the quantum-mechanical description of energy levels of many-electron systems. Though it is imperative to use samples having optical properties greatly different from the standard continuous spectrum of opaque objects ("black body" radiation) it is not always necessary to restrict the study to monatomic entities. Thus, spectral lines can be obtained (in absorption and in emission) from lanthanide compounds, containing from one to thirteen 4f electrons going from trivalent cerium to ytterbium, that are nearly as sharp as the ones from gaseous atoms. However, the presence of adjacent atoms modifies the simple picture of an isolated electron configuration, and in particular, it is possible to pump excited levels efficiently by energy transfer from species with intense absorption bands, such as the inter-shell transitions of other lanthanides and of thallium(I), lead(II) and bismuth(III) or the electron transfer bands of the uranyl ion or other complexes. On the other hand, it is possible to diminuish the multi-phonon relaxation (competing with sharp line luminescence) by selecting vitreous or crystalline materials with low phonon energies.

Obviously, one cannot circumvent the conservation of energy by lasers, but they may have unprecedented consequences for the future by allowing nuclear fusion of light elements, effects of non-linear optics and time-resolved spectroscopy, besides the more conventional applications of coherent light beams with negligible angular extension. In this book we attempt to provide suggestions for a variety of new laser materials and conceivable use of combined new properties.

Jerusalem and Geneva, January 1977

Renata Reisfeld
Christian K. Jørgensen

Contents

1. Analogies and Differences Between Monatomic Entities and Condensed Matter

(References to this Chapter are found p. 59).

The selective absorption or emission of narrow spectral lines in the visible and in the adjacent, near ultra-violet and near infra-red regions is normally thought to be a characteristic of transitions in individual atoms efficiently isolated from the surroundings by large distances to the nearest neighbour atoms. This phenomenon is of great importance for astrophysics and our conviction that the same elements exist in the Sun and the other stars as in the surface of the Earth is almost exclusively based on coincidences between positions of lines in stellar spectra and in light sources (such as flames, arcs, sparks and electric discharges in gases) available in the laboratory. This idea was established by Bunsen and Kirchhoff in 1860 and atomic light sources such as red neon, yellow sodium and blue-green mercury lamps (frequently combined with luminescent wall materials) have to a great extent replaced the continuous spectra emitted by incandescent solids in oil lamps and candles, and the electric carbon or tungsten filaments. Whereas Newton did not use sufficiently collimated light and a very narrow slit, Wollaston detected 1802 narrow, dark absorption lines in the Solar spectrum (also seen in the scattered light from the blue sky) and these were carefully studied by Fraunhofer since 1814. The line spectra of gaseous atoms and monatomic positive ions (such as M^+, M^{+2}, . . . called M II, M III, . . . by atomic spectroscopists because they emit the second, third, . . . spectrum of elements, M^0 representing M I) have been intensively studied for more than a century. The Ritz combination principle $h\nu = (E_2 - E_1)$ can be used to determine the energy levels E_1, E_2, . . . relative to the groundstate E_0 of the monatomic entity (only a minority of atoms have well-established long series of levels obeying the Rydberg formula and hence allowing the ionization energy $I = E_0\ (M^+) - E_0\ (M^0)$ to be determined with as high an accuracy as the other energy differences) and they are tabulated [1] by Charlotte Moore-Sitterly of the National Bureau of Standards in Washington, D. C. A very important empirical fact is that the individual levels (having the quantum number $J = 0, 1, 2, \ldots$ for an even and $J = 1/2, 3/2, 5/2, 7/2, \ldots$ for an odd number of electrons; combined with odd (ungerade) or even (gerade) parity) frequently can be classified in terms characterized by S and L in the Russell-Saunders coupling

$$J = S \oplus L = (S + L) \text{ or } (S + L - 1) \text{ or } (S + L - 2) \ldots \\ \text{or } (|S - L| + 1) \text{ or } |S - L| \qquad (1.1)$$

where $S = 0, 1, 2, \ldots$ for an even number of electrons are called singlet, triplet, quintet, . . . and $S = 1/2, 3/2, 5/2, \ldots$ for an odd number of electrons doublet, quartet, sextet, . . . It has no practical importance that q electrons at most can show $S = (q/2)$, already in the lithium atom, the first state with maximum S (= 3/2) occurs in the continuum about ten times higher than the first ionization energy. The quantum number L is a non-negative integer having traditional names

$$\begin{array}{llllllllllllll} L = & 0 & 1 & 2 & 3 & 4 & 5 & 6 & 7 & 8 & 9 & 10 & 11 & 12 \ldots \\ & S & P & D & F & G & H & I & K & L & M & N & O & Q \end{array} \quad (1.2)$$

and the individual level is denoted $^3H_6$, $^4I_{9/2}$, . . . where the left-hand superscript indicates the spin multiplicity $(2S + 1)$ equal to 3 ($S = 1$), 4 ($S = 3/2$), . . . and the right-hand subscript $J = 6, 9/2, \ldots$ Frequently, one adds a right-hand superscript small circle for *odd* levels, of which the energies are tabulated [1] with italicized numerals. Seen from the point of view of quantum mechanics [2, 3] each J-level corresponds to $(2J + 1)$ mutually orthogonal wave-functions Ψ and is said to contain $(2J + 1)$ *states.* A summation of $(2J + 1)$ for each level indicated by Eq. (1.1) gives $(2S + 1)(2L + 1)$ states in a term characterized by a given combination of S and L.

An even more important empirical fact is that the levels and terms can be successfully classified in *electron configurations* where each shell l [having the lower-case traditional names s, p, d, f, g, . . . in Eq. (1.2)] contains between 0 and $(4l + 2)$ electrons. l is a quantum number for one-electron wave-functions, *orbitals,* where the number [4] of angular node-planes (node-cones count for two angular nodes) is l and where the orbital can be written as the product of a radial function and an *angular function* which is a normalized linear combination of homogeneous polynomials $x^a y^b z^c / r^l$ in the Cartesian coordinates with the sum of the three non-negative integers $a + b + c = l$. The principal quantum number n (a positive integer not smaller than $l + 1$) is not strictly speaking a quantum number corresponding to a physical quantity (like the square of the momentum of orbital motion being $l(l + 1) h^2/4\pi^2$) though the radial function has $(n - l - 1)$ radial nodes for finite, positive distance r from origo of the coordinate system, where the nucleus is situated. There are no such radial nodes for 1s, 2p, 3d, 4f, . . . whereas other orbitals (such as 2s, 3s, 3p, . . .) have to be orthogonal on the previous shells with the same l by judicious choice of their radial functions. The groundstate of gaseous M^{+2} (where five exceptions are known, M = La, Gd, Lu, Ac and Th) and all known M^{+3} and M^{+4} belong to a configuration obtained by consecutive filling of the shells in the order

$$\begin{aligned} &1s \ll 2s < 2p \ll 3s < 3p \ll 3d < 4s < 4p \ll 4d < 5s < 5p \ll 4f < 5d < 6s \\ &< 6p \ll 5f < \ldots \end{aligned} \quad (1.3)$$

where the double inequality signs indicate configurations containing 2, 10, 18, 36, 54, 86, . . . electrons isoelectronic with the noble gases. Thus, Pr^{+3} has its $(Z - 3) = 56$ electrons distributed on the same closed shells [first line of Eq. (1.3)] containing 54 electrons as the xenon atom and two electrons in the 4f shell. This can be written [Xe]$4f^2$ where the number of electrons in each shell is written as a right-hand superscript (sometimes omitted when it is one; thus, the next-lowest configuration [5] of Pr^{+3} is [Xe]$4f^1 5d^1$ or [Xe]4f5d). Seen from this point of view, the behaviour of the neutral atoms is far less regular. A modified version of Eq. (1.3) putting $(n + 1)$ s orbitals before the nd shell has 20 exceptions, and frequently, several configurations partly coincide and it seems somewhat accidental exactly to which configuration the groundstate of the neutral gaseous atom belongs. Though the lowest level $^4I_{9/2}$ of the praseodymium atom belongs to [Xe]$4f^3 6s^2$ this configuration is represented by 41 levels distributed over more than 40,000 cm^{-1} or 5 eV, whereas the first level belonging to [Xe]$4f^2 5d 6s^2$ already occurs at 4,300 cm^{-1} above the groundstate.

It is worth noting the discrepancy [6, 7] between the *chemical* and the *spectroscopic* versions of the Periodic Table that neodymium is almost exclusively trivalent (written with Roman numerals in parenthesis Nd(III) by chemists, one unit lower than Nd IV without parenthesis used by atomic spectroscopists when speaking about Nd^{+3}) though it has only *two* valence electrons in the sense that the groundstate belongs to [Xe]$4f^4 6s^2$. This lack of relation between chemical behaviour and lowest configuration occurs frequently; the groundstate of the gaseous uranium atom belongs to [Rn]$5f^3 6d 7s^2$ though U(VI) and U(IV) are more common oxidation states than U(III). A completely extreme aspect of this discrepancy is, that helium and ytterbium are spectroscopic alkaline earths having the last occupied orbital *ns*, whereas the noble gases Ne, Ar, Kr and Xe terminate with a filled *np* shell.

A. The Configuration $4f^q$ as an Instance of Spherical Symmetry

Already in 1857, Gladstone noted that aqueous solutions containing praseodymium(III) and neodymium(III) (this mixture was believed to be an element, didymium, according to Mosander 1842, but was separated in 1885 by Auer von Welsbach by laborious techniques of fractional crystallization) show about ten narrow absorption bands in the visible. This observation is strikingly similar to the Fraunhofer lines and spectral lines in absorption present in monatomic vapours of metallic elements. Extensive studies of all kinds of coloured substances later showed that diatomic and polyatomic molecules in the gaseous state in many cases show band spectra with adjacent narrow lines organized in what is now recognized to be vibrational and rotational structure [8] of electronic transitions, whereas, a few sharp, irregularly distributed absorption bands in condensed matter (solutions, vitreous and crystalline solids) are an almost unique characteristic of the lanthanides. The 3d, 4d and 5d group compounds [6, 9] and the general dye-stuffs and other coloured organic molecules [10] show broad absorption bands, in the latter cases often exhibiting vibrational structure, but very rarely atomic-like sharp lines. After the advent of the "ligand field" theory [3, 11] the rare cases of d group complexes showing sharp lines (even known in emission from ruby $Al_{2-x}Cr_xO_3$ with a small amount of chromium(III) replacing colourless aluminium) in addition to broader bands have been explained as intra-sub-shell transitions where the excited state has almost the same electronic density in our three-dimensional space as the groundstate (however, the average reciprocal value $\langle r_{12}^{-1} \rangle$ of the distances between the electrons in the partly filled shell is different). According to Franck and Condon's principle no extensive co-excitation of vibrational states is expected when the equilibrium internuclear distances are the same in the excited electronic state as in the groundstate. It is very interesting that Ephraim and Mezener [12] argued that the presence of many narrow absorption bands in uranium(IV) compounds indicates a configuration [Rn]$5f^2$ in analogy with Pr(III) and confirming the suggestion by Goldschmidt that a new series of rare earths of predominantly quadrivalent elements starts with thorium. Of course, it was not known at that time [7] that $Z = 95$ to 100 are most stable in trivalent compounds, and that Md(II) and No(II) occur at the end of the 5f group.

Though all the elements with the atomic number Z between 57 (lanthanum) and 71 (lutetium) actually were discovered before 1907 (with exception of promethium ($Z = 61$) only having short-lived radioactive isotopes) a certain uncertainty remained about their definite nature until Moseley derived Z from X-ray spectra in 1913. Furthermore, the "Aufbau principle" of the Periodic Table such as Eq. (1.3) was not established before Stoner in 1924 suggested nl-shells containing at most $(4l + 2)$ electrons. The study of absorption spectra was somewhat delayed by the scarcity of high-purity samples obtained by tedious methods of separation, until Spedding developed the ion-exchange resin columns around 1942. In accordance with the present state of knowledge [13], all the M(III) with $q = (Z - 57)$ between 2 and 12 show a definite set of narrow absorption bands corresponding to the transitions from the lowest to a variety of higher J-levels of the configuration $[Xe]4f^q$. At this point, gadolinium ($Z = 64$) has a special position ($q = 7$) because the first excited levels occur at sufficiently high energy to correspond to absorption (and sometimes emission) bands in the ultra-violet, as first measured by Urbain. Ytterbium ($Z = 70$) was later shown to have a single group of closely adjacent, narrow bands in the near infra-red (close to 10,000 cm^{-1}) which can be identified as the transition from the groundstate $^2F_{7/2}$ to the only other level $^2F_{5/2}$ of $[Xe]4f^{13}$. One would expect cerium(III) to have a similar transition from $^2F_{5/2}$ to $^2F_{7/2}$ of [Xe]4f, but it occurs at a low wave-number close to 2,500 cm^{-1} with the result that it is difficult to detect, and is hidden by vibrational spectra in solvents or crystals containing hydrogen. Apart from the narrow bands due to internal transitions in $[Xe]4f^q$ the M(III) can show other broad bands due to excitations to another configuration $[Xe]4f^{q-1}5d$ or due to electron transfer, where an electron is lost from one or more reducing neighbour atoms and the lanthanide is reduced to M(II) by being $[Xe]4f^{q+1}$. Contrary to the case of d groups, the broad absorption bands are not very frequent below 50,000 cm^{-1} though M(II) generally have more prominent inter-shell and M(IV) electron transfer bands than M(III).

It is very important to note that the internal transitions in the partly filled 4f shells have positions almost independent of the ligands, the neighbour molecules or anions. We discuss below the detailed nature of these weak perturbations, but the main part of the problem has spherical symmetry. Though Sugar [5] found 12 of the 13 possible levels of $[Xe]4f^2$ in Pr^{+3} in 1965, it had been believed for many years that the narrow band spectra of M(III) are very similar to the excited levels of isolated M^{+3}. We have to make a few more remarks about the coupling schemes which can be represented by the operator $\oplus$ of Eq. (1.1). If a monatomic entity has a configuration with two partly filled shells, it is possible to *count* the number of terms having the desired combination of S and L by performing the operation $L = L_1 \oplus L_2$ and $S = S_1 \oplus S_2$ on each of the combinations (S_1, L_1) of the first and (S_2, L_2) of the second partly filled shell, using the general definition

$$Q_1 \oplus Q_2 = (Q_1 + Q_2) \text{ or } (Q_1 + Q_2 - 1) \text{ or } (Q_1 + Q_2 - 2) \ldots \text{ or } (|Q_1 - Q_2| + 1) \text{ or } |Q_1 - Q_2| \tag{1.4}$$

called vector-coupling by Hund [14]. Since this operation is associative (and commutative) the final result of $(Q_1 \oplus Q_2) \oplus Q_3$ is the same as of $Q_1 \oplus (Q_2 \oplus Q_3)$. As simple examples may be mentioned:

$$\begin{array}{ll} f^1d^1: & {}^1P, {}^3P, {}^1D, {}^3D, {}^1F, {}^3F, {}^1G, {}^3G, {}^1H, {}^3H \\ f^1d^1s^1: & \text{two } {}^2P, {}^4P, \text{two } {}^2D, {}^4D, \text{two } {}^2F, {}^4F, \text{two } {}^2G, {}^4G, \text{two } {}^2H, {}^4H \end{array} \quad (1.5)$$

where it is noted that the same combination of S and L can repeat itself twice or more in the same configuration. The operation (1.4) simply multiplies the number of states having the quantum number Q_1 with the number of states having Q_2. Hence the configurations in Eq. (1.5) comprise a number of states being the product of $(4l + 2)$ for each electron or $14 \cdot 10 = 140$ and $14 \cdot 10 \cdot 2 = 280$ in the two examples.

When a partly filled shell contains at least 2 and at most $4l$ electrons, it is not possible to apply Eq. (1.4) before knowing the terms belonging to l^q which is a complicated problem [2, 6] restricted by Pauli's exclusion principle though it is completely resolved in principle [15]. An explicit expression exists for

$$l^2: \ {}^1S, {}^3P, {}^1D, {}^3F, {}^1G, \ldots, {}^3(2l-1), {}^1(2l) \quad (1.6)$$

having all L-values from 0 to $2l$ alternatively combined with $S = 0$ or 1. The number of states in l^2 is $(2l + 1)(4l + 1)$ to be compared with two non-equivalent electrons $(nl)^1(n'l)^1$ having all the L values of Eq. (1.6) represented both as singlet and as triplet terms with the result that $(4l + 2)^2$ states subsist. In the partly filled shell l^q the number of states is the product

$$(4l+2) \cdot \frac{(4l+1)}{2} \cdot \frac{(4l)}{3} \cdot \frac{(4l-1)}{4} \cdots \frac{(4l+3-q)}{q} \quad (1.7)$$

which is the result of the permutation theory for the distribution of q *indiscernible* objects on $(4l + 2)$ sites. Equation (1.7) is identical for q and for $(4l + 2 - q)$ electrons in the partly filled l-shell. This identity goes much further in Pauli's theorem of *hole-equivalency* that the number of levels having a given value of J is the same in the two cases, as well as the number of terms having a definite combination of S and L. By the way, the ten terms given for f^1d^1 in Eq. (1.5) are the same for the three configurations f^1d^9, $f^{13}d^1$ and $f^{13}d^9$ related by hole-equivalency. The terms for three electrons or three holes are:

$$\begin{array}{ll} p^3: & {}^4S, {}^2P, {}^2D \\ d^3, d^7: & {}^4P, {}^4F, {}^2P, \text{two } {}^2D, {}^2F, {}^2G, {}^2H \\ f^3, f^{11}: & {}^4S, {}^4D, {}^4F, {}^4G, {}^4I, {}^2P, \text{two } {}^2D, \text{two } {}^2F, \text{two } {}^2G, \text{two } {}^2H, {}^2I, {}^2K, {}^2L \end{array} \quad (1.8)$$

comprising 20,120 and 364 states, respectively. There are no immediately evident regularities for a greater number of electrons in the shell, though it may be noted that among the 119 terms of f^7, the groundstate is ${}^8S$ and the sextet terms ${}^6P$, ${}^6D$, ${}^6F$, ${}^6G$, ${}^6H$ and ${}^6I$ having the same L values as f^2 in Eq. (1.6). One doublet $({}^2Q)$ has $L = 12$.

According to Hund [14] the lowest term of l^q has the highest L *compatible* with the highest $S = S_{max}$ of the configuration. $S_{max} = (q/2)$ in the first half of the shell, and $(4l + 2 - q)/2$ for $q > (2l + 1)$. The L value of the ground term is a symmetric function around maxima for *quarter* and three-quarter filled shells, and is:

$$\begin{array}{llll} f^0, f^{14}: {}^1S & f^1, f^{13}: {}^2F & f^2, f^{12}: {}^3H & f^3, f^{11}: {}^4I \\ f^7: \quad {}^8S & f^6, f^8: {}^7F & f^5, f^9: {}^6H & f^4, f^{10}: {}^5I \end{array} \qquad (1.9)$$

Obviously, such results are also symmetric around half-filled shells. This is true to a lesser degree for the relativistic effect usually called *spin-orbit coupling* where another of Hund's rules is that the lowest level has $J = L - S$ for $q \leqslant 2l$ but $J = L + S$ for the second half of the shell, $q > (2l + 1)$. In the case of more than one partly filled shell, Hund's rules do not always apply. Thus, the lowest level of the gaseous cerium atom is ${}^1G_4$ (and distinctly not ${}^3H_4$) belonging to $[Xe]4f5d6s^2$.Whereas, the lowest level of Ce^{+2} is ${}^3H_4$ (even parity) belonging to $[Xe]4f^2$ the configuration $[Xe]4f5d$ starts already at 3,277 cm^{-1} with ${}^1G_4$ (odd parity). The levels belonging to a given configuration have even or odd parity according to whether the sum of the l values of the electrons is even or odd. It is also calculated [6, 15] that the rules of Hund do not apply for all cases of l above 3, such as g^3.

It should be emphasized that Eqs. (1.5), (1.6) and (1.8) do not indicate the relative energies of the terms, and Hund's rules should not be construed to demand a systematic dependence of energy on L for given S value. The question of relative energies have three parts:

1. Group-theoretical conditions for the existence of definite (S, L)-terms and J-levels.
2. Algebraic relations between coefficients to parameters of interelectronic repulsion and spin-orbit coupling.
3. The numerical question of size of the parameters, either calculated from the radial function in Hartree-Fock approximation or considered as phenomenological parameters are fitted to obtain the best possible agreement with the experiment.

An important aspect of the group-theoretical (symmetry-determined) conditions is that the quantum numbers $S = L = 0$ characterizing a closed shell filled with $(4l + 2)$ electrons is the neutral element of the operation $\oplus$ in Eq. (1.4) always having $({}^1S) \oplus (Q_2) = Q_2$ like zero is the neutral element of addition and 1 the neutral element of multiplication. This means that we can speak about the configuration $4f^3$ of Pr^{+2} or Nd^{+3} without knowing anything about the detailed behaviour of the 54 electrons in closed shells. This remark has profound ramifications in chemistry. It is clear what we mean with *isoelectronic series* of monatomic entities such as the 54-electron series I^-, Xe, Cs^+, Ba^{+2}, La^{+3}, Ce^{+4}, Pr^{+5}, . . . though we must remember that the ground configuration [1] may change along an isoelectronic series. Thus, the 19-electron series is [Ar]4s for K and Ca^+ but [Ar]3d for Sc^{+2}, Ti^{+3}, V^{+4}, . . . and the 68-electron series $[Xe]4f^{12}6s^2$ for the gaseous erbium atom, $[Xe]4f^{13}6s$ for Tm^+ but the closed shells $[Xe]4f^{14}$ for Yb^{+2}, Lu^{+3}, Hf^{+4}, Ta^{+5} and undoubtedly many subsequent elements. Chemists [4] also speak about isoelectronic series such as Sb(–III), Te(–II), I(–I), Xe, Cs(I), Ba(II), La(III) and Ce(IV) whereas Pr(V) is far too oxidizing [16] to exist in compounds, even in fluorides and oxides and gaseous anions such as Sb^{-3} and Te^{-2} spontaneously lose electrons *in vacuo,* as all anions (including O^{-2} isoelectronic with neon) carrying more than one negative charge. The closed-shell segment of the 68-electron series Yb(II), Lu(III), Hf(IV), Ta(V), W(VI), Re(VII) and Os(VIII) is also familiar to chemists though they do by no means argue that the fractional atomic charge approaches + 8 (it seems to be closer to + 2 in the tetrahedral molecule OsO_4) and not even that the fractional atomic charge is a monotonic function of the oxidation state; it may easily be higher in lutetium(III) fluoride than in osmium(VIII) oxide. The oligoatomic molecules [6] are said to be isoelectronic if they contain the same nuclear distribution with essentially the same atomic core closed-shells, as the linear triatomic species OCO, NCN^{-2}, N_3^-, NNO, fulminate CNO^- and cyanate NCO^- among which the three former possess a centre of inversion and the point-group $D_{\infty h}$ whereas the symmetry of the three latter species is $C_{\infty v}$. The chemist speaks about pseudo-isoelectronic or *isologous* series of such species having a different number of electrons in closed inner shells but analogous distributions of the valence electrons. Remembering the classification of oxygen, sulphur, selenium and tellurium as chalkogens, species such as OCO, SCO, SeCO, SCS, SeCSe, NCO^-, NCS^- and $NCSe^-$ are isologous. The repetition of periods in the Periodic Table is a manifestation of isologous behaviour (though S may vary from high to low values in the d group compounds) and spectroscopists can argue that U(IV) is isologous with Pr(III) and U(III) and Np(IV) with Nd(III).

In Russell-Saunders coupling the individual (S, L) term is predicted [2] to follow the interval rule first proposed by Landé

$$E(J) = E(S, L) + \zeta\,[J(J+1) - \langle J(J+1)\rangle]/2 \tag{1.10}$$

where $E(S, L)$ is the average energy of the $(2S+1)(2L+1)$ states belonging to the term, and the average value is

$$\langle J(J+1)\rangle = S(S+1) + L(L+1) \tag{1.11}$$

as can be seen easily by induction and recursion. A corollary of Eq. (1.10) is that the distance between two levels, J and $(J-1)$, of the term is $J\zeta$. Further on, ζ can be evaluated from first principles [2] as a multiple $k\zeta_{nl}$ of the *Landé parameter* ζ_{nl} characterizing all the terms with one partly filled shell of the monatomic entity considered. It is empirically known [17] that ζ_{nl} tends to be proportional to $(z+1)^2Z^2$ where z is the ionic charge. The coefficient k is particularly simple for terms with S_{max}. In the first half of the shell, $S_{max} = q/2$ and $k = 1/q$. In the second half of the shell (q above $2l+1$), k is negative and equals $-1/2S_{max}$. There is a tradition among magnetochemists to call $k\zeta_{nl}$ for (the positive or negative) λ. This symbol is also used as a symmetry type in the linear symmetries [6] and we do not use it for the former purpose. Nearly all terms with S_{max} have $L \geqslant S_{max}$ in which case the total width of the term between the two extreme J-values $L + S_{max}$ and $L - S_{max}$ is $(L + 1/2)\zeta_{nl}$ exactly like it would be for a single nl-electron outside closed shells.

If there were no correlation effects [3] and no mixing of configurations, the interelectronic repulsion in one partly filled shell l^q can be treated [2] according to a theory first proposed by Slater 1929, where a series expansion (relative to polar coordinates with the nucleus at origo) is performed for the integrated average value

$\langle r_{12}^{-1} \rangle$ of reciprocal interelectronic distances. Fortunately enough, the triangular conditions of Gaunt restrict the number of parameters needed to describe term distances in l^q to l integrals, $F^2, F^4, F^6, \ldots, F^{2l}$ which are functions of the radial function of the partly filled shell. Actually, there is one more slightly larger parameter F^0 but it does not contribute to the term distances since its coefficient is $q(q-1)/2$ for all the terms. However, F^0 is a major contribution to the difference between the ionization energy and the electron affinity of the partly filled shell. Though the proportionality constant depends to a certain extent on the shape of the 4f radial function, the order of magnitude is $F^k = \varphi_k \langle r^{-1} \rangle$ with

$$\varphi_0 \sim 0.7 \qquad \varphi_2 \sim 0.5 \qquad \varphi_4 \sim 0.3 \qquad \varphi_6 \sim 0.2 \tag{1.12}$$

where the average reciprocal distance from the nucleus $\langle r^{-1} \rangle$ of the 4f electrons in hartree/bohr atomic units correspond to 14.4 eV/Å or 115,000 cm^{-1}/Å. It is also worth noting that the product $\langle r \rangle \langle r^{-1} \rangle$ which is $(2l+3)/(2l+2)$ for hydrogenic ($H, He^+, \ldots, Ar^{+17}, \ldots$) 1s, 2p, 3d, 4f, . . . radial functions without nodes is between 1.3 and 1.25 for most Hartree-Fock radial functions in many-electron atoms.

It is rather tedious [18] to apply the Slater-Condon-Shortley theory to f^q where it is also customary to introduce new, smaller parameters

$$F_2 = F^2/225 \qquad F_4 = F^4/1089 \qquad F_6 = F^6/7361.64 \tag{1.13}$$

in order to assure integers as coefficients f_k to the individual terms $(f_2F_2 + f_4F_4 + f_6F_6)$. It is very fortunate that Racah [19] introduced a much more practical method to obtain the same results as the F_k treatment, but with the new parameters

$$\begin{aligned} E^0 &= F^0 - 10F_2 - 33F_4 - 286F_6 \\ E^1 &= (70F_2 + 231F_4 + 2{,}002F_6)/9 \\ E^2 &= (F_2 - 3F_4 + 7F_6)/9 \\ E^3 &= (5F_2 + 6F_4 - 91F_6)/3 \end{aligned} \tag{1.14}$$

and at the same time, achieved a very subtle group-theoretical classification of the numerous terms showing the same combination of S and L by introducing new quantum numbers such as the *seniority number* v and other complicated quantities such as U and W. Since these fascinating results have minor connections with our subject, and since they have been described by Wybourne [20] and Judd [21] we do not consider them here, apart from the comment that the diagonal elements classified with the help of Casimir's group generally have rather large non-diagonal elements of interelectronic repulsion with the result that the new quantum numbers frequently are not particularly "good" when several terms have the same conbination of S and L.

One of the many advantages of Racah's parametrization $(e_0E^0 + e_1E^1 + e_2E^2 + e_3E^3)$ is that the energy differences between terms with S_{max} are definite multiples of E^3 alone, the coefficient e_3 then dependent on L:

$$\begin{array}{lllll} f^2, f^5, f^9, f^{12}: & H: -9 & F: \ 0 & P: +33 & \\ f^3, f^4, f^{10}, f^{11}: & I: -21 & S, F: \ 0 & G: +12 & D: +33 \end{array} \tag{1.15}$$

Both the Hartree-Fock radial functions [22] and the phenomenological values give E^1 very close to $10E^3$. The coefficient e_1 only depends [19] on S and the seniority number v (which is q for the large majority of terms in the first half of the shell, though it is 0 for ^1S of f^2 and 1 for one of the two ^2F of f^3 in Eq. (1.8), and 2 for seven terms of f^4) and is:

$$e_1 = \frac{9(q-v)}{2} + \frac{v(v+2)}{4} - S(S+1) \qquad (1.16)$$

This dependence on $S(S + 1)$ is an expression of a general result of the Slater-Condon-Shortley theory for l^q noted by one of us in 1956 that the energy difference between the *baricentre,* the average energy, of all the states having $S = S_0 - 1$ and the baricentre of all the states having $S = S_0$ turns out to be $2DS_0$ where D is a *spin-pairing parameter* constituting the same linear combination of F^k parameters for all q and S_0. Hence, one obtains a formal similarity to Eq. (1.10) by writing the average energy of all the states as containing a contribution of interelectronic repulsion

$$q(q-1)A_*/2 \qquad (1.17)$$

to which is added a contribution dependent on the total spin quantum number S:

$$D[\langle S(S+1)\rangle - S(S+1)] \qquad (1.18)$$

with the average value

$$\langle S(S+1)\rangle = \frac{3}{4}q\left\{1 - \frac{q-1}{4l+1}\right\} = \frac{3q(4l+2-q)}{16l+4} \qquad (1.19)$$

In the case of f-electrons ($l = 3$) the parameters are

$$\begin{aligned} A_* &= E^0 - (9E^1/13) \\ D &= 9E^1/8 \end{aligned} \qquad (1.20)$$

It may be noted that this treatment is aligned with the behaviour of the closed shell f^{14} showing the contribution $91A_*$ (which is smaller than $91F^0$ because of the systematic mutual avoidance of electrons in the anti-symmetrized Ψ in spite of the overall spherical symmetry) whereas, Racah [19] concentrated attention on the first half of the shell (q at most 7) with $e_1 = 0$ of Eq. (1.16) for S_{max}. This is no longer true in the second half (which one does not need to consider if one is exclusively interested in term distances within f^q) where $e_1 = 63$ for q = 14 as also discussed by Johnson [23]. It has been discussed in detail [4, 6] why the spin-pairing energy parameter D contains a factor $(2l + 3)/(2l + 2)$ expressing the average effect of decreased seniority number when S decreases. This is one of the several interesting algebraic results relating the differing coefficients to the parameters of interelectronic repulsion.

Obviously, the lowest term among several terms having a given baricentre has lower energy. This is why the distance $2DS_{max}$ (with D around 6,500 cm^{-1}or 0.8 eV) between the baricentres of states with $(S_{max} - 1)$ and S_{max} *is* in qualitative agreement with the spin-forbidden transitions moving towards the ultra-violet in the vicinity of $q = 7$ with $S_{max} = 7/2$ but ^5D and in particular its lowest level $^5D_0$ is very low in $4f^6$ [17,200 cm^{-1} in europium(III)] relative to the average quintet energy about 40,000 cm^{-1}. The lowest excited term ^5D is rather exceptional [24] by being the eigen-value of three diagonal elements in Racah's classification (of which one has $\nu = 4$) with highly differing energy, but at the same time unusually large non-diagonal elements of interelectronic repulsion. Further more, the coefficient $k\zeta_{4f}$ is enhanced with resulting large separations at least between the three first levels $^5D_0$, $^5D_1$ and $^5D_2$. Elliott, Judd and Runciman [25] calculated such coefficients k and f_2 in terms of a fixed value of F_2. The next quintet terms ^5L and ^5G of Eu(III) do not start before 25,000 cm^{-1}.

One should be careful not to filter off the mosquito and swallow the camel. There has been many attempts to refine Racah's treatment with additional parameters [20, 21] in addition to the E^1, E^2 and E^3 (with more or less pre-determined ratios) and ζ_{4f}. One consistent refinement was introduced by Trees [26, 27] using the $2l$ independent distances between the l^2 terms of Eq. (1.6). Curiously enough, this amelioration is far more important in 2p group atoms [4] and somewhat useful in the d groups, but less spectacular in the far more complicated configurations $4f^q$. Actually, the major problem is the *Watson effect* [3, 28] that *all* the parameters F^k of interelectronic repulsion derived from Hartree-Fock radial functions are roughly $(z + 3)/(z + 2)$ times the phenomenological values (where z is the ionic charge). A comparison within an isoelectronic series (such as $[Ar]3d^3$ Ti^+, V^{+2}, Cr^{+3}, Mn^{+4}, . . .) shows that the (almost invariant) deviation from the Hartree-Fock values tends to be more pronounced for low than for high L values, but one gets the general impression of a moderate dielectric constant decreasing all the term distances, probably [3] via two-electron substitutions to orbitals with comparable $\langle r \rangle$ and positive one-electron energies, i.e., configuration interaction with $4f^{q-2}$ (continuum g)^2 rather than with $4f^{q-2}5g^2$. There is no evidence for speaking in favour of the somewhat mythological "Coulomb hole" supplementing the Fermi hole produced by the requirements of anti-symmetrization. Actually, the distances between terms with S_{max} are, if anything, more decreased by the deviations from Hartree-Fock values than by the distances to terms with lower S. This state of affairs corresponds to E^3 being more diminished than E^1.

Recently, several authors [29–33] have pointed out that the identification of term distances with differences in $\langle r_{12}^{-1} \rangle$ may be somewhat oversimplified. Though the Slater theory is asymptotically valid for high ionic charges, there is indeed a tendency for small z to have excited terms with expanded radial functions, smaller attraction $Z\langle r^{-1} \rangle$ to the nucleus, and actually *decreased* over-all interelectronic repulsion. This is one of the many unexpected side-effects of the virial theorem [34] that the *kinetic energy* decreases when the total energy becomes less negative. In many ways, it is comparable to the idea discussed thoroughly by Ruedenberg [35] that covalent bonding is due to a decreased local contribution to the kinetic energy in the bond region between the two atoms considered. One should be careful not to think exclusively in terms of electrostatic potentials because the kinetic energy operator is the *only* effect preventing the implosion of atoms and molecules, since a stable electrostatic system is N times more stable if all the distances are divided by N. A closer analysis shows that it is still legitimate for the chemist [36] to ascribe the term distances (and hence the major part of the energy differences between 1 and 4 eV involved in ruby and in lanthanide lasers) to an originally more successful mutual avoidance between electrons in the partly filled shell in the groundstate (with high L and S_{max}) though the excited states have to compensate their more pronounced interelectronic repulsion in the partly filled shell by many minor modifications of the total wavefunctions in order to satisfy the virial theorem. In particular, there is perfectly observable evidence [37, 38] that the spin-pairing energy Eq. (1.18) is an actual stabilization in comparison to otherwise similar closed-shell systems.

In spite of the general consensus that M(III) have excited levels of essentially the same kind as the J-levels of $4f^q$ of M^{+3} the progress in the actual identification was rather slow. Ellis [39] suggested that the three bands in the blue of Pr(III) are transitions from $^3H_4$ to $^3P_0$, $^3P_1$ and $^3P_2$ within $4f^2$ but otherwise the field was opened up by Gobrecht [40] identifying transitions in the infra-red and the red with the J-levels predicted by Eq. (1.10) in Russell-Saunders coupling (which later turned out to be a fairly good approximation for the lowest term except for Er(III) and Tm(III) having q = 11 and 12). In order to avoid vibrational spectra of water or other hydrogen-containing materials, the rare earths were dissolved in molten borax beads. The atomic spectrum of Ce^{+3} studied by Lang indicated $\zeta_{4f} = 643\ cm^{-1}$ and the Yb(III) compounds clearly indicate $\zeta_{4f} = 2{,}950\ cm^{-1}$ and Gobrecht assumed (on the basis of the hydrogenic Sommerfeld formula which is rather unsound because it does not take the ionic charge into account) that ζ_{4f} varies smoothly between these two limits (q = 1 and 13). This is also accepted today, though the variation is not linear but somewhat curved. $[Xe]4f^2$ is a highly excited configuration of La^+ presumably subject to perturbations by adjacent levels belonging to other configurations, and several authors took over, somewhat uncritically, the exceedingly low $\varphi_6 = 0.04$ [cf. Eq. (1.12)] reported [2]. Hence, Bethe and Spedding [41] identified correctly the levels with $S_{max} = 1$ of thulium(III) and Satten [42] the levels with $S_{max} = 3/2$ of neodymium(III), whereas the decreased E^1 concomitant with a negligible F^6 according to Eq. (1.14) did not permit the levels of $S = 0$ of Pr(III) and Tm(III) [43] and $S = 1/2$ of Nd(III) [18] to be identified. Since 1955 the techniques of identifying J-levels have progressed enormously and have gone beyond the mere comparison between observed and calculated positions. An excellent catalogue of assignments are 5 papers by Carnall, Fields and Rajnak [44]. We are not going into detail to describe the method [20, 21] of determining J, which are to a certain extent based on the "ligand field" separations of the $(2J + 1)$ states in a certain number of sub-levels, but receive independent confirmation by the Judd-Ofelt parametrization of the band intensities. However, we may present a statistical survey of the number of J-levels of $4f^q$ calculated (with a definite choice of E^k and ζ_{4f}) to be below (including the groundstate) and above $25{,}000\ cm^{-1}$ (the beginning of the ultra-violet) compared with the firmly established assignments:

q =	M(III)	Below 25,000 cm^{-1} calculated	Below 25,000 cm^{-1} identified	Above 25,000 cm^{-1} calculated	Above 25,000 cm^{-1} identified
2	Pr	12	12	1	(1)
3	Nd	22	20	19	11
4	Pm	31	17	76	3
5	Sm	28	23	170	21
6	Eu	11	11	284	26
7	Gd	1	1	326	19
8	Tb	8	8	287	16
9	Dy	15	14	183	15
10	Ho	14	13	93	24
11	Er	11	11	30	17
12	Tm	7	7	6	5

(1.21)

The last level $^1S_0$ of Pr(III) is in the somewhat peculiar position of being detected only in the luminescence excitation spectrum [45, 46] and as fluorescent transitions to highly excited levels. It almost coincides with the broad absorption bands due to transitions to 4f5d.

Many atomic spectra [1] are not in as fine a shape as the M(III) spectra in Eq. (1.21). The main problems with M(III) are that the spin-forbidden transitions where S_{max} of the groundstate decreases by one (or two) units can be very weak and that they are crowded in the ultra-violet with the result that the identification is difficult. Already in the blue, Pr(III) has the weak transition to $^1I_6$ almost coinciding [43] with $^3P_1$. In actual practice, aqueous solutions are not transparent beyond 52,000 cm^{-1} and they need non-absorbing counter-ions such as perchlorate (both chloride and sulphate start broad, intense bands in the region above 50,000 cm^{-1} and may sometimes induce electron transfer bands, say of Eu(III) and Yb(III), at lower wavenumbers). Transparent, crystalline fluorides can be used up to 90,000 cm^{-1} (in *vacuo* or in spectrographs filled with helium or argon, since N_2 and O_2 absorb strongly above 55,000 cm^{-1}) as in a study [47] of Gd(III) in CaF_2. One of the astonishing aspects of Racah's classification [19] is that all the energy differences between ^8S and the six sextet terms of f^7 (which are not split to first-order by spin-orbit coupling) are exactly the opposite, expressed as $e_1E^1 + e_3E^3$ of the distances between the seven terms of f^2 from Eq. (1.6) corresponding to a large gap between the next-last ^3P and last ^1S of Pr(III).

The combination of the ζ_{5f} being about twice ζ_{4f} and the parameters of interelectronic repulsion E^k under equal circumstances 0.6 times as large in the 5f group as in the 4f group make the quantum number S much less defined in the transthorium elements, and the deviations from Russell-Saunders coupling necessitating the introduction of *intermediate coupling* (the opposite extreme of ascribing a definite value of $j = (l + 1/2)$ or $(l - 1/2)$ to each electron with positive l is only asymptotically valid for the atomic spectra of 6p group elements from thallium to radon, and for X-ray and photo-electron spectra involving inner shells) which are already perceptible in erbium(III) and thulium(III) are quite impressive in the 5f group. Thus, $[Rn]5f^7$ curium(III) has lost many of the half-filled shell properties [48, 49] of Gd(III), the squared amplitude [50] of ^{8}S character in the ($J = 7/2$) groundstate is only 80 percent, and the three next levels ($J = 7/2$) at 17,000, ($J = 5/2$) at 20,300 and ($J = 7/2$) at 21,900 cm^{-1}do not have pronounced ^{6}P character about a-half) in contrast to the three levels of gadolinium(III) between 32,000 and 34,000 cm^{-1}.

[Rn]5f^2 uranium(IV) aqua ions (of which neither the symmetry nor the coordination number N are known) have the last absorption band due to a ^{3}P sub-level at 23,400 cm^{-1} whereas $^1S_0$ was detected [51] as a weak, sharp band at 40,800 cm^{-1}. The most thorough studies of U(IV) are of octahedral hexahalide complexes [52, 53] such as UCl_6^{-2} and UBr_6^{-2} showing slightly smaller parameters of interelectronic repulsion and stronger "ligand field" splittings than the aqua ions, both phenomena indicating more pronounced covalent bonding when N is as low as 6 with concomitant shorter internuclear distances. However, covalent bonding can also be enhanced [54, 55] in compounds with high N but neighbour atoms of low electronegativity, such as $N = 12$ in the pink modification of $U(H_3BH)_4$ syncrystallized in hafnium(IV) boranate [54] and $N = 16$ in the octogonal [55] sandwich $U(C_8H_8)_2$ and in $U(C_5H_5)_3Cl$. Whereas, 21 of the 40 sub-levels of UCl_6^{-2} and UBr_6^{-2} have been identified [52] most other uranium(IV) compounds are in a rather preliminary stage since the energy differences between the 12 J-levels are comparable to the sub-level splitting of each J-level. The same problem occurs in neptunium(V) and plutonium(VI) compounds, whereas 5f^{4}Np(III), 5f^{5}Pu(III), 5f^{6}Am(III) and 5f^{7}Cm(III) show clear-cut band groups each corresponding to a definite J-level as in the lanthanides.

B. The "Ligand Field" as Minor Deviations from Spherical Symmetry

An important aspect of the atomic-like spectra of [Xe]4f^q in M(III) compounds is that the influence of the surrounding atoms is so remarkably weak. Though this is also true for [Rn]5f^6 Am(III) and subsequent elements, it is much less true for the 5f^2 systems U(IV), Np(V) and Pu(VI). From this point of view the earlier 5f elements approach to a certain extent the behaviour of the d groups, having the strong influence of the adjacent atoms on their excited states. The fact that octahedral cobalt(III) and nickel(II) complexes are known with green, blue, violet, purple, red, orange and yellow colours (the order mentioned corresponds to the complementary colour of a single absorption band moving towards higher wave-numbers through the visible from the red to the violet) actually represent a less complicated behaviour than one might have expected, since the variation as a function of the ligands [4, 9, 56] corresponds to two absorption bands of Co(III) moving regularly towards higher wave-numbers, whereas, octahedral Ni(II) has a somewhat varying spectrum of three spin-allowed and four spin-forbidden transitions below 30,000 cm^{-1}. However, it is clear that these excited states show no close similarity to the gaseous Co^{+3} and Ni^{+2}, in particular because the former ion has the ground term ^{5}D belonging to [Ar]3d^6. Though the great majority of isoelectronic iron(II) complexes and a rather unique case (CoF_6^{-3}) of cobalt(III) also have $S = 2$, the nickel(IV) complex NiF_6^{-2} and the very numerous cases of Co(III) all have $S = 0$ with the groundstate having a wave-function which is a linear combination of states belonging to ^{1}I, ^{1}G and ^{1}S. Though the excited levels of nickel(II) in a certain sense belong to ^{3}F, ^{1}D, ^{3}P, ^{1}G and ^{1}S of [Ar]3d^8 the distribution of the 45 states is quite different.

Around 1930 it was very fashionable [57] to calculate *Madelung potentials* as the sum of q_i/R_i in a given point (inside an atom) where all the other atoms i carrying (negative or positive) charges q_i occur at the distance R_i. If the nucleus of the atom considered is put at origo, it is possible to describe the Madelung potential inside the atom as the sum of a large component $V_0(r)$ possessing spherical symmetry and a small residual component showing angular dependence:

$$V_{Mad} = V_0(r) + V_{res}(x, y, z) \tag{1.22}$$

This separation makes it possible to calculate the electrostatic energy of a crystal consisting of spherically symmetric, non-overlapping ions as if the charges q_i were situated at points. A closed-shell ion or certain cases of half-filled shells such as the ^{6}S ground terms of 3d^5 manganese(II) or iron(III) and the ^{8}S ground terms of 4f^7 europium(II), gadolinium(III) or terbium(IV) are not influenced by the residual part of Eq. (1.22) whereas, the typical transition-group ion containing a partly filled shell has its lowest level favourably arranged with respect to its electronic density being stabilized by $V_{res}(x, y, z)$. Bethe [58] suggested to describe the energy levels of d and f group ions as differing perturbations of $V_{res}(x, y, z)$ of various of the $(2S + 1)$ $(2L + 1)$ states belonging to each of the Russell-Saunders terms in the former case, and to differing perturbation of the $(2J + 1)$ states of each J-level in the lanthanides, in recognition of the fact that the perturbations of the non-spherical part of the Madelung potential are much larger than ζ_{3d} in the iron group ions, but much smaller than ζ_{4f} in the lanthanides.

Classical electrostatic theory allows the series expansion of $V_{res}(x, y, z)$ in definite components having the same angular dependence as the angular parts A_l of the one-electron functions in spherical symmetry. This has allowed a wide range of interesting connections between applied group-theory [6, 59, 60] and the study of excited states of d^q and f^q in point-groups of highly different symmetry. A major condition for a given angular part A_l to be represented in the non-spherical part of the Madelung potential as a non-vanishing component is that it has *total symmetry* (like $l = 0$ in spherical symmetry) in the prevailing point-group (which may possibly not be one of the 32 point-groups [59] compatible with infinitely extended, repeated lattices). If we stay inside a given configuration l^q the electronic density has *even* parity. In systems containing a centre of inversion (centre de symétrie in French) all the total wave-functions have to choose between the two parities (g = gerade and u = ungerade in German):

$$\begin{aligned} &\text{even (g):} \quad \Psi(-x, -y, -z) = \Psi(x, y, z) \\ &\text{odd(u):} \quad \Psi(-x, -y, -z) = -\Psi(x, y, z) \end{aligned} \tag{1.23}$$

and the square representing the electronic density then has even parity. If we do not modify the l^q wave-functions appropriate for spherical symmetry, they retain this property. The perturbations of the A_l components on such electronic densities vanish unless l is even. Furthermore, the triangular conditions of Gaunt [2] applied to these spherical harmonics make the perturbation vanish if l is higher than twice l of the partly filled shell. The typical uniaxial point-groups have A_2 components of V_{res} whereas the point-group O_h present in regular octahedral MX_6 (or for that matter MX_8 in CaF_2 or tetrakaidecahedral MX_{12} on the K site of K_2PtCl_6, Sr site of the cubic perovskite $SrTiO_3$ and in cubic closed-packed metals) have V_{res} starting with A_4 proportional to $(5x^4 + 5y^4 + 5z^4 - 3r^4)$ having equivalent Cartesian axes (what is a definition of cubic point-groups). A sort of hyper-cubic point-group is the icosahedral K_h where one A_6 is the first totally symmetric A_l. Hence, it does not separate d^q energy levels. It is interesting that Judd [61] demonstrated from the J-sub-levels of magnesium double nitrates (which so frequently served for fractional crystallization) that the major perturbation is icosahedral, since the crystal structure [62] later was shown to consist of $[Mg(H_2O)_6^{++}]_3[M(O_2NO)_6^{-3}]_2, 6H_2O$ containing almost ideal icosahedral MO_{12} with the twelve oxygen neighbour atoms supplied by six bidentate nitrate ligands.

The electrostatic model of the "ligand field" frequently was applied to a set of charges q_i on a sphere with radius R considerably larger than the average radius of the partly filled shell $\langle r \rangle$. In such cases the energy differences produced by the A_2 perturbations in typical uniaxial cases are proportional to $\langle r^2 \rangle / R^3$, whereas, the octahedral point-group has energy differences proportional to $\langle r^4 \rangle / R^5$ and the icosahedral point-group $\langle r^6 \rangle / R^7$. This result was one of the many arguments which caused the collapse [6, 63] of the electrostatic model in 1956 for d group and [64] in 1963 for f group compounds. It had been a continuous source of worry that the *chromophore* MX_9 of symmetry D_{3h} (like a trigonal prism MX_6) known from crystalline $LaCl_3$ (each chloride anion is bound to three M) and from salts (such as bromates and ethylsulphates) of ennea-aqua ions $M(OH_2)_9^{+3}$ had J-sub-levels suggesting $\langle r^2 \rangle / R^3$ and $\langle r^6 \rangle / R^7$ to have the same order of magnitude, which is incompatible with an electrostatic model. Nevertheless, the many investigations of "ligand field" parameters were not a waste of time, because they turn out to be a linear transformation of one-electron energy differences of deeper physical significance.

Among the many other valuable heritages from the electrostatic model there is a closer understanding of the real (not complex) one-electron functions

$$\psi_{nl} = A_l \cdot R_{nl} / (2\sqrt{\pi} r) \tag{1.24}$$

which we choose to normalize on concentric shells (with the volume element $4\pi r^2 dr$) with the result that $\int_0^\infty (R_{nl})^2 \, dr = 1$ and the average value of $(A_l)^2$ on a spherical shell is 1. The sum of all $(2l + 1)$ different $(A_l)^2$ is invariantly $(2l + 1)$ as first pointed out by Unsöld. In agreement with crystallographic tradition we select the z-axis as the principal axis in the linear symmetries and other uniaxial point-groups. There is only one practical choice for A_0, A_1 and A_2:

$$
\begin{array}{llll}
\text{s:} & 1 & \text{d}\sigma\text{:} & \sqrt{5}(2z^2 - x^2 - y^2)/2r^2 \\
\text{p}\sigma\text{:} & \sqrt{3}z/r & \text{d}\pi\text{c:} & \sqrt{15}xz/r^2 \\
\text{p}\pi\text{c:} & \sqrt{3}x/r & \text{d}\pi\text{s:} & \sqrt{15}yz/r^2 \\
\text{p}\pi\text{s:} & \sqrt{3}y/r & \text{d}\delta\text{c:} & \sqrt{15}(x^2 - y^2)/2r^2 \\
 & & \text{d}\delta\text{s:} & \sqrt{15}xy/r^2
\end{array} \tag{1.25}
$$

The quantum number λ in the linear symmetries has trivial names chosen in analogy to Eq. (1.2), $\sigma(\lambda = 0)$, $\pi(\lambda = 1)$, $\delta(\lambda = 2)$, $\varphi(\lambda = 3)$, . . . and indicate the degree of the homogeneous polynomial in *two* of the Cartesian coordinates x and y entering the angular functions $x^a y^b / \rho^\lambda$ where the sum of the two non-negative integers $a + b = \lambda$ and the variable $\rho^2 = x^2 + y^2$. In the xy-plane, the angular function possess λ nodes (forming planes containing the z-axis, and having an angular separation $180°/\lambda$). The $(2l + 1)$ different A_l correspond to one σ, two π, two δ, . . . orbitals with λ at most l. In classical description [2] of the linear point-group C_∞ (lacking not only a centre of inversion like $C_{\infty v}$ characterizing hetero-atomic diatomic molecules, but also the symmetry of time-inversion, which is always present in univocal potentials) characterizing a linear magnetic field, the ψ are complex for positive λ, and λ is actually the $|m_l|$. We distinguish [6] the two A_l for a given λ by a new quantum number ç (the French letter c-cedille which is pronounced s) having the two possible values c and s (derived from "cosine" and "sine" in polar coordinates). For a given combination of l and λ, the λc and λs orbitals are transformed into each other by a rotation $90°/\lambda$ in the xy-plane. The λc orbital has a maximum at the x-axis, where λs vanishes. Seen from this point of view, $l\sigma$ (rotationally symmetric around the z axis) is c. A short definition of ç = s is that $\psi(x, -y, z) = -\psi(x, y, z)$ whereas c-orbitals are invariant for inversion of the sign of y. Organic chemists frequently talk about "σ" and "π" orbitals in planar molecules. These two values of a quantum number defined by

$$
\begin{array}{ll}
\text{“}\sigma\text{”:} & \psi(x, y, -z) = \psi(x, y, z) \\
\text{“}\pi\text{”:} & \psi(x, y, -z) = -\psi(x, y, z)
\end{array} \tag{1.26}
$$

if the molecular plane contain the x and y axes should not be confused with σ and π being the two first λ-values 0 and 1, but is rather an analogy to the parity (inverting all three coordinate signs) defined Eq. (1.23). Among the A_l, we have "σ" when $(l + \lambda)$ is an even integer, and "π" when $(l + \lambda)$ is odd. For a given l, we have $(l + 1)$ "σ" and l "π" orbitals.

The A_2 in Eq. (1.25) are, at the same time, adapted to linear and to cubic symmetry. In cubic point groups such as O_h the five d orbitals form two *sub-shells,* one consisting of dπc, dπs and dδs having angular functions concentrated in the regions between the Cartesian axes and are proportional to (xz), (yz) and (xy) obtained by cyclic permutations of the three coordinates. The other sub-shell consists of dσ and dδc concentrated closely around the Cartesian axes. It is less obvious that these two orbitals necessarily are equivalent in O_h but it can be demonstrated by rotation around the three-fold axes.

The seven Λ_3 adapted to linear symmetry are:

$$
\begin{aligned}
&f\sigma: && \sqrt{7}z(5z^2 - 3r^2)/2r^3 \\
&f\pi c: && \sqrt{42}x(5z^2 - r^2)/4r^3 \\
&f\pi s: && \sqrt{42}y(5z^2 - r^2)/4r^3 \\
&f\delta c: && \sqrt{105}z(x^2 - y^2)/2r^3 \\
&f\delta s: && \sqrt{105}xyz/r^3 \\
&f\varphi c: && \sqrt{70}x(3y^2 - x^2)/4r^3 \\
&f\varphi s: && \sqrt{70}y(3x^2 - y^2)/4r^3
\end{aligned}
\tag{1.27}
$$

These orbitals would be mixed in cubic symmetries, where the symmetry-adapted orbitals are:

$$
\begin{aligned}
&a_{2u}: && \sqrt{105}\, xyz/r^3 \\
&t_{1u}: && \sqrt{7}(5z^3 - 3zr^2)/2r^3 && \text{and two cyclic permutations} \\
&t_{2u}: && \sqrt{105}\, z(x^2 - y^2)/2r^3 && \text{and two cyclic permutations}
\end{aligned}
\tag{1.28}
$$

It may be noted that the shape of four of the d orbitals (except dσ) is the same and corresponds to the prototype of a quadrupole, whereas the f orbital of symmetry type a_{2u} in O_h is the typical octupole.

The first-order perturbation from an external set of electric charges q_i cannot separate one-electron energies nor energies of many-electron states to a finer grid than permitted by group theory in the point-group of the M site, and in actual practice, nearly all the manifold states which are allowed to separate in energy, do indeed distribute. This is the main reason why a large number of books published between the early treatise by Griffith [65] and the monograph by Sugano, Tanabe and Kamimura [66] (the two former authors [67] already systematized in 1954 the treatment of d^q in O_h assuming Russell-Saunders coupling) continue to exploit the electrostatic model with a slight touch of doubletalk in spite of the evidence which had already been accumulating since 1955 that the major part of the sub-shell energy differences in d^q is due to formation of *anti-bonding molecular orbitals* (MO) which can be described in the approximation of LCAO (linear combination of atomic orbitals) as a weak delocalization of the M orbital of Eq. (1.24) by adding a contribution with the opposite sign of the amplitude (producing a nodal surface surrounding the central atom) with exactly the same symmetry type (in the point-group of M) and hence the same number l of angular nodes, and consisting of a superposition of orbitals belonging to one or more ligating atoms. The LCAO model allows a continuous parametrization between the fully ionic cases (without any delocalization of the partly filled shell on the ligands) to pronounced covalent bonding with comparable amplitudes on the central atom and the ligands.

We are not discussing the problem at length, (very important for d^q) as it is of no direct importance for 4f group ions, i.e. that one may start the "ligand field" treatment with *strong-field diagonal elements* where each of the sub-shells [containing one, two or three of the $(2l + 1)$ orbitals] contain a definite number of electrons or with *weak-field diagonal elements* corresponding to Russell-Saunders S, L-terms from spherical symmetry of the isolated ion. In both cases one obtains identical results after taking the non-diagonal elements into account (consisting of parameters of interelectronic repulsion in the former case, and of "ligand field" parameters in the mixing of weak-field diagonal elements). The corresponding two-branched hyperbola when two sets of states show the same combination of S and the symmetry type Γ_n replacing L in the prevailing point-group and more complicated eigen-value curves, when the combination of S and Γ_n is represented three or more times, represent a competition between the sub-shell energy differences and term differences (due to interelectronic repulsion) of comparable magnitude. Thus, the diagrams of Tanabe and Sugano [67] illustrating the evolution of the energy levels of d^q in O_h has the ratio between the sub-shell difference Δ(called 10 Dq in early literature) and Racah's parameter of interelectronic repulsion $B = \frac{1}{49} F^2 - \frac{5}{441} F^4$ as the variable. The slopes of the tangents to the curves at $\Delta = 0$ indicate the weak-field diagonal elements, and the asymptotic behaviour for very large (Δ/B) correspond to the strong-field diagonal elements.

Herzberg [8] pointed out that "ligand field" theory is essentially MO theory applied to partly filled shells. This is perfectly true, but the predominant rôle of interelectronic repulsion in the d groups produce specific features, among which the strong stabilization of high S values according to Eq. (1.18) is the most surprising to chemists mainly working with elements outside the transition groups. Thus, organic chemists expect all stable molecules to have groundstates with $S = 0$ and hence be diamagnetic. Though the first excited level generally is a triplet ($S = 1$) it belongs to the same excited MO configuration as the second singlet level. Molecules or ions having positive S (which is always the case for an odd number of electrons) are called "free radicals" and are expected to dimerize rapidly or perform other reactions leading to proper singlet molecules. However, the situation that two or more orbitals have exactly (for group-theoretical reasons) or almost the same energy and contain at least two electrons (but at most 2 below the full occupation) producing a groundstate with $S = 1$ or higher values, can occur outside the transition groups. Faraday discovered that oxygen molecules O_2 are paramagnetic and Lennard-Jones explained this phenomenon on the basis of MO theory as two π_g orbitals (formed by a LCAO of one 2p orbital from each oxygen atom having parallel axes and one node-plane perpendicular on the linear axis) containing *two* electrons and having a groundstate with $S = 1$ and two excited levels (at 7,918 and 13,198 cm^{-1}) with $S = 0$. If the same two π_g orbitals (which are empty in N_2) contain three electrons, we have the superoxide anion O_2^- and if they contain one electron O_2^+ (which was previously known from molecular spectra of electric discharges, but Bartlett has now prepared salts of this cation). In both cases $S = 1/2$, whereas the peroxide anion O_2^{-2} with four π_g electrons is diamagnetic with $S = 0$ like the isoelectronic F_2. Calculations [68] on the chemically much less stable species C_2^+ show the groundstate to have $S = 3/2$ because two π_u and one σ_u electron have almost the same energy. This is the same mechanism producing $S = 3/2$ for the lowest level $^4H_{7/2}$ belonging to $[Xe]4f^26s$ observed in Ce^+ (though it is 1,472 cm^{-1} above the groundstate of Ce^+ belonging to [Xe]4f5d6s). A more clear-cut example is the groundstate $^4F_{3/2}$ of Zr^+ belonging to the Configuration $[Kr]4d^25s$. Interestingly enough two almost regular ^4F terms occur between 0 and 1,323 and between 2,572 and 3,758 cm^{-1}, the latter belonging to $[Kr]4d^3$.

Historically speaking, the electrostatic model did a tremendous service by allowing the complicated spectroscopic behaviour of partly filled shells in spherical symmetry to be smoothly incorporated in the general theory of hetero-atomic molecules and complex ions. There is not the slightest doubt that it would have been far less fruitful to start MO treatment from scratch, as one had to do for the excited states of electron transfer spectra of permanganate MnO_4^- and uranyl UO_2^{+2}. The description of heterocyclic organic molecules such as pyridine C_5H_5N goes back to Hückel assuming more negative diagonal elements of an effective one-electron operator for the more electronegative atoms, non-diagonal elements being the product of some average energy times the overlap integral between atomic orbitals on two adjacent atoms. This line of thought corresponds to a perturbation by substituting CH by a nitrogen atom in the otherwise comparable benzene C_6H_6. A model of hetero-atomic complexes suggested by Wolfsberg and Helmholz [69] became very popular [3, 70] around 1960. Qualitatively, this model is able to describe both internal transitions in the partly filled d shell and electron transfer spectra, though agreement with experience is obtained only [71, 72] by considerable fiddling with the semi-empirical parameters. Nevertheless, a general feature of this type of model has remained plausible, the non-diagonal elements proportional to the overlap integral S_{MX} between a linear combination of ligand (X) orbitals with the same symmetry type (in the point-group of M) as the M orbital considered, produce according to the formula of second-order perturbation theory (which is rendered slightly more complicated [3] by the presence of overlap integrals) energy differences proportional to $(S_{MX})^2$ and negative for the lower-energy bonding combination (this eigen-value has most of its electronic density concentrated on the X atoms, and has no additional nodal surface between M and X) and positive for the higher eigen-value (the electronic density of this anti-bonding orbital is to a great extent localized on M, and the orbital has a new nodal surface surrounding M). The diagonal sum-rule valid for secular determinants not involving overlap integrals would assure that the stabilization of the bonding and de-stabilization of the anti-bonding combination would be equal with opposite signs. This is distinctly not the case [3] when an overlap is included, and one finds that the anti-bonding orbital is twice or thrice as anti-bonding as the bonding orbital is bonding, in agreement with the general experience [8] in molecular spectroscopy.

The *angular overlap model* was introduced [64] for explaining the small one-electron energy differences between the seven 4f orbitals in lanthanide chromophores MX_N. It turns out that the squared overlap integrals $(S_{MX})^2$ can be written Ξ^2 times a diatomic contribution dependent on the nature of the atoms M and X and their distance, and that Ξ^2 can be calculated as the first-order perturbation of a singular *equiconsequential contact potential*

$$\Xi^2 = \sum_{i=1}^{N} [A_l(x_i, y_i, z_i)]^2 E_i \tag{1.29}$$

acting only at the positions of the N ligating atoms (where E_i is put equal to one if all atoms are identical, and at the same distances). The corresponding increase of one-electron energy is frequently written $\Xi^2\sigma^*$. If $E_i = 1$, the sum of Ξ^2 for all the $(2l + 1)$ orbitals belonging to the same shell is $N(2l + 1)$. Thus, $l = 2$ in octahedral symmetry have two equally anti-bonding orbitals with the result that their $\Xi^2 = 6 \cdot 5/2 = 15$. The three other orbitals are non-bonding in this approximation, with Ξ zero. The identification of the sub-shell energy difference Δ in MX_6 with $15\sigma^* = 15E_i$ allows the interesting conclusion (in good agreement with experience [9] but is entirely different from the electrostatic model predicting much larger contributions dependent on $\langle r^2 \rangle / R^3$ in contrast to Δ) that the dδc orbital from Eq. (1.25) in quadratic (the usual wording "square planar" is rather out of date, since squares normally are planar) MX_4 with the *same* distances is anti-bonding to the same extent $15E_i$, whereas the dσ orbital is a-third as anti-bonding, $\Xi^2 = 5$, and the three other d orbitals non-bonding. This degeneracy between the three orbitals is only of group-theoretical necessity in the point-group D_{4h} in the case of dπc and dπs. It may be noted that the zero-point of the angular overlap model corresponds to non-bonding orbitals and is situated NE_i *below* the baricenter energy of all the $(2l + 1)$ orbitals which constitutes the zero-point of the electrostatic model in the sense that it corresponds to the perturbation by the spherically symmetric component $V_0(r)$ of the Madelung potential in Eq. (1.22). It was already noted by Orgel [73] in 1955 that the consequences for "ligand field" stabilization are not essentially different in the two lines of thought, either counting the number of anti-bonding electrons or counting their deficit relative to the average value obtained for d^q by taking $(q/10)$ times the number of anti-bonding electrons for d^{10} (a concept which invites some doubt when comparing [74] with photo-electron spectra of d group compounds).

In fortunate situations (such as the point-groups D_{3h} or O_h) each of the seven orbitals in Eqs. (1.27) or (1.28) represents its own symmetry type (together with its degenerate counter-parts in the degenerate symmetry types, such as λ in linear symmetries). In such a case, one only needs to evaluate [64] the diagonal elements Eq. (1.29) of the equiconsequential contact term. The zircon ($ZrSiO_4$) type found in YPO_4 and YVO_4 have the point-group D_{2d} for the chromophore $Y(III)O_8$ where one non-diagonal element has to be introduced [75]. The non-diagonal element of the equiconsequential contact term is

$$\sum_{i=1}^{N} [A_l(x_i, y_i, z_i)][A'_l(x_i, y_i, z_i)]E_i \tag{1.30}$$

connecting the two orbitals A_l and A'_l. The Ξ^2 values are then the eigen-values of a secular determinant having the degree equal to the number of times the symmetry type is repeated among the $(2l + 1)$ orbitals. In the worst possible instance, the point-group C_1 having only an identity as a symmetry element (C_n axes only have a direction in space for n at least 2), we have $(2l + 1)$ diagonal and $(2l^2 + l)$ non-diagonal elements of the equiconsequential contact term. For f orbitals this makes 7 diagonal and 21 non-diagonal elements. The point-group C_s possessing a plane of symmetry divides the $(2l + 1)$ orbitals according to Eq. (1.26) in $(l + 1)$ orbitals being "σ" and l being "π". These two alternatives are generally written as single-prime and double-prime in the symbols for the symmetry types. Consequently, we have two secular determinants for f orbitals, one having 4 diagonal and 6 non-diagonal elements, and the other having 3 diagonal and 3 non-diagonal elements like the p orbitals in C_1. In D_{3h} we have 5 essentially different diagonal elements because fπc and fπs remain degenerate of group-theoretical necessity, and also fδc and fδs. It is worth noting that this is not true for fφc and fφs which can be separated by a perturbation having an angular dependence in the xy-plane periodic with the period 120°. Actually, one of the two fφ orbitals is non-bonding, having vanishing Ξ.

If one has a one-valued potential in each point, U(x,y,z), it is possible [6] to *generate symmetries* of an arbitrary point-group either by integration (spherical and linear symmetries) or by forming the average value in N_G points where N_G is the order of the point-group (48 for O_h, 24 for T_d, 16 for D_{4h}, 4n for D_{nh} and D_{nd}, 2n for D_n, S_{2n}, C_{nh} and C_{nv} and n for C_n *etc.*). A specially important generated symmetry is the *holohedrized* symmetry

$$U_{hol} = [U(x,y,z) + U(-x,-y,-z)]/2 \tag{1.31}$$

necessarily having a centre of inversion like the point-group C_i. The residual potential has *hemihedrized* symmetry

$$U_{hem} = U - U_{hol} = [U(x,y,z) - U(-x,-y,-z)]/2 \tag{1.32}$$

vanishing identically if U already had a centre of inversion. On the other hand, it is not always certain that holohedrization of the molecule possessing a nuclear skeleton realizing the point-group G restricts itself to produce the minimum result (the product of G with a centre of inversion, as O_h is formed from T_d characterizing a regular tetrahedron). This is particularly true for *orthoaxial chromophores* where three Cartesian axes can be brought to contain all the ligand nuclei. It must be stressed that orthoaxiality is not at all a point-group; an octahedral complex with six different monatomic ligands on the Cartesian axes has the point-group C_1 only. Whereas the bent water molecule has the point-group C_{2v} and the pyramidal NH_3 is C_{3v} having the minimum holohedrized symmetries D_{2h} (like a rectangle) and D_{3d} (a trigonal antiprism) the (slightly idealized) orthoaxial H_2S and PH_3 also have the symmetries C_{2v} and C_{3v} but holohedrized symmetries D_{4h} (like a square) and O_h (regular octahedron). The latter situation also occurs in octahedral *fac*-MX_3Y_3 with symmetry C_{3v} and holohedrized symmetry O_h. Besides a variety of arguments related to transition probabilities and band intensities, the main reason why these concepts are important for "ligand field" models is that only the holohedrized symmetry of the chromophore determines the energy levels of l^q. This is why the Tanabe-Sugano diagrams [67] orginally intended for MX_6 with the symmetry O_h can be applied equally well to regular tetrahedral MX_4 with the symmetry T_d (but the holohedrized symmetry O_h like a cube) as far as the energy levels go (though Δ is negative and the absorption bands are by one or two orders of magnitude more intense). Another interesting result is that both the first-order electrostatic perturbation $V_{res}(x,y,z)$ of Eq. (1.22) and the diagonal element Eq. (1.29) of the angular overlap model act on the electronic density proportional to the *square* $(A_l)^2$. This is one reason why the series expansion is allowed to stop at angular contributions of order $(2l)$ in Bethe's electrostatic model. It is possible [6] to discuss the *literal symmetry* (the adjective is added in order to avoid confusion with the site point-group) both of extended entities (sets of points, or one-valued functions of (x, y, z) such as one-electron wave-functions) and their *squares*. Thus, the cylindrical (rotational) symmetry of pσ, dσ, fσ, . . . makes the literal symmetry $D_{\infty h}$, whereas the common shape (rotated in space) of the four dπ and dδ orbitals has the literal symmetry D_{2h} (it is orthorhombic because of the alternating positive and negative signs in the quadrupole) but their squares have the literal symmetry D_{4h}. It is possible to write a given square $(A_l)^2$ as a linear combination of other A_l with $l = 0, 2, 4, \ldots, 2l$. According to Unsöld's theorem the spherically symmetric contribution ($l = 0$) is exactly 1, which is one way of obtaining the normalization constants. For instance, the square of the pσ orbital is $3z^2/r^2$ which is the sum of 1 and the dσ orbital multiplied by $2\sqrt{5}/5$ (it is customary to avoid square-roots in denominators). Such results are related to the conditions found by Kibler [76, 77] for ratios between parameters of the electrostatic model reproducing Ξ^2 values of Eq. (1.29).

Already in Bethe's paper [58] from 1929, the rather insidious question was discussed whether the small amount of electronic density of the partly filled shell close to the ligand nuclei (in the tail of the radial function seen from the point of view of the central atom) might conceivably be more perturbed (in the total energy) than the major part of the l shell by the rather small deviation of the Madelung potential from spherical symmetry. This idea is formalized in the equi-consequential contact term Eq. (1.29) introduced in 1963, but on the whole the suggestion has been completely neglected by "ligand field" theorists taking a certain pleasure in applied mathematics by manipulating the spherical harmonics. It must be emphasized that the reason covalent bonding with overlap integrals S_{MX} proportional to Ξ can be incorporated in an apparent first-order perturbation proportional to Ξ^2 is that the covalency in transition group complexes is sufficiently weak so that no cross-term and inter-ligand effects are perceptible. This statement might be doubted in d group complexes of sufficiently reducing and low-electronegativity ligands, but seems very plausible in f group compounds. The physical origin of the increase $\Xi^2 E_i$ of the one-electron energy is conceivably [6] the increased *local contribution to the kinetic energy* in the bond region, according to a suggestion by Ruedenberg [35]. One of the major effects of the kinetic energy operator is to prevent the implosion of atoms and molecules, and it is indeed quite consistent if the "ligand field" anti-bonding corresponds to the electronic density of the partly filled shell colliding with the electronic density of the ligands. In other words, the word "field" originally designating the external potential V_{Mad} slowly has changed its meaning to something like the pseudo-potential sometimes introduced in atomic calculations to represent the influence of orthogonalization of outer orbitals (in particular *via* their radial nodes) on the inner shells in the core. Ballhausen and Dahl [78] recently suggested that such a pseudo-potential originating in the ligands is a sort of minor addition to the conventional V_{res} but experimental evidence [6, 64] is rather the other way round, the purely electrostatic V_{res} being negligible relative to the covalent effects described by the angular overlap model. In a certain sense this model explains the differing anti-bonding effect on the $(2l + 1)$ orbitals of a given shell as a consequence of Pauli's exclusion principle, since the partly filled shell has to be orthogonal on all the ligand orbitals. Like shells with increasing n (for a given l) in a single atom, the kinetic energy operator is important for determining the energy.

The one-electron operator in Eqs. (1.29) and (1.30) represents weak covalent bonding of σ character, rotationally symmetric around the line segment M–X. This symbol means the same thing as the λ values used for classifying orbitals in linear symmetries. Hence, it is possible to discuss π bonding between orbitals having one node-plane containing M–X, *e.g.* between dπ of M and pπ of X. Actually, it is beyond doubt [6, 9] that the energy difference close to 14,000 cm^{-1} between the two empty dπ and the dδs orbital containing one electron in the $3d^1$ vanadyl ion $VO(H_2O)_4^{+2}$ and in the $4d^1$ molybdenum(V) complex $MoOCl_5^{-2}$ is due to unusually strong anti-bonding on the oxygen 2pπ. The same is true for the fluoride ligands [79, 96] in *trans*-$Cr(NH_3)_4F_2^+$. It is the general consensus [4, 6, 79] that the ratio between the negative π and the positive σ contribution to Δ in octahedral complexes varies between (–1):4 for certain anions to very small for ligands (such as ammonia) having only one lone-pair. At first, it would seem contradictory to seek an equiconsequential contact term for π orbitals, which by definition vanish at the ligand nuclei. However, Schäffer and one of us [80, 81] found that the π-anti-bonding effects can be introduced in the angular overlap model as the squares of the interaction of Kronecker dipoles at the ligand positions with the partial differential quotient of A_l in either the x- or y-direction perpendicular on the M–X axis. In octahedral MX_6 this makes the lower sub-shell (of symmetry type t_{2g}) consisting of the three orbitals dπc, dπs and dδs equally π-anti-bonding, whereas the σ-anti-bonding occurs in the upper sub-shell (e_g) consisting of dσ and dδc. In quadratic MX_4, dπc and dπs are half as π-anti-bonding as in MX_6 with the same distances, and dδs to the same extent as in the octahedral complex, as was first pointed out by Yamatera [82] based on the Wolfsberg-Helmholz model.

The treatment of π-anti-bonding effects can be extended [83, 84] to all the λ values at the most equal to l. The second differential quotients of A_l are perturbed by a Kronecker quadrupole, when expressing δ-anti-bonding and φ-anti-bonding (which might occur on f orbitals, but is expected to be exceedingly weak) by third differential quotients of A_l perturbed by a regular hexagon of alternating positive and negative signs. It was shown by Schäffer [83, 84] that including all $(2l + 1)$ parameters of $\sigma, \pi, \ldots, (\lambda = l)$ anti-bonding as perturbations of planar, regular 2λ-polygons perpendicular on the M–X axes assure a linear transformation of the conventional electrostatic parameters, with the result that one model is not better than the other if the free choice of phenomenological parameters is permitted. However, it is clear that the advantages of the angular overlap model are based on the predominance of σ effects with possible weak π effects added.

In view of the obvious biunique correspondence between the six independent energy differences between the f orbitals (in the case of vanishing non-diagonal elements) and the parameters of the electrostatic model, nobody doubts that one can give the same information about a chromophore in both models. The sense in which the angular overlap model in our opinion can be shown to be superior to the conventional electrostatic model is that if σ-anti-bonding is the predominant effect, the Eqs. (1.29) and (1.30) prescribe the relative energy differences when the angular distribution of the ligating atoms (assuming identical M–X distances R) is known. Hence, the validity of the angular overlap model can be shown by the *transferability* of the parameter σ^* from one chromophore MX_N to another with differing symmetry or different coordination number N. The value of σ^* is obtained from the best possible agreement between the calculated $\Xi^2\sigma^*$ and the observed one-electron energies relative to their common baricenter put at $N\sigma^*$. As seen in Table 1, it is very satisfactory that M(III) in ennea-aqua ions shows a sudden drop of σ^* from M = Ce to Pr and then a weak, smooth decrease toward M = Tm, much like the evolution of Δ in the 3d or the 4d group. Another similarity with the d groups is that σ^* for M(III) substituted in $LaCl_3$ is about two-thirds the values for aqua ions. It must be admitted that the agreement with $\Xi^2\sigma^*$ is not perfect, though it is statistically very impressive. There are several conceivable reasons for the minor disagreements: the angular positions of the oxygen atoms have a certain experimental uncertainty in the crystal structures of bromates and ethylsulphates of the aqua ions; the three equatorial ligands have longer distances than the six forming the trigonal prism (adaptation of smaller E_i values in Eq. (1.29) for the longer distances ameliorates [64] the agreement sensibly) and we may have weak effects of π-anti-bonding (see below). Recently, Linares and Louat [85] continued the study of Er(III) [75] in YPO_4 and YVO_4 by a careful comparison of σ^* for Eu(III) in these and related lattices. As seen in Table 2, the agreement is remarkably good in spite of the conceivable π-anti-bonding in oxides. The nine-coordinated $M(OH)_3$ (M = Tb, Dy, Ho) also agree very well [86] with a pure σ-anti-bonding description, and are included in Table 1.

Table 1. Parameters σ^* of the angular overlap model for Crystalline ennea-aqua salts of ethylsulphate and for ions incorporated in anhydrous lanthanum trichloride. The parameters Ξ^2 of anti-bonding character of the 4f orbitals are given, as well as the values (Ξ^2 – 9) relative to the baricenter of type 1 (equal influence of all nine neighbour atoms) and (Ξ^2 – 12.6) of type 2 (60 percent larger influence of the six atoms in the trigonal prism compared with the three atoms in the equatorial plane). The one-electron energies [64] are in cm^{-1}. calculated (Ξ^2 – 9.72) for $Er(OH)_3$ are given [86] for comparison

	fσ	fπ	fδ	fφc	fφs	σ*
Ξ^2 (Type 1)	1.31	12.80	9.84	16.41	0.00	–
(Ξ^2 – 9)	– 7.69	+ 3.80	+ 0.84	+ 7.41	– 9.00	–
Ξ^2 (Type 2)	2.1	18.1	15.7	18.4	0.0	–
(Ξ^2 – 12.6)	– 10.5	+ 5.5	+ 3.1	+ 5.8	– 12.6	–
(Ξ^2 – 9.72)($Er(OH)_3$)	– 9.35	+ 6.47	+ 0.05	+ 6.04	– 9.72	–
$Ce(H_2O)_9(C_2H_5SO_4)_3$	–416	+250	– 35	+411	–425	45
$Pr(H_2O)_9(C_2H_5SO_4)_3$	–268	+113	+ 87	+195	–323	27
$Pr_xLa_{1-x}Cl_3$	–164	+112	+ 39	+ 64	–200	15
$Nd(H_2O)_9(C_2H_5SO_4)_3$	–255	+147	+ 61	+144	–304	25
$Sm_xLa_{1-x}Cl_3$	–154	+149	– 10	+ 98	–220	18
$Tb(H_2O)_9(C_2H_5SO_4)_3$	–177	+121	+ 89	+ 52	–294	19
$Dy(H_2O)_9(C_2H_5SO_4)_3$	–165	+118	+ 99	+ 49	–317	19
$Dy_xLa_{1-x}Cl_3$	– 96	+ 90	+ 42	+ 12	–180	11
$Ho(H_2O)_9(C_2H_5SO_4)_3$	–160	+115	+100	+ 12	–280	16
$Er(H_2O)_9(C_2H_5SO_4)_3$	–167	+119	+103	+ 9	–277	16
$Tm(H_2O)_9(C_2H_5SO_4)_3$	– 76	+115	+ 79	+ 28	–294	16

One of the first questions which were asked about the angular overlap model was the dependence of σ^* on R. This can be directly investigated by hydrostatic pressure applied to Tm(II) in cubic CaF_2. In such a chromophore MX_8, the a_{2u} orbital of Eq. (1.28) is strongly σ-anti-bonding (pointing toward the eight neighbour atoms) with $\Xi^2 = (280/9) = 31.111$, whereas each of the three t_{1u} orbitals have $\Xi^2 = (224/27) = 8.296$ giving the sum $8 \cdot 7 = 56$. The observed absorption spectra are consistent with the ratio (15/4) of anti-bonding character of these two sub-shells and give σ^* close to 43 cm^{-1} which increases 8 percent for each percent decrease of R, according to Burns [87]. Taken at its face value, this measurement might suggest a dependence on R^{-8} like in the old electrostatic model, but it was already discussed by Smith [88, 89] that an exponential decrease with R may have the same differential quotient, when $(S_{MX})^2$ is evaluated [90]. This has particularly been verified [88] for Δ of NaCl-type NiO where all the atoms are at special (and not general) positions, and the Ni–O distance exactly half the unit cell parameter a_0 which can be measured precisely, even under high pressure. Octahedral f group chromophores should have $\Xi^2 = 14$ for each of the three t_{1u} orbitals of Eq. (1.28) whereas, t_{2u} and a_{2u} should remain non-bonding. In the 5f group complexes [91, 92] PaX_6^{-2}, UX_6^- and NpF_6 one finds a certain separation between t_{2u} and a_{2u} but it is almost certainly due to π-anti-bonding, and it becomes progressively less important for the lower oxidation states Pa(IV) and U(IV) where σ^* is close to 200 cm^{-1} in the hexahalide complexes, less than 1,360 cm^{-1} for $4d^6RhCl_6^{-3}$ but much larger than the σ^* values for 4f group compounds given in Tables 1 and 2.

Table 2. Parameters σ^* (in cm^{-1}) for zircon type [85] phosphate, arsenate and vanadate (with 4 short and 4 long distances to the oxygen nuclei) and for nine-coordinated hydroxides [86]

$Nd_xY_{1-x}VO_4$	(four 2,291 Å): 63	(four 2,433 Å): 39
$Eu_xGd_{1-x}AsO_4$	(four 2,346 Å): 37	(four 2,446 Å): 26
$Eu_xGd_{1-x}VO_4$	(four 2,336 Å): 43	(four 2,462 Å): 28
$Eu_xY_{1-x}PO_4$	(four 2,313 Å): 42	(four 2,374 Å): 34
$Eu_xY_{1-x}AsO_4$	(four 2,300 Å): 41	(four 2,412 Å): 28
$Eu_xY_{1-x}VO_4$	(four 2,291 Å): 42	(four 2,433 Å): 26
$Eu_xLu_{1-x}PO_4$	(four 2,264 Å): 44	(four 2,346 Å): 33
$Eu_xLu_{1-x}AsO_4$	(four 2,248 Å): 43	(four 2,385 Å): 27
$Eu_xLu_{1-x}VO_4$	(four 2,242 Å): 44	(four 2,408 Å): 25
$Er_xY_{1-x}PO_4$	(four 2,313 Å): 43	(four 2,374 Å): 35
$Er_xY_{1-x}AsO_4$	(four 2,300 Å): 41	(four 2,412 Å): 28
$Er_xY_{1-x}VO_4$	(four 2,291 Å): 50	(four 2,433 Å): 31
$Tm_xY_{1-x}VO_4$	(four 2,291 Å): 45	(four 2,433 Å): 28
$Tb_xY_{1-x}(OH)_3$	27.3	
$Tb(OH)_3$	29.3	
$Dy(OH)_3$	26.3	
$Ho_xY_{1-x}(OH)_3$	24.6	
$Er_xY_{1-x}(OH)_3$	23.8	
$Er(OH)_3$	23.3	
$Er_xTb_{1-x}(OH)_3$	25.3	
$Yb_xY_{1-x}(OH)_3$	23.7	

In Copenhagen, Schäffer and Harnung [83, 84, 93] have developed the angular overlap model in a variety of interesting investigations. In the meantime, Newman [94] developed the superposition model, discussing the additivity of "ligand field" parameters from each M–X diatomic grouping in a chromophore. In view of what was said above, this has almost the same consequence as the angular overlap model. Recently, Schäffer [95, 96] made a careful analysis of the additive (like Ξ^2) and non-additive "ligand field" models. Since there are many reasons (discussed in Chapter 3) to believe that the partly filled 4f shell is far less delocalized than 3d in iron group complexes, it is clear that these higher-order effects of covalent bonding are less important in the lanthanides.

However, we remain with a serious problem. The one-electron energy differences (interpreted as $\Xi^2\sigma^*$ in the angular overlap model) are not directly observable in any lanthanide chromophore (not even f^1 or f^{13} because of the strong spin-orbit coupling). Seen from a practical point of view, nearly all the σ^* values obtained until now have been calculated from the electrostatic parameters evaluated by fitting the observed spectra to the lengthy expressions summarized in the books by Wybourne [20], Judd [21] and Dieke and Crosswhite [97]. It would be extremely valuable if a theorist could devise practical formulae for the energy differences between sub-levels of a given J-level (even with the possible restriction of Russell-Saunders coupling). What is technically the MO configurations (having exactly zero, one or two electrons in each of the seven f-like orbitals) are generally highly mixed in a given sub-level, quite in contrast to the strong-field diagonal elements, which are frequently remarkably good approximations to the energy levels of d^q especially in octahedral symmetry. The general spreading of one-electron energies is rarely above $3N\sigma^*$ but the separation of a given J-level in sub-levels is usually smaller in spite of the fact that well-defined MO configurations from f^3 to f^{11} would be able to change their sum of one-electron energies three times the total spreading for f^1. We need the distinction between an *even* and an *odd* number of f electrons for the following group-theoretical comments:

q = 2, 4, 6, 8, 10, 12. By the same token as L and J in spherical symmetry take the same values (non-negative integers), the symmetry types Γ_n and Γ_J are also the same in the finite point-groups. In the linear symmetries L and J are replaced by Λ and Ω (both series of non-negative integers). It can be seen from the treatise of Griffith [65] that the number of sub-levels is fairly low in the two most frequent among the seven cubic groups:

$J =$	0	1	2	3	4	5	6	7	8 . . .
O_h	1	1	2	3	4	4	6	6	7 . . .
K_h	1	1	1	2	2	3	4	4	4 . . .

(1.33)

but since the number of J levels is comparatively high, it cannot usually be argued that the sub-levels correspond to definite MO configurations. In a series of pioneer papers in 1948, Hellwege [98] introduced the *crystal quantum number* μ, being the residue [0, 1, 2, . . . , (n – 1)] by division of M_J with the highest C_n in uniaxial chromophores. It is argued by several authors [21] that μ is not satisfactory because two sub-levels with different symmetry type Γ_J sometimes may have the same μ. Nevertheless, there are many reasons why this treatment of uniaxial point-groups is a helpful approximation. Though two groups G_1 and G_2 can be in the relation that G_1 is a *super-group* to G_2 (and G_2 a *sub-group* to G_1) the point-groups do not form a well-ordered hierarchy [6], it is not always meaningful to say that G_3 or G_4 is the most "distinguished" group. For instance, the two linear point-groups $D_{\infty h}$ and $C_{\infty v}$ (the three others D_∞, $C_{\infty h}$ and C_∞ cannot characterize a one-valued function [6] of x, y, z) have an infinite number of symmetry types and free rotation with any C_n about *one* axis (this can only be done about one or *all* axes, in the latter case in spherical symmetry R_{3i}) whereas, O_h has only the order 48, ten different Γ_n (five combined with even or odd parity) *but* all three Cartesian axes equivalent and four of the symmetry types Γ_n having the *degeneracy number* 3. The linear symmetries have only two Cartesian axes equivalent, and the positive Λ have only the degeneracy number 2. By the way, the icosahedral group [61] with the centre of inversion K_h has the highest order (120) among the point-groups not exceeding C_{30} axes, it combines C_5 axes with three equivalent Cartesian axes (and is hence cubic) and has the degeneracy numbers 1, 3, 4 and 5 for its ten Γ_n, (the only higher degeneracy numbers are $(2L + 1)$ in spherical symmetry) *but* it is not a super-group of O_h. In the non-cubic point-groups one cannot have more than one C_n with $n = 3, 4, \ldots$ and those having such a principal axis are called *uniaxial,* as well as the dihedral group C_2. A low-quality imitation of cubic behaviour is found in the orthorhombic groups D_{2h} and D_2 with three orthogonal C_2 axes, between which one cannot be selected as principal axis. There is an obvious interest for chemists in the net of super-group relations, where sub-groups can be obtained by removing one or more elements of symmetry, perhaps by a weak perturbation, such as the disposal of next-nearest neighbour atoms, or by weak distortions. It is a necessary, but not sufficient condition for G_1 being a super-group to G_2 that the ratio between the orders of G_1 and G_2 is an integer, at least 2.

The general result of the fact that all uniaxial groups are sub-groups of $D_{\infty h}$ (which is not the case for the cubic groups) has specific corollaries of great interest. In the linear symmetries, a given J-level splits in $(J + 1)$ sub-levels, one state having $\Omega = 0$ [actually, it has an additional quantum number added as a right-hand superscript plus or minus, 0^+ or 0^-, according to whether the more-electron wave-function Ψ is invariant or changes signs by reflection Eq. (1.26)], in an *arbitrary* plane containing the linear axis, 0^- is obtained by the combination of even J with odd parity or of odd J with even parity and two states in each of the $\Omega = |M_J| = 1, 2, \ldots, J$. The uniaxial point-groups having the principal axis C_n $(n = 3, 4, 5, \ldots)$ conserve certain of these group-theoretical degeneracies. When n is an odd integer only the Ω values which are multiples of n have their two states separated in energy, whereas the others have two states having the same energy like the Ω sub-level in linear symmetry. When n is an even integer, the Ω values which are multiples of $(n/2)$ are the only ones to split. The reason for this distinction between odd and even n is related to the fact that the square $(A_{l\lambda})^2$ has 2λ equivalent maxima in the xy-plane, and that it is invariant for a rotation $180°/\lambda$ in this plane perpendicular on the principal axes. When n is even the operation C_n is a rotation $360°/n$ in the xy-plane with the result that λ shows a periodicity compatible with a non-vanishing perturbation when $2\lambda = n$ or a multiple of n. When n is odd, the holohedrized perturbation according to Eq.

(1.31) possesses a C_{2n} axis, when U(x, y, z) itself has C_n as principal axis. Correspondingly, the condition for non-vanishing perturbation is $2\lambda = 2n$. Hence, the C_3 axis present in D_{3h} has the result shown in Table 1 that the only λ value separated in two different one-electron energies is 3(φ). On the other hand, in quadratic chromophores MX_4 possessing a four-fold axis C_4 in D_{4h} already separate $\lambda = 2$ such as dδc and dδs (as mentioned above) or fδc and fδs, but not the two fφ.

It is of value to have correspondence tables of *descent in symmetry.* These have been given for Γ_J in a very complete form by Prather [99], though this author concentrates his attention on the electrostatic model alone. For the chemist it is important [6] to know the effects of a chromophore *almost* having a high symmetry, but being degraded to a sub-group, perhaps sofar as to C_1.

q = 1, 3, 5, 7, 9, 11, 13. For an odd number of electrons in the f shell, one has to consider *double-group* quantum numbers Γ_J differing from the symmetry types Γ_n characterizing energy levels the same way as *L* in spherical and Λ in linear symmetries. The subject of double-groups has been magistrally treated by Griffith [65], they are based on a formalism where a two-valued property changes sign by rotation 360° and remains invariant by twice as large a rotation 720°. This mechanism of providing double-groups with twice as high an order as the conventional group, successfully incorporates the relativistic effects of the spin quantum number of the individual electrons which is known to have the two alternatives $m_s = +1/2$ or $-1/2$, once a linear axis has been selected. The notation for Γ_J in double-groups has not been as unanimously and clearly established as for Γ_n and we tend to use Bethe's original proposal [58] such as Γ_6, Γ_7,and Γ_8 (strictly speaking combined with even or odd parity, Γ_{6g}, Γ_{6u}, . . .) in O_h (with the degeneracy numbers 2, 2 and 4) subsequent to Γ_n (n = 1, 2, 3, 4, 5) now called A_1, A_2, E, T_1 and T_2 (the letters A and B correspond to the degeneracy number 1, E to 2, T (unfrequently called F) to 3, U to 4 and V to 5; Griffith [65] calls Γ_J E, U or W according to degeneracy numbers 2, 4 or 6) whereas, Prather [99] constructs a nomenclature based on the lowest *J*-value for which some (or all) of the $(2J + 1)$ states have Γ_J considered.

Already in 1930, Kramers demonstrated that all energy levels of a system containing an *odd* number of electrons necessarily [65] contain an *even number of states* two and two related by the operation of *time-reversal* (we are speaking here about stationary states) in the sense that only an external perturbation not invariant for time-reversal (the factually important example is a magnetic field, C_∞ remembering the suggestion by Ampère that all magnetic fields are due to moving electric charges) is able to separate the two states forming, what is now called a *Kramers doublet.*

With exception of a few Γ_J in cubic groups (such as Γ_8 in O_h being the symmetry type of all four states of $J = 3/2$, and the Γ_J = W in icosahedral symmetry characterizing all six states of $J = 5/2$) all other Γ_J are individual Kramers doublets. This is particularly true for Ω in the linear symmetries having the possible values 1/2, 3/2, 5/2, . . . up to *J* for a given *J*-level. In the uniaxial point-groups with principal C_n axis, Ω values having the same absolute residue by division with n (in many cases, especially for even n, this quantum number [98, 99] is identical for (n – k) and (n + k) where k is an odd integer divided by 2) are able to mix in a sub-level with the same Γ_J. If this analysis is carried out for the (complex) Ψ having $M_J = +\Omega$ or $-\Omega$, non-diagonal elements of the "ligand field" can mix M_J values differing by a multiple of n. Thus, the $4f^3$ levels given in Table 3 classified according to three alternatives of μ originate in a site of the point-group D_{3h} having one C_6 axis in the holohedrized symmetry Eq. (1.31). It may be noted that each *J*-level in all non-cubic point-groups is divided

Table 3. Sub-levels of $4f^3$ of neodymium (III) in lanthanum chloride measured at 4 °K. Energy levels (in cm^{-1}) in parentheses are calculated [100] from a Hamiltonian containing 20 adjustable parameters producing a mean error of 8 cm^{-1} in comparison with the 101 identified sub-levels. The quantum number is μ

$^4I_{9/2}$	5/2: 0.00	1/2: 115.39	3/2: 123.21	5/2: 244.4	3/2: 249.4
$^4I_{11/2}$	3/2: 1,973.85	1/2: 2,012.58	5/2: 2,026.90	1/2: 2,044.19	3/2: 2,051.60
	5/2: 2,058.90				
$^4I_{13/2}$	1/2: 3,931.84	3/2: 3,974.88	1/2: 3,998.89	5/2: 4,012.92	1/2: 4,031.86
	3/2: 4,042.08	5/2: 4,083			
$^4I_{15/2}$	1/2: 5,869.3	1/2: 5,942.4	3/2: 5,992.4	5/2: (6,052)	1/2: 6,079.5
	3/2: 6,154.2	3/2: (6,194)	5/2: (6,257)		
$^4F_{3/2}$	1/2: 11,423.90	3/2: 11,453.91			
$^4F_{5/2}$	1/2: 12,458.37	3/2: 12,480.65	5/2: 12,487.67		
$^2H_{9/2}$	5/2: 12,536.14	3/2: 12,557.73	3/2: (12,640)	1/2: (12,648)	5/2: (12,702)
$^4F_{7/2}$	1/2: 13,396.05	5/2: 13,400.02	3/2: 13,474.53	5/2: 13,488.08	
$^4S_{3/2}$	1/2: 13,527.22	3/2: 13,531.23			
$^4F_{9/2}$	5/2: 14,705.73	5/2: 14,710.70	3/2: 14,715.30	1/2: 14,721.87	3/2: 14,759.25
$^2H_{11/2}$	3/2: 15,907.08	5/2: 15,923.91	1/2: 15,948.12	3/2: 15,953	1/2: 15,960.77
	5/2: (15,973)				
$^4G_{5/2}$	1/2: 17,095.12	5/2: 17,099 (?)	3/2: 17,165.17		
$^2G_{7/2}$	3/2: 17,228.75	5/2: (17,254)	5/2: (17,294)	1/2: 17,297.44	
$^4G_{7/2}$	3/2: 18,993.59	5/2: 19,012.88	1/2: 19,042.24	5/2: 19,078.02	
$^4G_{9/2}$	1/2: 19,430.93	3/2: 19,434.75	5/2: (19,437)	3/2: 19,454.60	5/2: 19,458.72
$^2K_{13/2}$	1/2: 19,546.65	1/2: 19,556.10	3/2: (19,571)	5/2: (19,573)	3/2: 19,653.67
	5/2: (19,618)	1/2: (19,673)			
$^2G_{9/2}$	1/2: 21,029.08	3/2: 21,042.33	5/2: (21,049)	3/2: (21,062)	5/2: (21,069)
$^2D_{3/2}$	1/2: 21,161.71	3/2: 21,187.72			
$^4G_{11/2}$	1/2: 21,369.56	3/2: 21,393.35	1/2: 21,407.90	3/2: 21,465.95	5/2: (21,477)
	5/2: 21,527.55				
$^2K_{15/2}$	1/2: (21,534)	1/2: 21,554.38	3/2: 21,572.64	3/2: (21,578)	5/2: (21,608)
	5/2: (21,624)	3/2: (21,643)	1/2: 21,651.16		
$^2P_{1/2}$	1/2: 23,214.93				
$^2D_{5/2}$	1/2: 23,759.73	3/2: 23,778.42	5/2: (23,804)		
$^2P_{3/2}$	1/2: 26,134.85	3/2: (26.166)			
$^4D_{3/2}$	3/2: 27,972.88	1/2: 27,975.68			
$^4D_{5/2}$	3/2: (28,118)	5/2: (28,129)	1/2: (28,183)	observed: 28,105–28,210	
$^4D_{1/2}$	1/2: 28,514.52				
$^2I_{11/2}$	1/2: 29,218.34	1/2: 29,314.53	3/2: 29,320.19	3/2: 29,327.76	5/2: (29,349)
	5/2: (29,357)				
$^2L_{15/2}$	3/2: 30,042.85	5/2: (30,104)	3/2: 30,110.82	1/2: (30,144)	1/2: 30,165
	5/2: (30,209)	3/2: 30,252	1/2: 30,289.90		
$^4D_{7/2}$	5/2: (30,181)	5/2: (30,258)	1/2: (30,268)	3/2: (30,271)	
$^2I_{13/2}$	1/2: 30,603	5/2: (30,696)	3/2: (30,712)	1/2: (30,717)	5/2: (30,732)
	1/2: (30,742)	3/2: (30,745)			
$^2L_{17/2}$	5/2: (31,504)	3/2: 31,608.9	3/2: (30,615)	1/2: 31,629	5/2: (30,622)
	1/2: (30,640)	5/2: (30,700)	3/2: (30,729)	1/2: (30,751)	
$^2H_{9/2}$	3/2: 32,683	5/2: (32,712)	3/2: 32,754,8	1/2: 32,785.5	5/2: (32,810)

in $(J + 1/2)$ sub-levels each consisting of one Kramers doublet. As an example, we give in Table 3 the sub-levels of $4f^3$ in Nd(III)Cl_9 of symmetry D_{3h} in $Nd_xLa_{1-x}Cl_3$ according to Dieke and Crosswhite [97] later extended [100] to 101 among the 182 Kramers doublets.

There is one aspect of the sub-level distribution that may be worthwhile discussing in view of our interest in stimulated emission, that is, the *half-width* of the absorption lines. In atomic spectra the two main sources of line width is Doppler effect due to thermal motion (the average velocity v is proportional to the square-root of the ratio between the absolute temperature and the atomic weight; a typical value of $(v/c) = 10^{-6}$ produces a width of a few hundredth cm^{-1}) and Heisenberg's uncertainty principle (the one-sided half-width δ of the corresponding Lorentzian curve is 1.8 cm^{-1} divided by the half-life $t_{1/2}$ of the excited state in the unit 10^{-12} sec; since $t_{1/2}$ normally is longer than 10^{-9} sec, this produces widths of at most some thousandth cm^{-1}). Both of these effects persist by extrapolation to very low gas pressures. Furthermore, at higher pressures broadening may occur due to interatomic perturbations, as a weak beginning of the general effects in condensed matter, where for instance Ammeter and Schlosnagle [101] have studied the line spectra of aluminium and gallium atoms trapped in solid noble-gas matrices at 4 °K. The question of intrinsic width in atomic spectral lines is of great importance for metrology, since a relative value of $(\lambda/\delta) = 10^6$ puts a higher limit of a million wave-lengths, $10^6\ \lambda$, of a train of coherent waves producing interference in Fabry-Perot etalons, corresponding to 60 cm for orange light. The new definition of the meter is 1,650,763.73 λ in vacuo of the emission from the next-lowest level (with $J = 1$) of $4p^5 6d$ to the lowest level (also with $J = 1$) of the configuration terminating $4p^5 5p$ of krypton 86, which is not so easy to recognize (if we return to primitive conditions of society) as the red cadmium line previously used.

It is quite obvious that even at 4 °K absorption and emission lines in crystals or vitreous solids are far broader than atomic lines and very rarely below 1 cm^{-1}. In the d group compounds most transitions [9] are broad structures with δ between 500 and 2,000 cm^{-1}, usually with shapes as Gaussian error-curves. These half-widths, large compared with kT (Boltzmann's constant k is 0.7 cm^{-1}) are readily explained by Franck and Condon's principle demanding a projection of the vibrational function of the groundstate on the potential hypersurface of the excited electronic state keeping the internuclear distances invariant. This explanation has been quantitatively verified [102, 103] for $3d^5$ manganese(II) complexes, where some excited states have almost the same equilibrium internuclear distances as the groundstate [because they have the same sub-shell configuration $(t_{2g})^3(e_g)^2$ with two anti-bonding electrons] with concomitant small δ, whereas the two first excited levels (rather unusually) have shorter equilibrium distances and other excited states longer equilibrium distances. In both cases, the δ are large and can be calculated from the variation of the Tanabe-Sugano diagram [67] assuming the sub-shell energy difference Δ to be proportional to R^{-5}. The only curious feature of the Franck-Condon broading is that the co-excitation of vibrations does not (usually) produce a distinct structure of equidistant bands as in most gaseous polyatomic molecules [8]. However, the ruby [104] and salts [105] of $Ni(H_2O)_6^{+2}$ show vibrational structure by cooling below 100 °K, as well as intra-sub-shell transitions in $5d^3(t_{2g})^3$ rhenium(IV) [106] and $5d^4$

$(t_{2g})^4$ osmium(IV) [107] hexahalide complexes. In the latter case very weak co-excited vibrations of the solvent can be perceived, as previously demonstrated [108] by deuterium substitution on the O–H and O–D stretching frequency co-excited in the intra-sub-shell [56] transition $^3A_{2g} \rightarrow {}^1E_g$ in $Ni(H_2O)_6^{+2}$. With exception of the relatively narrow intra-sub-shell transitions, the "electronic origin" between the two electronic states without additional vibrational excitation is not generally observable in the weak foot at the beginning of the d group absorption bands. At room temperature this region is obscured by "hot bands" due to excitation from vibrationally excited ground-state situations.

It is not generally recognized that the behaviour of the 4f group chromophores is entirely different, with exception of the rather unique case [109, 110] of hexahalides MX_6^{-3} having weak transitions accompanied by excitation (or at higher T, de-excitation) of the three fundamental frequences (below 250 cm^{-1}) of odd parity, but lacking the electronic origins (at least in certain crystals) except when the transitions are allowed as magnetic dipole radiation. Actually, many cases are known of crystal absorption spectra with a certain amount of co-excited vibrations, as when Spedding [111] studied $Pr_2(SO_4)_3$, $8H_2O$, but the major features are always the pure electronic transitions. A naive explanation is that the potential hypersurfaces (having $3N$-6 spatial variables for non-linear species containing N nuclei) run *exactly* parallel with constant distances but it is difficult to believe that Franck and Condon's principle is so effective in the 4f group, when co-excitation of vibrations is still so conspicuous [106, 107] in ReX_6^{-2} and OsX_6^{-2}. Anyhow, the actual electronic transitions show a finite δ far in excess of atomic spectral lines. We return in Chapter 3 to the conceivable mechanismus for a weak change of the wave-number in slightly non-equivalent sites. It is well-known that the thermal motion cannot be totally suppressed (even at 0 °K) and a Franck-Condon projection of the vibrational ground-state on a potential hypersurface showing a perceptible slope because of a slightly varying nephelauxetic effect combined with slightly varying sub-level, energy differences induced by the "ligand field" may be the most plausible explanation of δ in the 1 to 10 cm^{-1} range. However, very low librational frequences (replacing rotations in solids) then need to be involved.

Since we are interested in stimulated emission which may discriminate in favour of very sharp wave-number intervals (such as in tunable dye lasers) the *nuclear magnetic hyperfine structure* may be important. It is well known from the atomic spectra of praseodymium 141 that the hyperfine levels having the quantum number $F = J \oplus I$ obtained by coupling the $I = 5/2$ of the nucleus with J according to Eq. (1.4) have a spreading of several cm^{-1}. Hellwege, Hüfner and Pelzl [112] succeeded in detecting the hyperfine structure ($I = 7/2$) of holmium 165 in $[Ho(H_2O)_9](C_2H_5SO_4)_3$. In the lanthanides the isotopic change of the Rydberg constant (the reciprocal value of 1,822 times the atomic weight) hardly gets outside the seventh decimal, but other sources of *isotopic shifts* related to the finite volume of the nucleus may yield a few tenths of a cm^{-1}. The shape of the nucleus, such as the energy differences measured directly by *nuclear quadrupole resonance* [113] have effects of the same order of magnitude. It is clear that all of these nuclear effects (specific for each isotope) generally are masked by the standard δ in condensed matter, but they might get involved in sophisticated mechanisms of energy transfer.

C. Inter-shell Transitions to Empty 5d and 6s Orbitals

With exception of exceedingly rare magnetic dipole and pressure-induced electric quadrupole transitions the atomic line spectra consist of *electric dipole transitions,* having the following absolutely strict, and other approximately less severe *selection rules:*

1. The two energy levels E_2 and E_1 have *opposite parity.* Either the lower level E_1 is even and the upper E_2 odd, or the other way round. Such transitions are called *Laporte-allowed* and may be forbidden for other reasons, such as the strict selection rule 2, or the other rules given below. Transitions between levels of same parity are Laporte-forbidden.
2. J changes at most one unit. In absorption, J_2 must have one of three values $(J_1 - 1)$, J_1 or $(J_1 + 1)$. In emission, J_1 must have one of the three values $(J_2 - 1)$, J_2 or $(J_2 + 1)$. Furthermore, transitions between $J_1 = J_2 = 0$ are absolutely forbidden, not only for dipole, but for any kind of multipole radiation.
3. In the approximation of *well-defined configurations* the configurations of E_1 and E_2 differ in the combination of n and l for exactly one electron which can be said to perform the quantum jump, and l_2 is either $(l_1 - 1)$ or $(l_1 + 1)$. This condition is an additional constraint on Laporte's rule (no. 1) since the parity is even or odd according to whether the sum of l values of all the electrons in the configuration is even or odd. The parity of closed shells is even, which is the neutral element of the operation combining parities.
4. In the approximation of the *Russell-Saunders coupling,* S is invariant $(S_1 = S_2)$ for *spin-allowed transitions* (it is characteristic for cases of intermediate coupling that two levels with the same J mix their two differing S values to a certain extent, by non-diagonal elements of the spin-orbit coupling [2], and hence, spin-forbidden transitions become perceptible, obeying rule 2. in monatomic entities).
5. In the approximation of the Russell-Saunders coupling, L changes by one unit at the most.

These selection rules are closely related to the symmetry type of electric dipoles being the same (in all point-groups, not only the spherical R_{3i}) as of the three p orbitals, or as the three orthogonal vectors defined in a Cartesian coordinate system. The group-theoretical product $Q_1 \oplus (l = 1)$ formed in spherical symmetry according to Eq. (1.4) has to contain Q_2 (perhaps in addition to other results) or, what is the same thing expressed in a more symmetric and systematic manner, the product $Q_1 \oplus (l = 1) \oplus Q_2$ has to contain the totally symmetric Γ_1 of the point-group considered. For instance, in the linear symmetry $D_{\infty h}$ (conserving rule 1, like all other point-groups involving a centre of inversion) the p orbitals have $\Lambda = 0$ and 1, and correspondingly, rule 2 is replaced by Ω changing at the most by one unit and rule 5 by Λ changing at the most by one unit also. Transitions from even Σ to odd Δ or Φ levels are symmetry-forbidden, in spite of the necessary condition of the opposite parity being satisfied.

The *oscillator strength* P (called f by many authors) is given expressed in atomic units

$$\frac{1}{3} < \Psi_1 \, |\bar{r}| \, \Psi_2 >^2 (E_2 - E_1) \tag{1.34}$$

where electric dipole moment $\bar{r}$ in the product $\Psi_1 \Psi_2$ is measured in units of 2.54 Debye, the product of the electronic charge and a bohr unit 0.5292 Å, and the transition energy $(E_2 - E_1)$ in hartree = 2 rydberg = 219,475 cm^{-1} = 27.2116 eV. If Ψ_1 and Ψ_2 have the same parity, $\bar{r}$ vanishes. If we consider one-electron jumps in spherical symmetry according to selection rule no. 3, only products $A_l A_{l-1}$ and $A_l A_{l+1}$ are able to possess dipole moments. The numerical value of $\bar{r}$ is then determined by the specific radial functions. It is a *sum rule* in quantum mechanics [2] that a system containing q electrons has the sum of P for all its transitions equal to q. Most of the oscillator strength of most systems is concentrated in the interval 10 to 1,000 eV, in the vacuo ultra-violet and soft X-ray regions. It is very rare for a spectral line to have P above 1.5; the yellow [Ne]3s → [Ne]3p lines of the sodium atom have their P values very close to 0.33 (excited level $J = 1/2$) and 0.66 ($J = 3/2$). It is a fairly good approximation [114] to divide P in contributions $(4l + 2)$ to each filled nl-shell, and 1 to each external electron but most of the excitation [115, 116] tends to take place to continuum orbitals with positive one-electron energy.

In condensed matter we generally have *molar extinction coefficients* ϵ following Beer's law

$$\log_{10}(I/I_0) = \epsilon c l \tag{1.35}$$

that the decadic logarithm of the ratio between I, the beam intensity of monochromatic light leaving the sample after traversing a thickness l cm, and I_0 entering the sample is proportional to the concentration c in moles/litre (1 litre is now 1,000 cm^3) with a quantity ϵ for a single light-absorbing species. The fundamental equation for *spectrochemical analysis* is that ϵc of several coloured species simultaneously present in homogeneous distribution usually are superposed without mutual interference (exceptions are supposed to indicate chemical interactions)

$$\log_{10}(I/I_0) = (\epsilon_1 c_1 + \epsilon_2 c_2 + \epsilon_3 c_3 + \ldots) l \tag{1.36}$$

allowing (in principle) the concentrations of n species to be measured at (advisedly somewhat more than) n wave-lengths, if unfortunate degeneracies do not occur by one spectrum being very close to a linear combination of the other spectra. Actually, the absorption spectrum is ϵ as a function of the *wave-number* σ in cm^{-1} (many authors use the symbol ν which we reserve for the frequency in sec^{-1} (Hertz) involving the product σc with the velocity of light in vacuo). At a certain time, 1 cm^{-1} was called 1K = 1 kayser, and 1,000 cm^{-1} are called 1kK = 1 kilokayser. The great majority of commercial spectrophotometers for use in the near infra-red, visible and ultraviolet are calibrated in wave-length λ (which has a more obvious reason when optical gratings are used for the dispersion instead of prisms) in Å (10^{-10}m) or the ten times larger unit nanometer (nm, previously mμ). It is to be remembered that (dis regarding subtleties such as the Doppler effect) it is the frequency which is the invariant property of a transition, since the wave-length is shortened by the same factor n (the refractive index in isotropic materials) as the velocity of light (c/n). Hence, the values of λ or σ given in spectroscopic tables are the quantities extrapolated to vacuo with n = 1. The oscillator strength of a given absorption band is

$P = 4.32 \cdot 10^{-9} \int \epsilon \, d\sigma$

which is proportional with the product of (one-sided) half-width δ and maximum ϵ_{max} for bands of comparable shape. It is usually a good approximation to write a Gaussian error-curve

$$\epsilon = \epsilon_{max} \cdot 2^{-(\sigma-\sigma_0)^2/\delta^2} \tag{1.37}$$

in which case

$$P = 9.20 \cdot 10^{-9} \, \epsilon_{max} \delta \tag{1.38}$$

where δ is measured in cm^{-1}. Values of P have been tabulated for many internal transitions [9] in the partly filled d shell, typically between 10^{-5} and 10^{-3}. The origin of intensity in these Laporte-forbidden transitions will be discussed further in Chapter 3. Strong electron transfer bands such as the transition in the green of MnO_4^- have $P = 0.03$, and the strongest known transition in the visible of a transition-group complex is the first electron transfer band of PtI_6^{-2} with $P = 0.18$.

In view of the definitions given above, all observable atomic spectral lines are inter-shell transitions; and a good approximation according to selection rule no. 3, there is one electron jumping from one shell to another. This is one among several reasons why it was no simple task for Sugar [5] to identify the excited levels of the lowest configuration [Xe] $4f^2$ of gaseous Pr^{+3} since they are only indirectly available via Ritz's combination principle applied to transitions between configurations of opposite parity, *e.g.* to the next-lowest configuration [Xe]4f5d. The situation is far more complicated [117] in the gaseous praseodymium atom where the levels of the two lowest, odd and even, configurations [Xe]$4f^3 6s^2$ and [Xe]$4f^2 5d6s^2$ have to be extracted from the myriads of spectral lines in the visible due to excited configurations such as [Xe]$4f^3$6s6p and [Xe]$4f^2$5d6s6p, respectively.

In the absorption spectra of gaseous molecules [8] one frequently finds characteristic series of bands in the interval starting some 30,000 cm^{-1} below the first ionization energy I of the molecule (for non-transition group compounds I varies from some 70,000 cm^{-1} for conjugated hydrocarbons such as C_6H_6 to some 120,000 cm^{-1} for fluorides). Relative to I, these series can sometimes be written

$$E = I - (z + 1)^2 \text{ rydberg}/(n - d)^2 \tag{1.39}$$

and are then called *Rydberg transitions* in reminiscence of monatomic entities containing one electron outside closed shells, such as Li, Na, K, Rb, Cs or the isoelectronic Be^+, Mg^+, Ca^+, Sr^+ and Ba^+ (having the ionic charge $z = 1$) where the fractional value of the Rydberg defect d remains roughly invariant for a given value of l, but rapidly decreases from a large value for s electrons to almost zero for $l = 3$ or at least 4. The behaviour of the Rydberg defect in the lanthanides has recently been discussed [118].

In condensed matter one would not generally expect Rydberg transitions to occur because of the difficulties of orthogonalizing orbitals with large average radii inherent in high values of $(n - d)$ in Eq. (1.39) on the filled orbitals of the adjacent atoms.

Nevertheless, several categories of inter-shell transitions are known in inorganic compounds [4, 9, 104]. The most striking is perhaps ionic iodides I^- both in salts of colourless cations and in sufficiently transparent solvents (such as water, alcohols and acetonitrile) having two strong bands close to 45,000 and 52,000 cm^{-1}. The first band has $\epsilon_{max} = 13{,}500$ and $P = 0.25$ not exhausting a large part of the contribution 6 allotted to the loosest bound, filled shell $5p^6$. The detailed nature of these excited levels has been intensively discussed [119] and they are also observed [120] in organic iodides such as CH_3I and C_2H_5I. Franck and Scheibe pointed out that the energy differences between the two maxima are close to the value 7,603 cm^{-1} separating the first excited level $^2P_{1/2}$ from the groundstate $^2P_{3/2}$ of the iodine atom having a configuration terminating $5p^5$. By the way, this magnetic dipolar transition can be used in lasers [121]. In view of recent evidence from cooled films of solid alkaline metal iodides [122, 123] there is no longer any doubt that the optical excitations (following Franck and Condon's principle) to a good approximation are the transitions to the two levels with $J = 1$ (vide supra) of $5p^56s$ separated by $3/2\ \zeta_{5p}$ like $5p^5$, and followed at slightly higher wave-numbers by transitions to various levels of $5p^55d$. This is exactly the same behaviour found, not only in gaseous xenon [1] but also in xenon atoms incorporated in solidified argon or krypton, or in solid films of pure xenon [119]. When Jortner classifies these transitions as "charge transfer to the solvent" this may be perfectly true for the subsequent evolution of the excited configuration $5p^56s$ by modification of the internuclear distances, like in any other photochemical reaction [124]. The inter-shell transitions in the elements iodine, xenon and caesium preceeding the lanthanides have recently been discussed [125].

Certain post-transition group complexes have very strong inter-shell transitions in the ultra-violet with P close to 1. This is particularly true for the isoelectronic series (starting with the mercury atom) thallium(I), lead(II) and bismuth(III). In condensed matter one observes weak perturbations of the surrounding atoms [126] on the mercury resonance line at 39,412 cm^{-1} much in the same way as for xenon discussed above. The general shape of these spectra are a strong and a very strong transition separated by 9,000 to 15,000 cm^{-1}. If the energy levels of the post-transition group compounds can be described as a good approximation in spherical symmetry (the ostensive absence of a centre of inversion in many crystal structures may cast some doubt [6] on this hypothesis) and if we write $[Xe]4f^{14}5d^{10}$ as [78], the groundstate is the unique state $^1S_0$ of $[78]6s^2$. The first excited configuration [78]6s6p contains the four levels $^3P_0$, $^3P_1$, $^3P_2$ and $^1P_1$ in Russell-Saunders coupling but the intermediate coupling is so pronounced that the two levels with $J = 1$ have comparable amounts of triplet and singlet character. It is exactly the two transitions from the groundstate to these two ($J = 1$) which are allowed by selection rule no. 2, as first pointed out by Seitz [127] in the case of thallium(I) substituted in alkaline metal halide crystals. A large number of complexes in solution [128] show the transitions at lower wave-numbers than the corresponding gaseous ions Tl^+, Pb^{+2} and Bi^{+3}, as discussed in several places in this book, and roughly comparable to the mercury atom, or slightly lower. The lowest transition $^1S_0 \rightarrow {}^3P_0$ is severely forbidden and seems not to have been observed in absorption, presumably because it is indeed very weak and hidden in the rather broad band due to $^3P_1$. It has the observable consequence [129, 130] that at sufficiently low temperatures the luminescence of bismuth(III) in certain crystals shows a superposition of two exponential decay curves, the slower of which corresponding to the thermal heating of the $^3P_0$ to the closely adjacent (with or without strongly modified internuclear distances) $^3P_1$ which then emits violet light.

In the d groups inter-shell transitions are not frequent. One of the best established cases [128] is a weak band ($\epsilon = 18$) at 40,500 cm^{-1} of $Fe(H_2O)_6^{++}$ almost certainly belonging to the lowest quintet state $^5A_{1g}$ (unless the spin-forbidden transition to $^7A_{1g}$ has acquired sufficient intensity) of the configuration [Ar]$3d^54s$ representing an oxidation to the conditional oxidation state [4] iron[III]. In conjugated ligands having low-lying empty MO such as 2,2′-dipyridyl, 1,10-phenanthroline, acetylacetonate [131] and picolinate, it is well-known that reducing central atoms containing partly filled d-shells such as Ti(III), V(II), Cr(II), Fe(II), Mo(III), Ru(II), Os(II) and Pt(II) or filled d-shells such as Cu(I) show strong absorption bands in the visible or near ultra-violet. Such *inverted electron transfer* transitions have a certain similarity with inter-shell transitions and were systematically studied [132] in a large number of iridium(III) pyridine complexes prepared by Delépine. In the case of iron(II), the conjugated ligands do not need to be cyclic [133] in order to show inverted electron transfer bands in the visible. However, other species such as $Cu(NH_3)_2^+$ have their transitions to $3d^94s$ at considerably higher energy than gaseous Cu^+. There is no doubt that the strong absorption bands [9, 128] above 40,000 cm^{-1} in d^{10} cyanide complexes such as $Cu(CN)_2^-$ and $Au(CN)_2^-$ are due to inverted electron transfer in view of their absence in the isoelectronic zinc(II) and mercury(II) cyanides. Comparable transitions [134] have recently been studied in isonitrile complexes of the strongly reducing central atoms $4d^8$ rhodium(I) and $5d^8$ iridium(I). In view of the very high ionization energies of the d shell measured [16, 74] by photo-electron spectra, it does no longer seem plausible that the strong bands close to 50,000 cm^{-1} observed [135, 136] for tetrahedral MX_4^{-2} (M = Zn, Cd, Hg) and TlX_4^- are due to nd → $(n + 1)$s excitation, but rather to a mixture (indifferently compatible with MO theory) of electron transfer for X_{np} to Ms (as is most definitely [137] present in $SnCl_6^{-2}$, $SbCl_6^-$ and yellow $PbCl_6^{-2}$) with internal X intra-shell transitions $np \rightarrow (n+1)$s. By the way, the strong iodide 5p → 6s transitions seem [137] to persist to a certain extent at higher wave-numbers than the genuine electron transfer to the central atom 5d shell in OsI_6^{-2} and PtI_6^{-2}.

Already in 1931, Freed [138] pointed out that cerium(III) aqua ions incorporated in $La(H_2O)_9(C_2H_5SO_4)_3$ have four strong, relatively narrow bands ($\delta \sim 800$ cm^{-1}) at the same positions as the aqueous perchlorate solution [139] at 39,600, 41,700, 45,100 and 47,400 cm^{-1} (with ϵ = 710, 600, 380 and 270, respectively). Freed identified these transitions with 4f → 5d transitions of the single electron outside the xenon core. If the aqua ion is a definite species, the 5d shell provides five Kramers doublets (ζ_{5d} is very close to 1,000 cm^{-1} since Ce^{+3} has $^2D_{3/2}$ at 49,737 and $^2D_{5/2}$ at 52,226 cm^{-1}) and it is pertinent whether the weak band ($\epsilon = 16$) at 33,700 cm^{-1} is a fifth band of the *same* aqua ion (for some reason having very low intensity) or is a strong band of an aqua ion with different composition, say $Ce(H_2O)_8^{+3}$, occurring in very low concentration in Eq. (1.36). Actually, the weak band at 33,700 cm^{-1} has rather peculiar characteristics [140] and is sensitive to variations of temperature, to ion-pair formation, and even to the choice of D_2O as solvent. Brinen and one of us [141] found a *sixth* band of the cerium(III) aqua ion(s) at 50,000 cm^{-1} ($\epsilon = 170$). If it is the same species, the most plausible assignment is 4f → 6s which is against the approximate selection rule no. 3, but such an otherwise Laporte-allowed transition may intensify in non-spherical surroundings. It may be noted that the sum of oscillator strength P for the cerium(III) aqua ion is 0.02, an order of magnitude smaller than 6s → 6p in post-transition group species.

It is possible to study a variety of ligands bound to Ce(III) [139]. There are two experimental difficulties: one should avoid oxidation by air oxygen to cerium(IV) which is favoured at higher pH (these yellow complexes have broad electron transfer bands in the near ultra-violet) and the solutions should be perfectly limpid. The Tyndall effect of colloidal solutions and the turbidity of even a very small amount of a dispersed precipitate produce Rayleigh scattering acting as an effective background ϵ in Eq. (1.36), increasing rapidly with increasing σ (approximately the fourth power) and displacing the maxima (defined as for an analytical function) toward lower σ than the values without the background.

It is difficult to predict how large the one-electron energy differences will be in the empty 5d shell studied as excitation energies of cerium(III) complexes compared with the sub-shell energy difference Δ in the $5d^6$ iridium(III) complexes (known [4, 128] to be 23,100 cm^{-1} for $IrBr_6^{-3}$, 25,000 cm^{-1} for $IrCl_6^{-3}$ and 41,200 cm^{-1} for $Ir(NH_3)_6^{+3}$). Though one can no longer defend the dependence on $\langle r^4 \rangle / R^5$ in the old electrostatic model, there are two counter-acting effects in the larger average radius of the 5d orbitals in Ce^{+3} compared with $Ir^{+3}$and, at the same time, the longer internuclear distances M–X for M = Ce(III) than Ir(III). There are various indications that the order of magnitude of $\sigma^* (= \Delta/15)$ is roughly the same. Loh [142] studied $Ce(III)F_8$ cubal chromophores incorporated (with charge-compensating defects) in CaF_2 and the isotypic SrF_2 and BaF_2 . Here, the narrow band at lower energy (32,500, 33,600 and 34,200 cm^{-1}, respectively) can be assigned to the lower sub-shell ($d\sigma$ and $d\delta c$) of symmetry type e_g which is approximately non-bonding (by avoiding the 8 neighbour atoms) whereas, we believe that the three broad transitions scattered some 4,000 cm^{-1} close to 51,000 cm^{-1}, correspond to the sub-shell t_{2g} being split to some extent by the lower symmetry induced by adjacent defects or by the mismatch of the ionic radii and also to a smaller extent by spin-orbit coupling. In the angular overlap model Eq. (1.29) the sub-shell energy difference is 13.33 σ^* and according to our interpretation, the average of the three upper components are situated 19,000 cm^{-1} above the lower sub-shell in CaF_2, to be compared with 17,200 cm^{-1} in SrF_2 and 17,500 cm^{-1} in BaF_2. The most clear-cut case would be octahedral [109] $CeCl_6^{-3}$ and $CeBr_6^{-3}$ in acetonitrile having the transitions to the lower sub-shell (now t_{2g}) at 30,300 and 29,150 cm^{-1}, respectively. The band is somewhat asymmetric, probably due to the spin-orbit coupling expected to be $3/2\, \zeta_{5d}$ or 1,500 cm^{-1}. Unfortunately, it has not been possible to evaluate Δ from the detection of the next band [109], though a safe lower limit is 15,000 cm^{-1}. Hence, we expect σ^* values of the order of 1,000 to 2,000 cm^{-1} for the 5d shell of cerium(III), about 30 times larger than σ^* for the 4f shell given in Tables 1 and 2.

If the angular overlap model is applied [64] to the chromophore $M(III)O_9$ of symmetry D_{3h} with the positions of the oxygen nuclei known from crystal structures of salts of the ennea-aqua ion, the 5d shell falls in three sub-shells with Ξ^2 calculated in the first line for identical M–O distances, and in the second line corrected for differing distances by multiplying the contributions from the three oxygen atoms in the plane by 0.8:

$d\sigma$:	5.625	$d\pi$:	11.25	$d\delta$:	8.4375
modified:	4.875		11.25		7.3125

(1.40)

The C_3 axis has no common prime-factor with $l = 2$, and hence, the degeneracies between λc and λs in linear symmetries are not removed, in contrast to the behaviour of φc and φs for $l = 3$ in Table 1. Of course, the order of energy in Eq. (1.40) is not the same as in a linear chromophore XMX. If the angular overlap treatment is approximately valid in this case, the only conceivable identification considers the five first bands, the weak band at 33,700 cm^{-1} as $d\sigma$, 39,600 and 41,700 cm^{-1} $d\delta$; 45,100 and 47,400 cm^{-1} $d\pi$; and 50,000 cm^{-1} perhaps 6s. This is not as arbitrary as it may look at first, because the over-all spreading of spin-orbit coupling is severely restricted. At the first approximation in linear symmetry, the two ω components of a given set of two λ orbitals are separated to the extent $\lambda\zeta_{nd}$. Hence, $d\pi$ should split about 1,000 cm^{-1} and $d\delta$ about 2,000 cm^{-1}. Several smaller second-order effects would be superposed in practice. If this identification is accepted, σ^* is close to 2,500 cm^{-1} and an important corollary is that the weak band at 33,700 cm^{-1} indeed belongs to the majority of the aqua ions, and not to a minor concentration of an aqua ion with differing constitution. The low intensity may be due to some accidental interference. Thus, in gaseous Ce^{+3}, the transition [2] from $^2F_{5/2}$ to $^2D_{3/2}$ is 14 times stronger than to $^2D_{5/2}$.

Frequently, cerium(III) in solids (crystals or glasses) fluoresce from the lowest 5d Kramers doublet (with a very short life-time between 10^{-8} and 10^{-7}sec) to the groundstate $^2F_{5/2}$ or to the other 4f level $^2F_{7/2}$ situated about 2,000 cm^{-1} above the groundstate. It is possible to measure the *excitation spectrum* by plotting the relative yield of light at a given output wave-number σ_0 (usually chosen as one of the maxima in the luminescence spectrum) as a function of the incoming monochromatic σ. Since the mechanism of radiationless transitions from the higher electronic states to the luminescent state generally is fairly effective and normally does not vary in a dramatic fashion with σ, the shape of the excitation spectrum tends to be rather similar to the absorption spectrum. However, it may be showing finer details, for instance, when other absorbing species contribute in Eq. (1.36), and the excitation spectrum can be more informative in the case of powdered samples, where the $\log_{10}(I/I_0)$ of the reflection spectrum at best is a monotonic (but by no means a proportional) function of the molar extinction coefficient ϵ, as theoretically treated by Kubelka and Munk.

Blasse and Bril [143] compared the positions of emission bands and of bands in the excitation spectrum (shoulders in parentheses) of compounds, where 1 to 2 percent of the yttrium or lanthanum has been replaced by cerium(III). A few examples are:

	Emission	Excitation
YBO_3	23,800, 25,400	27,400, (29,000), 40,800, (43,500)
$LaBO_3$	26,600, 28,400	30,800, 36,900, 41,500
YPO_4	28,300, 30,000	32,800, (34,200), 39,600
$LaPO_4$	29,800, 31,500	36,200, 38,800, 41,600
$YAl_3B_4O_{12}$	27,200, 29,100	31,000, 36,600, 39,200
$Y_3Al_5O_{12}$	18,200, 27,800	22,000, 29,400, (37,000), (44,000)

(1.41)

It may be noted that the first 5d level of yttrium aluminium garnet occurs already at 22,000 cm^{-1} (and this cubic crystal is an exceptional case, where a second 5d level fluoresces). However, other oxides (such as stoichiometric Ce_2O_3 which is olive-yellow in contrast [165] to dark blue CeO_{2-x}) have low-lying 5d absorption bands, such as the perovskite [143] $La_{0.99}Ce_{0.01}AlO_3$ at 24,200 and 31,200 cm^{-1}. If Eq. (1.40) is taken seriously for the ennea-aqua ion, the non-bonding zero-point for 5d orbitals would also occur at 21,000 cm^{-1}. It is difficult to understand why this value is 9,000 cm^{-1} below the first 5d level of $CeCl_6^{-3}$ though it can be argued that the lower sub-shell is π-anti-bonding in the latter complex.

Borate glasses [144] show absorption maxima due to 4f → 5d transitions in Ce(III) at 31,950, 33,450, 39,200, 45,100 and 47,200 cm^{-1} and phosphate glasses at 33,500, 40,750, 44,000, 47,200 and 51,300 cm^{-1}. The reproducibility of these spectra suggest the presence of a dominant chromophore. The only energy difference susceptible of being due mainly to spin-orbit coupling is 1,500 cm^{-1} between the two first bands in the borate glass. The sum of the oscillator strength P is 0.016 in the borate but only 0.006 in the phosphate glass. As discussed further in Chapter 4, such transitions are important for the energy transfer [151] to other lanthanides, such as terbium(III) and thulium(III) emitting narrow bands of luminescence due to internal transitions in the configurations $4f^8$ and $4f^{12}$.

The only other lanthanide M(III) aqua ions to show 4f → 5d inter-shell transitions below the solvent cut-off at 52,000 cm^{-1} is praseodymium(III) and terbium(III). The aqueous Pr(III) perchlorate solution [141] has a band at 46,800 cm^{-1} followed by a 16 times stronger band at 52,900 cm^{-1}. In gaseous Pr^{+3}, Sugar [5] found all 20 J-levels of [Xe]4f5d distributed between 61,171 and 78,777 cm^{-1}. It is noted that both Ce(III) and Pr(III) complexes have their 4f → 5d transitions some 10,000 to 20,000 cm^{-1} below the wave-numbers in the corresponding gaseous ions. This situation which is also known when comparing 6s → 6p transitions in thallium(I) and lead(II) [128, 144] with Tl^+ and Pb^{+2} is called the nephelauxetic effect by many authors in spite of the fact that it involves one-electron energy differences to a larger extent than parameters of interelectronic repulsion. However, there is the valid analogy that both such inter-shell transitions and the usual phenomenological parameters derived from term distances serve as useful indicators [4, 6, 28, 36] for the fractional atomic charge of the central atom, usually above + 1 in disagreement with Pauling's electroneutrality principle, but always below the oxidation state showing deviations from purely electrovalent bonding. For comparison, it may be mentioned that $^2D_{3/2}$ is the groundstate of La^{+2} 7,195 cm^{-1} *below* the odd $^2F_{5/2}$ which becomes the lowest level of the isoelectronic Ce^{+3}, and that the lowest level of [Xe]4f5d occurs in Ce^{+2} 3,277 cm^{-1} above the groundstate belonging to $[Xe]4f^2$ like Pr^{+3}. Though a linear dependence of the inter-shell transitions is a definite oversimplification, the evidence available for M(III) complexes suggests an effective fractional atomic charge between 2.7 and 2.8.

It was discovered by Stewart and Kato [145] that terbium(III) aqua ions possess a strong band (ϵ = 374) at 45,900 cm^{-1} in contrast [146] to Dy(III), Ho(III) and Er(III). However, this transition does not go to the lowest term ^9D of $4f^75d$ but is rather spin-allowed, going to ^7D. This can be more clearly seen in the case [109] of $TbCl_6^{-3}$ in acetonitrile having a weak (ϵ = 28) band at 36,800 cm^{-1} due to ^9D and a strong (ϵ = 1,500) at 42,750 cm^{-1} going to ^7D. The difference close to 8,000 cm^{-1} between the two terms ^9D and ^7D belonging to the lowest configuration [Xe]$4f^7 5d$ is well-known [147] in isoelectronic Gd^{+2}. The other terms have considerably higher energy because they are based on the fractional parentage of ^6P, ^6I, . . . of $4f^7$ rather than ^8S. In the chromophores containing terbium(III), one expects the "ligand field" to separate the five 5d-like orbital energies, each time combined with a good approximation of ^8S, exclusively. However, no clear-cut separations have been reported. The terbium(III) aqua ions show a mildly undulating structure close to 38,000 cm^{-1}, probably representing transitions to the odd ^9D rather than the spin-forbidden transitions to quintet terms of $4f^8$, predicted to be very numerous [25, 44] in the same region. Recently, Spector and Sugar [233] studied Tb^{+3} and found the five levels $^9D_2$ to $^9D_6$ of $4f^7 5d$ between 51,404 and 54,882 cm^{-1} and $^7D_5$ to $^7D_1$ between 62,681 and 64,312 cm^{-1} above the $4f^8$ groundlevel $^7F_6$.

Loh [148] succeeded in measuring the $4f^q \rightarrow 4f^{q-1}5d$ transitions of all the M(III) from M = Ce to Yb (excepting Pm and Gd) in low concentration in fluorite CaF_2 which is transparent well beyond 80,000 cm^{-1}. The variation of the lowest band as a function of q is a very characteristic double zig-zag curve (with comparable wave-numbers for (q + 7) and q electrons in the 4f shell, and the lowest for q = 1 and 8) as rationalized in the refined spin-pairing energy theory discussed below in section 1D. These wave-numbers are compiled in Table 4.

In view of the behaviour of gaseous M^{+2} having much lower energies of the first level of [Xe]$4f^{q-1}5d$ than M^{+3} (12,847 cm^{-1} in Pr^{+2}, 33,856 cm^{-1} recently determined [149] in Eu^{+2}, 22,897 cm^{-1} in Tm^{+2} and 33,386 cm^{-1} in Yb^{+2}) it is not surprising that M(II) have inter-shell transitions in the visible and frequently in the near infra-red. All M(II) aqua ions are thermodynamically unstable towards the evolution of hydrogen, but Butement [150] measured the first transition of the dark red

Table 4. Distances (in cm^{-1}) from the lowest level of $4f^q$ (groundstate if the value is positive) to the lowest level of $4f^{q-1}$ 5d in gaseous M^{+2} and M^{+3} and in M (II) [152] and in M (III) [148] incorporated in calcium fluoride. Values in parentheses are calculated from the refined spin-pairing energy treatment

q =	M =	M^{+2}	M(II)CaF_2	M =	M^{+3}	M(III)CaF_2
1	La	− 7,195	negative	Ce	49,737	32,500
2	Ce	+ 3,277	− 7,100	Pr	61,171	45,600
3	Pr	+ 12,847	4,000	Nd	(70,100)	55,900
4	Nd	(+ 16,000)	7,000	Pm	(73,300)	–
5	Pm	(+ 17,100)	–	Sm	(74,400)	59,500
6	Sm	+ 26,283	15,000	Eu	(81,800)	68,500
7	Eu	+ 33,856	25,000	Gd	(91,200)	> 78,000
8	Gd	− 2,381	negative	Tb	51,404	46,500
9	Tb	+ 8,972	negative?	Dy	(66,900)	58,900
10	Dy	(+ 17,200)	10,400	Ho	(74,500)	64,100
11	Ho	(+ 18,100)	11,100	Er	(75,400)	64,200
12	Er	+ 16,976	10,900	Tm	(74,400)	64,000
13	Tm	+ 22,897	17,000	Yb	(80,200)	70,700
14	Yb	+ 33,386	27,500	Lu	90,432	> 80,000

$4f^6$ samarium(II) at 17,900 cm^{-1} and in the ultra-violet of $4f^7$ europium(II) at 31,200 cm^{-1} and of the closed-shell $4f^{14}$ ytterbium(II) at 28,400 cm^{-1}. M(II) can be substituted in CaF_2, SrF_2 and BaF_2 without charge-compensating constituents, and even the most reducing (M = Pr and Nd) still remaining as $4f^q$ systems can be maintained in CaF_2 without forming "F colour centres" which are anion vacancies occupied by one electron, as known from blue rock salt $NaCl_{1-x}$. Feofilov and Kaplyanskii have made very thorough studies of M(II) in fluorite-type lattices described in a long series of papers in the journal "Optics and Spectroscopy", but for our purposes the systematic comparison by McClure and Kiss [152] has been tabulated in Table 4. We must make an exception for M = La, Ce, Gd and Tb calculated to have $4f^{q-1}5d$ ground levels (q = 0, 1, 7 and 8) and where the exact origin of the broad, apparently Laporte-allowed, absorption bands is somewhat uncertain. Though it has been argued [153] that Ce(II) in CaF_2 shows transitions from 4f5d to $4f^2$ it must also be noted that the strongly coloured terbium-containing crystals [154] seem to contain Tb(III) connected directly with a colour centre. This behaviour is reminiscent of the fact [155, 156] that undiluted compounds (such as stoichiometric MS, MSe, MTe and MI_2) which are calculated to have $4f^{q-1}5d$ groundstates actually are metallic, containing M[III] and one conduction electron per M, whereas, the non-metallic (semi-conducting) compounds have magnetic and other physical properties appropriate for $4f^q$ and M(II). This classification even applies to the metallic elements (we except the highly complicated behaviour of some of the allotropic modifications of cerium) where europium and ytterbium are barium-like M[II] with comparatively low boiling-points and unusually large molar volumes, whereas, the others are M[III]. We discuss below in Chapter 3 the recent evidence from photo-electron spectra for the simultaneous presence of Tm[II] and Tm[III] on an instantaneous picture of TmTe.

For our purposes it is important to note the influence of the neighbour atoms on the position of the lowest level of $4f^{q-1}5d$ in M(II) just above or just below the wave-number of a potentially luminescent *J*-level of $4f^q$. For instance, the seven lowest *J*-levels belonging to the term ^7F of $4f^6$ have within a few-tenths of a percent the same positions [157] in Sm^{+2} and in a large number of crystals containing Sm(II), illustrating the negligible chemical influence on ζ_{4f}, the level $^5D_0$ close to 17,200 cm^{-1} in the isoelectronic Eu(III) occurs not far from 14,500 cm^{-1} in Sm(II) but it is *above* the lowest level of $4f^5 5d$ in CaF_2 and most compounds, and *below* in SrF_2, BaFCl, $LaCl_3$, $LaBr_3$ and $SrCl_2$ in order of decreasing wave-numbers.

The major importance of this distinction is that only in the latter cases can one hope to observe narrow-line luminescence from $^5D_0$ to $^7F_0$, $^7F_1$, $^7F_2$, . . . of the same configuration. By the same token, the absorption spectrum of Eu(II) aqua ions reported by Butement [150] (and repeated on a Cary spectrophotometer by one of us) shows small wiggles close to 31,300 cm^{-1} corresponding to narrow, weak transitions from ^8S to ^6I within $4f^7$ superposed the broad, strong transitions to $4f^6 5d$. In view of the influence of the ligands on the inter-shell transitions discussed above, it is not surprising that it is a few double fluorides [158] such as $SrAlF_5$ and $BaAlF_5$ which place the narrow transitions to ^6P at 27,800 cm^{-1}, below the first broad bands due to $4f^6 5d$. $^6P_{7/2}$ occurs in Eu^{+2} at [149] the marginally higher energy 28,200 cm^{-1}. The violet luminescence of Eu(II) in CaF_2 is a *Stokes shift* towards lower wave-numbers (in agreement with Franck and Condon's principle) of the lowest level of $4f^6 5d$ having a common "electronic origin" in emission and absorption at 24,200 cm^{-1}. In silicates such as $SrBe_2Si_2O_7$ and $BaBe_2Si_2O_7$ it is possible [159] to observe superposed broad-band and narrow-line (^6P) emission of Eu(II).

Johnson and Sandoe [160, 161] discuss the absorption spectra of Sm(II), Eu(II) and Yb(II) in various cubic crystals and in molten salts. These authors suggest a coupling scheme where a definite *J*-level of $4f^{q-1}$ constitutes the main part of the fractional parentage together with a "ligand field" selected Kramers doublet formed by the five 5d-like orbitals. Sub-shell energy differences of the order of magnitude 10,000 cm^{-1} are evaluated in this way. When metallic europium is dissolved in liquid ammonia, the absorption spectrum [162] is a superposition of the band in the near infra-red giving the usual blue colour of solvated electrons [163] and the same inter-shell transitions in the near ultra-violet as seen in a solution of EuI_2 in liquid ammonia, so the diagnosis is europium(II) phlogistonide.

The broad, intense $5f^3 \rightarrow 5f^2 6d$ transitions are quite prominent [164] in the near ultra-violet of the grey uranium(III) aqua ions and in the visible of the red chloro complexes. As reviewed elsewhere [4], these inter-shell transitions move to higher wave-numbers in Np(III), Pu(III) and Am(III) and have even been detected [170] as $5f^8 \rightarrow 5f^7 6d$ in the berkelium(III) complex $BkCl_6^{-3}$. They have also been found [166] in the M(IV) complexes UX_6^{-2}, NpX_6^{-2} and PuX_6^{-2} showing only a moderate decrease when X^- is varied from chloride over bromide to iodide, in contrast to the electron transfer bands. Seen from a theoretical point of view, the simplest system is $5f^1$ protactinium(IV) but with the exception [92] of octahedral PaX_6^{-2} the symmetry and even the stoichiometry of most Pa(IV) complexes in solution is not known. It is striking [167] that Pa(IV) aqua ions (which can be prepared in acidic solutions like Th(IV), U(IV), Np(IV) and Pu(IV) but unlike all other quadrivalent elements) has four strong $5f \rightarrow 6d$ transitions very similar to the four strongest bands of cerium(III) aqua ions. Shiloh, Givon and Marcus [168] have combined ion-exchange studies of americium(III) halide complexes with investigations of the $5f^6$ and $5f^5 6d$ excited levels.

Miles [169] argued that the linear variation (with the correct slope 8,065 cm^{-1} per volt) of the wave-numbers of $4f \rightarrow 5d$ and $5f \rightarrow 6d$ transitions with the standard oxidation potentials E^0 of the aqua ions demonstrates that we see inverted electron transfer, whereas it is now well-known [170] that the normal electron transfer bands move in the *opposite direction,* of M(IV) as a linear function of E^0 of the corresponding M(III) species, and of M(III) complexes as a linear function of E^0 of M(II). However, it can only be argued that the inter-shell transitions decrease the number of f electrons by one, but it cannot be maintained that it escapes at great distance. Actually, it remains trapped in a 5d (or 6d) orbital with rather low average radius.

Contrary to the inverted electron transfer spectra of iron(II), copper(I) and other d group complexes (or for that matter the transfer of an iodide 5p electron to the empty orbitals of a quaternized alkylpyridinium ion used by Kosower [171] as definition of a characteristic parameter for the polarity of the solvent) there are not any established cases where $4f \rightarrow 5d$ transitions merge with inverted electron transfer. This is partly due to the reluctance of lanthanides to form complexes with conjugated ligands having low electronegativity, but it must also be admitted [172] that the green colour of $Yb(C_5H_5)_3$ is due to normal electron transfer *from* the cyclopentadienide ligands. Other cases of such electron transfer bands are known in $M(C_5H_5)_3$ for M = Sm, Eu and Tm. On the other hand, the yellow colour [173] of M = Ce and the colourless M = Tb clearly show that the inter-shell transitions at most have wavenumbers 15,000 cm^{-1} below the corresponding M(III) aqua ions.

D. Electron Transfer Bands, Including the Uranyl Ion

In organic chemistry, one frequently speaks about "charge-transfer complexes". For reasons too lengthy to give here, we are not fully convinced about the theoretical considerations behind this concept. Anyhow, it is frequently the case that a new absorption band occurs in the visible or near ultra-violet of such adducts (usually between two neutral molecules) which can be ascribed to the transfer of an electron from the reducing species to empty low-lying MO of the oxidizing species. Thus, many inorganic molecules or hydrocarbons can transfer an electron to the iodine molecule I_2 and linear relations [174] can be found between the wave-number of the electron transfer band and the ionization energy of the reducing constituent, frequently with the theoretical slope 8,065 cm^{-1}/eV. Many authors also argue that the linear asymmetric I_3^- having very strong absorption bands (in aqueous solution ϵ = 26,400 at 28,350 and 40,000 at 34,800 cm^{-1}) is also such an adduct of I^- with the much less coloured I_2, though the fact that I_3^- is symmetric in salts of sufficiently large cations and (according to the Raman spectrum) in nitromethane solution also allows a classification [4] as a di-iodo complex of linear iodine(I) known from ICl_2^- and the pyridine complex $I(NC_5H_5)_2^+$. Perkampus [175] wrote a book about the adducts of aromatic hydrocarbons and compiles the new electron transfer bands found in the adducts with $TiCl_4$, $VOCl_3$ and WCl_6.

Based on the LCAO approximation in MO theory, many authors argue that one should speak about "charge transfer bands" in a chromophore MX_N rather than electron transfer bands, because the electron comes from an orbital which may be delocalized mainly on some of the N ligating atoms X, but generally has a non-vanishing amplitude on M, to an empty or partly filled shell centered on M but delocalized to a smaller or larger extent on the X atoms. Hence, the fractional atomic charge of the central atom is changed by less than one unit, *i.e.* one cannot say that exactly one electron has been transferred. We completely agree that this statement is entirely correct (actually, the adducts of reducing and oxidizing molecules are closer to electron transfer in this sense than almost all other cases) but the problem is that many other transitions *also* involve a partial transfer of charge. Both the inter-shell transitions discussed in section 1C and excitations within d^q from a roughly non-bonding sub-shell to a higher, anti-bonding sub-shell (which are the main subject for "ligand field" theory [4, 6]) tend to increase the fractional charge of M, as would also be the case for "inverted electron transfer" to conjugated ligands from partly or fully occupied d shells. Rather, we argue [4] that the genuine characteristic for electron transfer transitions is to *decrease* the oxidation state of M (characterized by spectroscopic and magnetic properties) by one unit, and for inverted electron transfer to *increase* the oxidation state of M by one unit. Thus, the first strong absorption band of $FeCl_4^-$, $Fe(NCS)(H_2O)_5^{+2}$ and $Fe(CN)_6^{-3}$ is electron transfer going from $3d^5$Fe(III) to $3d^6$ Fe(II), whereas red Fe dip_3^{+2} (2,2′-dipyridyl) and purple iron(II) croconate FeC_5O_5 shows inverted electron transfer from $3d^6$Fe(II) to $3d^5$Fe(III). In other words, exactly *one* electron changes the orbital type in the *preponderant electron configuration* [4] classifying correctly the low-lying energy levels and the groundstate of the system, independently of the unavoidable correlation effects [3]. With "orbital type" in the preponderant electron configuration, we mean MO concentrated in M *or* mainly on the X atoms. In relatively rare cases it is not possible [4] to define the oxidation state of M, and certain strong absorption bands of the system are not necessarily electron transfer bands, and in other cases the oxidation state can be debated, partly using stereochemical arguments.

It is not generally realized that Fromherz [176] until 1932 worked on electron transfer spectra of many complexes in aqueous solution. It was recognized early that partly filled d shells such as $3d^5$ iron(III), $3d^9$ copper(II) and $5d^6$ platinum(IV) readily show electron transfer bands in the visible with sufficiently reducing ligands, whereas d^{10} systems have their absorption bands at much higher wave-numbers. Unfortunately, the argument was somewhat confused by the intense 6s → 6p transitions in thallium(I) and lead(II) also studied by Fromherz. Franck and Scheibe suggested that the two *J*-levels of the excited configuration terminating $5p^5 6s$ of ionic iodides in solution correspond to electron transfer to the *solvent*, and Fromherz [176] suspected that the electron transfer bands of post-transition group halide complexes are modified versions of such transitions, the empty s orbital of the central atom taking over more of the electronic density than the solvent would do. These relations between the lowest inter-shell transition in iodide [119] and the electron transfer bands were later taken up by Katzin [177] and very subtle studies of the effect of bromide and iodide spin-orbit coupling on electron transfer energy levels have been performed [119, 137, 178, 179].

Rabinowitch [180] discussed the electron transfer spectra of the iron(III) complexes $Fe(H_2O)_5X^{+2}$ (also $Fe(H_2O)_6^{+3}$ has an electron transfer band at 42,000 cm^{-1}) and Linhard and Weigel [181] of the robust (kinetically slowly reacting) cobalt(III) complexes $Co(NH_3)_5X^{+2}$. In both cases a dramatic decrease of the wave-numbers was observed as a function of increasing reducing character of the ligand, $(F^-) < H_2O < < NH_3 < Cl^- < Br^- < I^-$. This variation agrees with chemical intuition, but not with the ionization energies I_0 of the gaseous ions X^- which are now measured from spectra in the near ultra-violet [182, 183] and turn out to have the low values in eV to be compared with the ionization energies I_1 of the gaseous atoms:

$$\begin{array}{lllllllllll} I_0: & H^- & 0.75 & F^- & 3.40 & Cl^- & 3.61 & Br^- & 3.36 & I^- & 3.06 \\ I_1: & H & 13.60 & F & 17.42 & Cl & 12.97 & Br & 11.81 & I & 10.45 \end{array} \qquad (1.42)$$

There are essentially two possible explanations of the discrepancy between the negligible variation of I_0 for the four halides and the variation close to 7 eV of the electron transfer bands from fluoride to iodide for a given central atom in a definite oxidation state. Either the spherical part of the Madelung potential of Eq. (1.22) is sufficiently large (its numerical value varies in the alkaline metal halides from 12.5 eV in LiF to

6.4 eV in CsI and is 15.3 eV in CaF_2 if the crystals are fully electrovalent) to affect the observed variation from fluoro to iodo complexes (the counterpart in solution is the stabilization $z^2/2r$ in a perfect dielectric, corresponding to 7.2 eV, divided with the effective ionic radius r in Å) or otherwise, a kind of electroneutrality principle operates, where the covalent bonding in the complex is sufficiently strong to make the halide ligands comparable to neutral halogen atoms in Eq. (1.42) as far as the ionization energy I of the loosest bound MO, mainly localized on X goes. Today, very much is known from the photo-electron spectra of gaseous molecules and solids [16, 74, 184, 185] and the fact that I is slightly below I_1 of the neutral atoms seems to have *both* reasons in a variable mixture, but producing almost the same results. Though the effect of quasi-stationary positive potentials (typically between 1 and 5 volt) maintained on insulating samples during the bombardment with X-rays in a photo-electron spectrometer has been rather difficult to evaluate [186], it is established today [184, 187] that I in solids tends to be 1 to 4 eV lower than in the corresponding gaseous molecules, due to interatomic relaxation effects [188]. Thus, the observed I of the loosest bound halide np orbitals is slightly lower than the values predicted from the Madelung potential, such as 3.4 + 12.5 = 15.9 eV for LiF and 3.1 + 6.4 = 9.5 eV for CsI. We return in Chapter 3 to the comparison of I values of the ligands and of the partly filled d or f shell in transition-group compounds.

It is striking when comparing large material of hexahalide MX_6^{+z-6} absorption spectra [9, 137] that the wave-number variation as a function of X ($F \gg Cl > Br > I$) is almost the same, independent of the central atom. One of us [3] introduced the *optical electronegativity* x_{opt} of the ligands which can be brought to have the Pauling values of electronegativity

$$F^- \; 3.9 \qquad Cl^- \; 3.0 \qquad Br^- \; 2.8 \qquad I^- \; 2.5 \tag{1.43}$$

if the definition is that the wave-number $\sigma_{e.t.}$ of the first Laporte-allowed electron transfer band is written

$$\sigma_{e.t.} = 30{,}000 \text{ cm}^{-1} \, [x_{opt}(X) - x_{uncorr}(M)] \tag{1.44}$$

where x_{uncorr} is the *uncorrected* optical electronegativity of the central atom. The meaning of this statement is that one has to subtract the large sub-shell energy difference Δ (typically 20,000 to 40,000 cm^{-1} in the 4d and 5d groups) from the observed $\sigma_{e.t.}$ if the electron transfer is to the upper sub-shell (this cannot be avoided in $4d^6$ Rh(III) and Pd(IV) and in $5d^6$ Ir(III) and Pt(IV), nor to the empty anti-bonding dδc orbital in quadratic $MX_4{}^{+z-4}$ formed by $4d^8$Pd(II) and $5d^8$Pt(II) and Au(III) halide complexes). This type of correction is not needed in the 4f group chromophores with negligible "ligand field" effects, but the other type of correction obtaining $x_{opt}(M)$ from $x_{uncorr}(M)$ introduces the change from q to (q + 1) electrons in the partly filled shell of the spin-pairing energy Eq. (1.18) proportional to the parameter D obtained as a known linear combination of phenomenological parameters of interelectronic repulsion. The correction for spin-pairing energy is considerably larger in the 4f than in the 4d and 5d groups. However, we are not normally very interested in evaluating the corrected $x_{opt}(M)$ for the 4f group (where the groundstate always has S_{max}), unless we want to stress the linear increase [119] as a function of q = 0 to 13. As we see below, there is a tremendous difference between the optical electronegativities of M(II), M(III) and M(IV) in the 4f group, whereas, the smooth variation of x_{opt} in the d group is much more moderate. Thus, osmium has 1.95 for Os(III), 2.2 for Os(IV) and 2.6 for Os(VI) known in OsF_6 whereas, iridium has x_{opt} = 2.25 for Ir(III), 2.4 for Ir(IV) and 2.9 for Ir(VI).

Once the optical electronegativities were established for a large number of central atoms in differing oxidation states, it was possible to extend Eq. (1.43) to other ligands. Water has the high value 3.5, sulphate SO_4^{-2} 3.2, ammonia NH_3 3.3, but the nitrogen-end of the ambidentate ligand thiocyanate NCS^- 2.6, azide N_3^- 2.8, acetylacetonate 2.7, the sulphur-bound ligands $(C_2H_5O)_2PS_2^-$ 2.7 and $(C_2H_5)_2NCS_2^-$ 2.6. The main reason why there is not a definite order O > N > S of the non-halide ligands seems to be the presence of one *or* more lone-pairs on the ligating atom. This is not too unreasonable in view of the fact [189] that the σ orbitals have about 12,000 cm^{-1} higher excitation energy than the π orbitals in bromo complexes and about 15,000 cm^{-1} in chloro complexes, corresponding to x_{opt} = 3.2 and 3.5 in the two cases. Ammonia has only one lone-pair behaving as a σ orbital. The same is true for the hydride ligand known from the rhenium(VII) complex ReH_9^{-2} where an uncertain value ~3.2 can be estimated [189]. This ennea-complex has the same symmetry as the 4f M(III) aqua ions. The electron transfer spectra (in particular of the d groups) and the optical electronegativities have been reviewed [189] in 1970. By the way, the *I* of gaseous halides [185] are generally close to

$$I = (1 + 3.7\, x_{opt})\ \mathrm{eV} \tag{1.45}$$

where the numerical constant 3.7 eV is the same as the 30,000 cm^{-1} in Eq. (1.44).

Ion-pairs have weaker electron transfer bands at higher wave-numbers than the directly coordinated complexes. Linhard [181] found that the ion-pair $Co(NH_3)_6^{+3}$, I^- has a band at 36,700 cm^{-1} to be compared with the first electron transfer band of $Co(NH_3)_5I^{+2}$ at 26,100 cm^{-1}. Similar comparisons [190] can be made between brown $Ru(NH_3)_6^{+3}$, I^- at 25,000 cm^{-1} and violet $Ru(NH_3)_5I^{+2}$ at 18,600 cm^{-1}. In many ways it is surprising enough that the electric dipole moment Eq. (1.34) of the transitions at larger distance remains as relatively strong because the overlap decreases exponentially. That the wave-number increases can be seen qualitatively [3] from the difference between the ionization energy of the ligand and the electron affinity of the partly filled shell being diminished with what is colloquially speaking the attraction between the hole on the ligand orbitals and the new electron on M. Actually, the shift observed is considerably smaller than the theoretical value of the correction (being roughly 14.4 eV divided by the distance in Å between the hole and the origin at the M nucleus).

Other examples of electron transfer through a fairly long distance is the deviations from the rule of additivity of ionic colours found [191] in silver(I) and thallium(I) salts such as blue $AgMnO_4$, carmine-red Ag_2CrO_4, orange Ag_2ReCl_6 (other salts of $ReCl_6^{-2}$ are pale green), black Tl_2OsBr_6 ($OsBr_6^{-2}$ is tomato-red), dark green Tl_2IrCl_6 and blue (highly ephemeric, discovered by Delépine) Ag_2IrCl_6 to be compared with orange-brown $IrCl_6^{-2}$. It is not perfectly clear whether the new absorption band at lower wave-numbers is due to direct transfer of one Ag4d or Tl6s electron to the partly filled $5d(t_{2g})$ sub-shell or rather due to a modification of the loosest bound MO (say πt_{1g}) by partial delocalization on silver or thallium. This question may perhaps be answered if it were known whether Ag_2IrF_6 is strongly coloured or not. The idealized formula [192] for Prussian blue is $K[Fe^{II}(CN)_6Fe^{III}]$ and it might be argued that the strong band at 14,000 cm^{-1} is due to the simultaneous presence of two oxidation states of the same element [193] like many cases recently studied in solution [194, 195] but in our opinion, the electron transfer takes place between the chromophore $Fe(II)C_6$ with $S = 0$ and $Fe(III)N_6$ with $S = 5/2$ which are so different that the transition is similar to the other cases of distant atoms of metallic elements. The same situation occurs in the purple [193] $K[Ru^{II}(CN)_6Fe^{III}]$ and fox-red [196] $K_2[Fe^{II}(CN)_6Cu^{II}]$ and $K_2[Fe^{II}(CN)_6UO_2]$ (previously used for toning daguerrotypes) where the ferrocyanide constituent loses an electron to the oxidizing elementbound to the nitrogen end of the ambidentate cyanide ligands.

Whereas, most common rocks are grey or black due to the simultaneous presence of Fe(II) and Fe(III) it is not perfectly certain whether electron transfer bands due to mixed oxidation states occur in the 4f group, though rather convincing indications are the blue [165] to black colours of sub-stoichiometric CeO_{2-x} and the intensely black colour of Pr_6O_{11} compared with red-brown PrO_2. On the other hand, Tb_4O_7 is only marginally deeper brown than TbO_2. Various mixed oxides containing Eu(II) and Eu(III) are conspicuously dark. Hofmann and Höschele discovered 1915 the strong and unexpected blue colour of the fluorite-type $Ce_{1-x}U_xO_2$. It was considered [4] in 1969 to have electron transfer from $5f^2U(IV)$ to $5f^0Ce(IV)$ producing excited states where each atom contains one f electron. However, recent evidence from photo-electron spectra [197] suggests that the groundstate may contain a certain amount of Ce(III) and U(V). Like the partial air oxidation of $Fe(OH)_2$ produces a very dark green colour, the precipitate of pure $U(OH)_4$ is apple-green but becomes brown by a slight oxidation. Air-oxidized $Ce(OH)_3$ turns very pale purple [165].

The detailed similarity of the internal transitions in the partly filled 4f shell with the *J*-levels predicted for spherical symmetry, and the successful identification of $4f \rightarrow 5d$ inter-shell transitions in Ce(III) and many M(II) created the impression for many physicists that lanthanide absorption and emission spectra are essentially atomic spectra. Because of this mental blocking, it was not presumed to be likely to observe electron transfer spectra. The only reason why such transitions would not be observed in the 4f group, when they are so frequent in the d groups [189], might be that the 4f shell is so protected against overlap with the ligand orbitals that the transition dipole moment Eq. (1.34) necessarily vanishes. An argument in favour of this possibility was at one time that the first electron transfer bands of the uranyl ion [132] (to be discussed in more detail below) have ϵ below 10, but it was inevitable to ascribe the orange or yellow colours of cerium(IV) compounds to electron transfer. Though the stoichiometry of the complexes is not very well established, it was beyond doubt [198] that solutions of Ce(IV) in aqueous sulphuric acid have broad electron transfer bands in the near ultra-violet with ϵ not much below 5,000. It might then be argued that Ce(IV) is not a genuine lanthanide, or at least that its empty 4f shell has a larger average radius than usual. A plausible indication is that the well-established octahedral species [109] yellow $CeCl_6^{-2}$ has electron transfer bands at 26,600 cm^{-1},

ϵ = 5,200 and P = 0.14 and at 39,200 cm^{-1}, ϵ = 13,800 and P = 0.4 and (the somewhat unstable) purple $CeBr_6^{-2}$ at 19,200 cm^{-1} with ϵ about 5,700. These are values typical for d group hexahalides [9, 189] and show that the major problem when attempting to observe 4f group electron transfer spectra, is to combine a sufficiently oxidizing M(III) with sufficiently reducing (and not too strongly coloured) ligands. We return below to the specific questions of mixed oxides containing Pr(IV) and Tb(IV).

So far we know the first reference to electron transfer transitions in 4f group M(III) is the statement [199] that the absorption edge of molten chlorides rises particularly early in the near ultra-violet in the two cases of M = Eu and Yb which, are most readily reduced to M(II). One of us [200] made a systematic study of the 0.002 molar solutions of anhydrous bromides in almost anhydrous ethanol of all M (excepting Pm) from Ce to Lu. It is possible to make MBr_3 by heating a finely powdered mixture of M_2O_3 and NH_4Br till most of the ammonium bromide has sublimed away. If too much humidity is present, insoluble MOBr is formed. As seen in Table 5, the electron transfer bands move toward higher wave-numbers

$$4f^6\,Eu(III) < 4f^{13}\,Yb(III) < 4f^5\,Sm(III) < 4f^{12}\,Tm(III) < \ldots \qquad (1.46)$$

in a way which was rationalized [200] with the refined spin-pairing-energy theory [4, 13]. The ϵ values vary between 300 for Sm and 40 for Yb, an order of magnitude lower than for typical d group bromo complexes [9]. As a by-product it was shown [200] that the 4f → 5d transitions have lower wave-numbers for M = Ce, Pr and Tb than for the aqua ions. The ethanolic solutions containing M = (Nd), Gd, Dy, Ho, Er and Lu showed no comparably strong bands before the inter-shell transitions of bromide (having the excited configuration terminating $4p^5 5s$) rising slightly before the absorption edge of C_2H_5OH. The weak point in this experimentation is that the constitution of the complex is not known, though the major part is probably solvated MBr^{+2}. The corresponding experiments with iodide were without significant results because the strongly absorbing I_3^- is formed in trace amounts.

Later Ryan [109] succeeded in preparing salts of octahedral MCl_6^{-3} and MBr_6^{-3} which can also be studied without solvolysis in a mixture of acetonitrile and succinonitrile. Table 5 gives the electron transfer bands of these well-defined species together with results [110] for MI_6^{-3} which are even more difficult to handle. Here, the visible colours are strongly influenced; EuI_6^{-3} is dark green and YbI_6^{-3} purple. The ϵ values tend to decrease from some 1,000 to slightly above 100 going from M = Sm toward M = Yb. It is noted that the shift 8,000 cm^{-1} from the hexachloro to the hexabromo complex and 10,000 cm^{-1} from the latter to the hexa-iodo complex are in marked excess of the values 6,000 and 9,000 cm^{-1} given in Eq. (1.43). The difference in Eq. (1.44) works in such a way that one may either say that the ligands in the iodo complex are unusually reducing (having anomalously low x_{opt}) or that the central atom is more oxidizing (higher x_{uncorr}) than usual. The two effects [6] which seem to collaborate, are that the M–I distances are shorter than in other complexes with higher coordination number N and that a considerable amount of repulsion between adjacent ligands decreases the ionization energy of the loosest bound MO with four angular nodes and symmetry type πt_{1g}. It is well-known from

Table 5. Electron transfer bands (cm^{-1}) of M(III) complexes in solution and measured in reflection spectra of crystals

	Sm(III)	Eu(III)	Tm(III)	Yb(III)
MI_6^{-3} in CH_3CN	24,900, 32,800	14,800, 22,200, 26,700	28,000	17,850, 22,400, 27,000
MBr_6^{-3} in CH_3CN	35,000	24,500, 32,400, 37,000	38,600	29,200, 41,700
MBr_3, anhydrous	34,500	26,000	–	28,200
MBr^{+2} in C_2H_5OH	40,200	31,200	44,500	35,500
MCl_6^{-3} in CH_3CN	43,100	33,200, 42,600	–	36,700
MCl^{+2} in C_2H_5OH	45,700	36,200	–	41,000
MSO_4^+ in H_2O	48,100	41,700	–	44,500
$M_2(SO_4)_3$, $8H_2O$	50,300	41,700	–	47,200
$M_2(SO_4)_3$, anhydrous	50,800	42,200	–	48,500
MPO_4, anhydrous	52,000	43,500	–	48,400
$M_2(CO_3)_3$, $3H_2O$	52,100	42,400	–	47,800
M(III) aqua ions	–	53,200	–	–

4d and 5d group hexahalides that inter-ligand repulsion has this effect, since measurements by Drickamer [201] at high pressure (typically up to 100,000 atm.) show that the first electron transfer band not only broadens, but also moves towards *lower* wave-numbers. Seen from a thermodynamical point of view, this is a proof that the equilibrium internuclear distances of the excited state are *shorter* than in the ground-state, in spite of all arguments [4] based on a loss of a part of the Madelung stabilization. The same conclusion is drawn from the shift towards lower wave-numbers [203] of the many narrow electron transfer bands of $IrBr_6^{-2}$ (an enormous amount of details [204, 205] are known about the various excited states) going from water to a long series of organic solvents. By the way, roughly the same order of solvents is obtained as in Kosower's series [171] of solvent polarity. A related problem is that $4fMX_6^{-3}$ have electron transfer bands at considerably lower wave-numbers than solvated MX^{+2}. Such an observation seems quite normal to the d group chemist [128, 174] because $Rh(H_2O)_5Cl^{+2}$ has its first electron transfer band at much higher energy than $RhCl_6^{-3}$. However, this is due to increasing sub-shell energy difference Δ when chloride is replaced by water (in blatant disagreement with any electrostatic "ligand field" model). If electron transfer to the lower, roughly non-bonding sub-shell is considered, as in the $4d^5$ ruthenium(III) chloro complexes $Ru(H_2O)_{6-n}Cl_n^{+3-n}$ the electron transfer band [206] remains roughly constant at 29,000 cm^{-1}. Using Eq. (1.44). $EuBr_6^{-3}$ indicates x_{uncorr} of Eu(III) to be 1.98 and $EuBr^{+2}$ in ethanol the lower value 1.76. As we said above, it would perhaps be more consistent to change x_{opt} of bromide, but the general tradition is to modify the optical electronegativity of the central atom. Since Eu(III) is the most oxidizing M(III) in the 4f group, it may be interesting to compare with the *least* oxidizing M(IV), cerium having $x_{uncorr} = 2.11$ in the hexachloro and 2.16 in the hexabromo complex. Blasse [202] has reviewed the influence of the internuclear distances and various parameters characterizing the neighbour atoms and the lattice type an positions of the electron transfer spectra of Eu(III), Pr(IV) and Tb(IV) and their relations to luminescent behaviour.

One would generally expect a continuously increasing stability of the electrons in the partly filled shell when keeping the oxidation state constant and increasing atomic number Z, for the simple reason that the increased nuclear charge adding $(-1/r)$ in atomic units to the central field cannot be completely screened by the additional electron. Nevertheless, this is not the impression one obtains [207] from the seemingly irregular variation of the standard oxidation potential E^0 of $M(H_2O)_6^{+2}$ to $M(H_2O)_6^{+3}$ in the 3d group. One contribution to this fluctuation is the "ligand field" stabilization [73, 128] evaluated from observed sub-shell energy differences Δ (though one is somewhat cautious [74] when comparing with d^{10} systems) and another the spin-pairing energy Eq. (1.18) combined with an effect comparable to Eq. (1.15) that F terms (with $L = 3$) having S_{max} are stabilized $-9B/2$ relative to the baricentre of all states having S_{max} determined by the position of the P term at $15B$ higher energy, where B is a parameter of interelectronic repulsion introduced by Racah, varying [4] in the 3d group between 400 and 1,000 cm^{-1}. Whereas, the "ligand field" stabilization is entirely negligible in the 4f group, the spin-pairing effects are more important than in the 3d group and explain the rather alarming fact that it is easier to oxidize $4f^{14}$Yb(II) to $4f^{13}$Yb(III) than $4f^7$Eu(II) to $4f^6$Eu(III). If such a thing had happened in the 3d group, it would have been easier to oxidize Zn(II) to (the actually unknown) Zn(III) than to oxidize Mn(II) to Mn(III). Combining Eqs. (1.15) and (1.18) the lowest (S, L) term of f^q is situated below the baricenter of *all* the f^q states

$$\begin{array}{lll}
f^2, f^{12} & {}^3H: & -8/13\,D - 9E^3 \\
f^3, f^{11} & {}^4I: & -24/13\,D - 21E^3 \\
f^4, f^{10} & {}^5I: & -48/13\,D - 21E^3 \\
f^5, f^9 & {}^6H: & -80/13\,D - 9E^3 \\
f^6, f^8 & {}^7F: & -120/13\,D \\
f^7 & {}^8S: & -168/13\,D
\end{array} \qquad (1.47)$$

to which can be added the first-order stabilization of the lowest J-level due to spin-orbit coupling, $-(L+1)\zeta_{nf}/2$ for q = 1 to 6, and $-L\zeta_{nf}/2$ for q = 8 to 13. Eq. (1.47) would vanish for q = 0, 1, 13 and 14. It may be noted that the coefficient to D is $-S_{max}(S_{max} - 1/2)16/13$. If we consider a process where an electron is added to $4f^q$ forming $4f^{q+1}$ (such as an electron transfer transition) it needs an energy we write as a linear function $W - q(E-A)$ to which is added the difference between Eq. (1.47) for (q + 1) and q, for instance:

$$\begin{array}{ll}
q = 5: & W - 5(E-A) - 40/13\,D + 9E^3 + \zeta_{nf} \\
q = 6: & W - 6(E-A) - 48/13\,D + 2\zeta_{nf} \\
q = 7: & W - 7(E-A) + 48/13\,D - 3/2\zeta_{nf} \\
q = 12: & W - 12(E-A) + 8/13\,D + 9E^3 + \zeta_{nf} \\
q = 13: & W - 13(E-A) + 3/2\zeta_{nf}
\end{array} \qquad (1.48)$$

In this model, the comparable wave-number of the electron transfer bands for q = 6 and 13 corresponds to 7(E–A) being closely similar to $48D/13$. If $D = 6{,}500\ cm^{-1}$,

(E–A) is then about 3,400 cm^{-1}. Such a relation assures comparable wave-numbers for (q + 7) and q electrons in the 4f shell of the groundstate. Actually, there is a weak trend as a function of the ligands to increase (E–A) from 2,900 cm^{-1} in the aqua ions to slightly higher values in more covalent complexes, for instance 3,300 cm^{-1} for the hexa-iodo complexes.

This linear treatment can also be extended [200] to cases where an electron is removed from the 4f shell, going from $4f^q$ to $4f^{q-1}$. In such a case the similarity of wave-numbers of the first 4d → 5d transition for q = 2 and 8 corresponds to

$$
\begin{aligned}
q = 2: &\quad W + 2(E\!-\!A) + 8/13\,D + 9E^3 + \zeta_{nf} \\
q = 8: &\quad W + 8(E\!-\!A) - 48/13\,D + 3/2\,\zeta_{nf}
\end{aligned}
\qquad (1.49)
$$

where (E–A) is somewhat larger, about 5,300 cm^{-1}. As pointed out by McClure and Kiss [152] studying M(II) and Loh [148] M(III) in CaF_2 this increase reflects the non-negligible repulsion between 4f and 5d electrons. Other instances where a 4f electron is removed are the standard oxidation potentials E^0 of M(II) aqua ions which have been systematized by Nugent, Baybarz, Burnett and Ryan [170], also predicting the very high E^0 for M(III) aqua ions. When E^0 of europium(II) aqua ions is −0.35 V and of ytterbium(II) −1.15 V, this is equivalent to 6,400 cm^{-1} lower $\sigma_{e.t.}$ of Eu(III) than of Yb(III) which would give (E–A) = 2,800 cm^{-1} by insertion in Eq. (1.48). It is also possible to describe the ionization energies [208] I_3 of M^{+2} and I_4 of M^{+3}. We may for instance compare the half-filled (q = 7) with filled (q = 14) shells:

$$
\begin{aligned}
&Eu^{+2}\ I_3 = 24.70\,eV \quad Yb^{+2}\ I_3 = 25.03\,eV \\
&Gd^{+3}\ I_4 = 44.01 \qquad\ \ Lu^{+3}\ I_4 = 45.19
\end{aligned}
\qquad (1.50)
$$

It would perhaps be more consistent to apply a somewhat larger D (say 1.0 eV), for M^{+3} than for M^{+2}. Then (E–A) is approximately 0.2 eV larger in the former case, corresponding to an actual increase of I_4 in Eq. (1.50). As discussed in Chapter 3, it is also possible to relate the I(4f) values measured via photo-electron spectra [13, 16] of the metallic elements and solid compounds to the refined spin-pairing energy treatment.

It is a characteristic feature for the differences Eqs. (1.48) and (1.49) to form what is colloquially called the *double zig-zag curve* as a function of q. When a 4f electron is *added* these curves increase from q = 0 to 6, suddenly drop back and increase in the same manner from q = 7 to 13. Both segments have a roughly horizontal plateau (for q = 2, 3, 4 and 9, 10, 11) due to the approximate equality of $12E^3$ and $8D/13$ in Eq. (1.47). By the same token, the curve for *removing* one 4f electron from $4f^q$ decreases from q = 1 to 7, suddenly jumps up again and decrease in the same manner from q = 8 to 14. This time the plateaux occur for q = 3, 4, 5 and 10, 11, 12. Tables 5 and 4 illustrate these two trends. In atomic spectroscopy it has become customary [13, 118] to speak about the *system difference,* the energy difference between the lowest J-level of a configuration $4f^{q-1}5d$ (with or without additional 6s electrons) and the lowest J-level of $4f^q$ (adding the same number, 0, 1 or 2, of 6s electrons). The variation of the system differences with q also follows the refined spin-pairing energy theory, even when they are negative. It was previously noted that the condition for metallicity of certain M(II) compounds [155, 156] acquiring the conditional [4] oxidation state M[III] was to have the system difference negative (with a consistent choice of (E–A) and W) and hence spontaneously losing a 4f electron.

One may discuss the neglect of differential changes of the large parameters (E–A) and D going from one configuration to another. There is not the slightest doubt that percentage-wise very small changes of expressions such as $q(q-1) A_*/2$ in Eq. (1.17) might completely wipe out the double zig-zag curves. However, the good agreement of electron transfer spectra, 4f → 5d transitions, straightforward ionization of monatomic entities, photo-electron spectra and E^0 values of the same oxidation state as a *function of* q strongly suggests that all differential changes are successfully incorporated in the linear parameters W and (E–A). This fortunate situation is already known from experience in the other transition groups [4, 7].

The composite symbol chosen for the parameter (E–A) indicates that it is the difference between an increased attraction E by the central field going from one element to the next, and increased interelectronic repulsion A in the partly filled shell. One of the major contributions to A is the change qA_* of Eq. (1.17) going from q to (q–1). However, all the differential changes of other parameters may contribute negative or positive contributions to A. Seen from an empirical point of view, it is one of the characteristic properties of partly filled shells with small average radii (with concomitant large differences between the ionization energy and the electron affinity [125], or more generally, between consecutive ionization energies I_z and I_{z+1}) to have small (E–A). Thus, the 3d group aqua ions [207] have E^0 values suggesting (E–A) = 8,000 cm^{-1}, and the 5f group [170, 209] has larger (E–A) than the 4f group. It has been argued [209] that the typical behaviour of almost invariant oxidation state occurs for shells having $n = l + 1$ (and no radial nodes) and accentuates in the direction 3d < 4f < 5g whereas, the typical behaviour [210] in the other transition groups (4d, 5d, 5f, . . .) of high and readily varying oxidation states in the beginning, but *lower* oxidation states at the end (compare silver(I) with copper(II), or mendelevium(II) and nobelium(II) with thulium(III) and ytterbium(III)) is connected with intrinsically larger (E–A).

Vander Sluis and Nugent [211] argue that (E–A) decreases from q = 1 to 14 because A necessarily increases. This suggestion is compatible with a slightly different form of the plateaux q = 10, 11, 12 and 3, 4, 5 but seems related to the somewhat artificial problem of Johnson [23] that Racah's coefficient e_1 in the second half of the shell (and S_{max}) has the value 9 (q–7) and not zero as for q below 8. It is not possible *a priori* to know what variation of the parameters of interelectronic repulsion can be incorporated in a linear parameter such as (E–A), and with just a few more degrees of liberty, we are in the situation described by Herakleitos that everything flows. It is perhaps important to note that non-linear (and specifically parabolic [4, 7] behaviour is clearly observed of *total energies,* for instance, of configuration baricentres as a function of occupation numbers of the individual shells, whereas, ionization or excitation energies going across a transition series from q = 1 to q = (4l + 2) vary in a linear manner to a remarkable precision after corrections for spin-pairing energy etc.

At the same time as the chloro and bromo complexes of the 4f group were shown to have electron transfer spectra, it was attempted to find other suitable ligands.

Besides the chemical problem that trivalent lanthanides are hard in Pearson's classification [116] and shows great affinity to ligating oxygen atoms in humidity or solvents, most reducing ligands have strong internal transitions in the ultra-violet, masking the weaker electron transfer bands. The absorption spectra of many d group complexes of sulphur-containing ligands have been measured [212], and Delépine prepared many dithiocarbamates containing bidentate $R_2NCS_2^-$ in 1908. We confirmed [200] his observation that the neodymium(III) compound $Nd(S_2CN(C_4H_9)_2)_3$ is sky-blue. This is also the colour of Nd_2O_3, whereas other compounds or glasses containing Nd(III) show various lilac, mauve or pink shades. The sky-blue colour is due to a combination of two effects discussed in Chapter 3, the nephelauxetic effect shifting the *J*-levels slightly toward lower energy, and the intensification of the hypersensitive pseudoquadrupolar transition $^4I_{9/2} \rightarrow {}^4G_{5/2}$ in the yellow. When the absorption spectrum of samarium(III) dithiocarbamate was measured after extraction into 1,2-dichloroethane (perhaps together with some $(C_4H_9)_2NH_2^+$) only an edge slightly earlier in the near ultra-violet than found for the other dithiocarbamates, was marginal evidence for electron transfer bands. Compared with this example, there is not the slightest doubt that the orange colour of europium(III) dithiocarbamate (having [200] a broad band at 22,200 cm^{-1}) and the lemon-yellow colour of ytterbium(III) dithiocarbamate are due to electron transfer bands. The former complex, as also [109] $EuBr_6^{-3}$, show the very narrow, but weak transitions to $^5D_2$ of $4f^6$ superposed on the much stronger electron transfer band. By the way, europium is a unique element, where also another oxidation state, Eu(II) aqua ions [150] show weak, sharp transitions (to^6 P levels of $4f^7$) superposed the much stronger inter-shell transition.

Barnes and Day [213, 214] studied the absorption spectra of aqueous solutions of M(III) with a variety of oxygen-containing anions. The adherence to Eq. (1.46) showed clearly that sulphate, selenate, phosphate, . . . indeed produce electron transfer bands, at least of Eu(III) and Yb(III). The optical electronegativities x_{opt} of such ligands are governed by Franck and Condon's principle, since the optical transitions are too rapid to allow the internuclear distances to change. For instance, hypophosphite $H_2PO_2^-$ (which can be considered [4] as a mixed hydrido-oxo complex of P(V) in analogy to FPO_3^{-2} and $F_2PO_2^-$) is chemically far more reducing than sulphate, but as ligand for Eu(III) it has x_{opt} slightly above SO_4^{-2}. The chemical redox reactions such as

$$2Cu_{aq}^{+2} + 4\, I^- = 2CuI + I_2 \qquad (1.51)$$

are *adiabatic* (allowing the internuclear distances to vary) like all thermodynamical studies [116] of standard oxidation potentials and complex formation constants. It can be extrapolated from Eq. (1.43) and the observation that purple $CuBr_4^{-2}$ has the first electron transfer bands at 16,200 and 18,600 cm^{-1} that the hypothetical species CuI_4^{-2} would have the first such band at 9,000 cm^{-1} (where OsI_6^{-2} actually has a narrow band [179]) but this prediction does not by itself tell that Eq. (1.51) is a very rapid reaction, favoured both kinetically and from the point of view of free energy.

Barnes and Pincott [215] continued the studies of oxygen-containing anions by measuring the reflection spectra of various carbonates, sulphates etc. It was recognized early [202, 216, 217] that the excitation spectrum of the europium(III) luminescence of $^5D_0$ to ^7F levels shows maxima in mixed oxides and in phosphates due to electron transfer bands. Borate, phosphate, silicate and germanate glasses [144, 218, 219] also show electron transfer to Eu(III) between 38,500 and 43,700 cm^{-1} to be compared with the interval 41,000 to 43,000 cm^{-1} for the first broad maximum of excitation spectra of typical mixed oxides [202, 217] with exception of 38,100 cm^{-1} for Eu(III) in the disordered fluorite $Gd_{0.5}Zr_{0.5}O_{1.75}$ and 32,300 cm^{-1} in the perovskite $LaAlO_3$.

Barnes and Day [213, 214] repeated the experiments with anhydrous bromides [200] in ethanol, using chlorides. The electron transfer spectra of such solutions presumably containing solvated MCl^{+2} are given in Table 5. These authors also found a surprisingly strong effect of the relative proportion of ethanol and water in the solvent on the positions of the electron transfer band of europium(III) thiocyanate. This solvent shift may be connected with unusually strong variations of the Eu-N distances or of the angle EuNC (known to vary a lot in crystal structures of other thiocyanates) whereas, it is less probable that the ambidentate alternative EuSCN is realized. Measurements of electric conductivities and classical physico-chemical methods of determining ion activities do not allow a ready distinction between inner-sphere (unidentate or bidentate) sulphate complexes and outer-sphere ion-pairs. It was suggested [220] that variations in the relative concentrations of the inner-sphere sulphate complexes by measuring the ultra-violet absorption spectra could be detected since the ion-pairs are expected to have vanishing transition dipole moments Eq. (1.34) and considerably higher wave-numbers. Anyhow, it is beyond doubt [139] that 0.5 to 5 molar hydrochloric acid does not modify the $4f \rightarrow 5d$ transitions of cerium(III) nor the internal $4f^3$ transitions of neodymium(III) demonstrating the unperturbed persistence of the inner sphere of the aqua ions, in spite of other physico-chemical techniques indicating chloride ion-pairs to form with the first formation constant close to 1 litre/mole. Actually, Malkova, Shutova and Yatsimirskii [221] showed that a definite inner-sphere monochloro complex $Nd(H_2O)_xCl^{+2}$ is formed between 6 and 10 molar HCl.

It is very difficult to propose a definite x_{opt} for ligated *oxide.* For instance, [222] the four tetrahedral complexes $CrO_3(OH)^-$, CrO_3F^-, CrO_3Cl^- and CrO_3Br^- are all orange and have closely similar electron transfer spectra. This dilemma has been very much discussed [223] and it would appear today that a plausible x_{opt} is 3.1, slightly below sulphate and much below water. However, the major problem is that the π-antibonding effects on the d orbitals are exceptionally strong, and hence, even protonation of yellow CrO_4^{-2} to $CrO_3(OH)^-$ produces a shift of (at least) the first electron transfer band because the central atom already seems more oxidizing with one oxo ligand less. Hence, the $x_{opt} = 3.1 - 0.9 = 2.2$ and $3.1 - 0.6 = 2.5$, one would calculate for Cr(VI) and Mn(VII) from the band positions of chromate, and permanganate have no common measure with the much higher values they would have in oxygen-free halide complexes. For instance, the first electron transfer band at 43,200 cm^{-1} of MoO_4^{--} gives $x_{opt} = 3.1 - 1.4 = 1.7$ for Mo(VI), whereas the band at 54,000 cm^{-1} of MoF_6 indicates $3.9 - 1.8 = 2.1$.

At this point, the rare earths may provide a valuable argument. It is known from many mixed oxides [224] that praseodymium(IV) and terbium(IV) are readily formed by calcination in air, whereas these oxidation states are not known in aqueous solution. They produce roughly the same colour (in agreement with Eq. (1.47) for q = 1 and 8) in a given oxide, but it depends dramatically on the kind of oxide. Both $Pr_xTh_{1-x}O_2$ and $Tb_xTh_{1-x}O_2$ are deep purple (even for $x = 10^{-3}$) with a broad band in the reflection spectrum at 19,000 cm^{-1}. However, the corresponding isotypic (fluorite) $Pr_xCe_{1-x}O_2$ and $Tb_xCe_{1-x}O_2$ are chamois, and there was a time when the pale lemon-yellow colour of CeO_2 was suspected to be due to impurities. It has played a rôle in the history of the discovery of the rare earths [13] that colourless Y_2O_3 is coloured intensely orange by traces of terbium (or praseodymium). This is again due to electron transfer from oxygen 2p to the 4f shell of M(IV), since the orange colour disappears by heating in hydrogen and reappears reversibly by heating in air. It would be difficult to ascribe these shifts to a highly varying Madelung stabilization of the oxide. Contrary to the case of the halides it is the general consensus to consider Eq. (1.44) with varying x_{opt} of the ligand. These studies have been continued by Hoefdraad [225] finding colours up to orange and yellow of Pr(IV) and Tb(IV) in $BaZrO_3$ and $BaThO_3$[N of Pr(IV) is 6] and $ZrSiO_4$ and $ZrGeO_4$(N = 8), whereas $ThGeO_4$ (also N = 8) is coloured violet.

Both PrF_4, TbF_4 and various double fluorides are colourless. It is very interesting that Asprey and Varga [226] found an intense electron transfer band close to 26,000 cm^{-1} of the only known $4f^2$Nd(IV) and $4f^8$Dy(IV) compounds, orange Cs_3NdF_7 and Cs_3DyF_7. Hence, x_{uncorr} of both Nd(IV) and Dy(IV) is 3.05, in agreement with Eq. (1.47). Whereas, the electron transfer spectra are known of the most oxidizing M(III) in Eq. (1.46) it is clear that we only know the electron transfer spectra of the *least* oxidizing M(IV) because the other M(IV) undergo spontaneous and rapid reduction by the ligands in analogy to Eq. (1.51). The x_{uncorr} derived [109] for $CeCl_6^{-2}$ is probably rather on the high side, and 2.05 seems to be a reasonable estimate for Ce(IV) in oxides. Then, one would interpolate values close to 2.5 for Pr(IV) and Tb(IV). If these values are accepted, the oxide in ThO_2 has x_{opt} above 3.1 and in Y_2O_3 above 3.3. Such optical electronegativities are not significantly influenced by π-anti-bonding of f-orbitals but certain doubts may be expressed as to whether the treatment does not have an intrinsic uncertainty which is less pronounced in the halide complexes.

The word *thermoluminescence* is generally reserved for the light emission by the heating of samples which have previously stocked energy by being exposed to X-ray or other radiation. Hence, we use the word *candoluminescence* for the emission of mixed oxides in a gas or hydrogen flame. This subject was reviewed by Ivey [227] also discussing the classical invention of Auer von Welsbach, the *mantle* consisting of the fluorite-type $Ce_{0.01}Th_{0.99}O_2$ in the form of an inorganic textile obtained by cautious ignition of a handkerchief-type material saturated with a strong aqueous solution of the nitrates of the elements to form the mixed oxide. As discussed in section 3C, we have recently started experimentation with the internal $4f^q$ transitions in candoluminescence of Nd(III), Ho(III), Er(III) and Tm(III) in mantles consisting of a variety of mixed oxides [228, 229], but there is little doubt that the broad, continuous spectrum in the visible from the greenish white light from ThO_2 containing small amounts of Ce(IV) or yellowish light of Pr(IV) is related to the electron transfer spectrum of the chromophore $M(IV)O_8$, broadened (and probably somewhat shifted) at higher temperature.

The luminescence of the *uranyl ion,* OUO^{+2} having the symmetry $D_{\infty h}$ has attracted attention since Brewster's study (1833), and was used to establish the *Stokes shift.* Like many organic molecules, both the absorption and emission spectra of UO_2^{+2} in many solids and solvents exhibit a vibrational structure essentially consisting of a progression of roughly equidistant components. Stokes discovered that one sharp feature (the electronic origin) is common for both spectra, and generalized this observation to the rule that the luminescence has lower or identical wave-numbers (at a time when Planck had not yet proposed the photon energy $h\nu$) compared with the absorption bands forming the excitation spectrum. The uranyl ion also served to disprove the absolute distinction between supposed instantaneous fluorescence and long-lived phosphorescence (though this word derives etymologically from the white modification of phosphorus, the latter element produces rather chemoluminescence by its slow oxidation) when Becquerel measured its life-time close to 10^{-4} sec. Based on the idea that when the X-rays discovered by Röntgen makes fluorescent compounds like $Ba[Pt(CN)_4]$ emit light, it might be conceivable that fluorescence is accompanied by the emission of X-rays, Becquerel discovered radioactivity in 1896. This curious incident had a tremendous impact on human history, and it may still be debated whether the discovery of uranium by Klaproth in 1789 had a greater lasting influence than another event that year.

The luminescence, and more generally the excited states, of the uranyl ion has been treated in a book [230] by Rabinowitch and Linn Belford and in a review [231] by Burrows and Kemp. It is obvious that the first excited levels are electron transfer states in the sense defined above. Several authors have discussed whether they are singlet or triplet states, and proposed various alternatives of MO configurations. In our opinion [232] this question has no meaningful answer. The uranyl ion is a rather special case because of *large spin-orbit coupling in the empty 5f shell* (the order of magnitude of ζ_{5f} is 2,000 cm^{-1}) and because the one-electron energy differences between some of the orbitals (in particular the two $5f\varphi$ and the two $5f\delta$) are so small as to be negligible compared with the coupling between the "hole" mainly delocalized on the two oxo ligands and the 5f electron almost entirely localized on the uranium atom. This behaviour is not observed [119, 189] in the other electron transfer spectra, for instance of the 4d and 5d group hexahalide complexes. Thus, the visible spectra of osmium(IV) and iridium(IV) complexes [137, 204] even with mixed halides [178] are strikingly similar in spite of the fact that the transfer of an electron from one of three sets of MO (πt_{1g}, $(\pi + \sigma)t_{1u}$ and πt_{2u}) to the sub-shell $(5d\ t_{2g})^5$ of Ir(IV) produces 6 mutually orthogonal states, whereas Os(IV) being reduced to Os(III) in the excited state by changing from $(5d\ t_{2g})^4$ to $(5d\ t_{2g})^5$ produces 36 independent states. In other words, there is no perceptible separation in energy of these 36 states compared with the three Kramers doublets of the former system. In the uranyl ion, there is little doubt that the exceptionally low band intensity is partly connected with the transitions to the *even* states being Laporte-forbidden, the loosest bound MO having lost one of its four electrons $(\pi_u)^3$ and the 5f shell having odd parity. This behaviour is different from the linear molecule OCO having the loosest bound MO of symmetry type π_g (but they are non-bonding, whereas π_u forms a bonding combination with the carbon $2p\pi$ orbitals) but it can be rationalized by strong bonding of the next-highest filled MO of symmetry type π_g with $U6d\pi$. We argue [232] that the fluorescent state of even parity has $\Omega = 4$ and is the lowest eigen-value of three diagonal elements having $\Omega = 4$ representing six among the $4 \cdot 14 = 56$ states of the manifold $(\pi_u)^3 5f^1$. One diagonal element corresponds to the transfer of a $\pi_u(\omega = 1/2)$ electron to $5f\varphi(\omega = 7/2)$ and two other diagonal elements to the transfer of a $\pi_u(\omega = 3/2)$ electron (of marginally higher energy) to the two *different* (mutually orthogonal mixtures [132] of comparable squared amplitudes of $5f\varphi$ and $5f\delta$) $\omega = 5/2$. On the condition that $5f\varphi$ and $5f\delta$ have almost the same energy, this lowest eigen-value cannot avoid being stabilized by spin-orbit coupling almost to the same extent, $-2\zeta_{5f}$, as $^2F_{5/2}$ in spherical symmetry. This level with $\Omega = 4$ has neither well-defined S nor Λ. Another candidate would have been the unique case of $\Omega = 5$ corresponding to a transfer of a $\pi_u(\omega = 3/2)$ electron to $5f\varphi(\omega = 7/2)$ possessing not only a well-defined MO configuration but also the definite values of $S = 1$ and $\Lambda = 4$. However, its contribution from spin-orbit coupling is $+3/2\zeta_{5f}$ like $^2F_{7/2}$.

Actually, the uranyl absorption spectrum having the electronic origin in aqueous solution 20,500 cm^{-1} shows another set of weak bands at 7,900 cm^{-1} higher wave-numbers. It was suggested [132] that this energy difference roughly represents the separation of the two J-levels of ^2F. This argument is independent of the question whether $5f\pi$ has much higher energy than $5f\delta$ and $5f\varphi$, whereas it is almost certain that $5f\sigma$ has several thousand cm^{-1} higher one-electron energy. This problem can be directly studied in the absorption spectrum of the neptunyl ion $5f^1 NpO_2^{+2}$ where the sharp, high band at 8,200 cm^{-1} may be due to the transition from the $\Omega = 5/2$ ground-state to the lowest sub-level of $^2F_{7/2}$ having $\Omega = 7/2$. At higher wave-numbers [132, 230] the uranyl ion develops a monotonically increasing, very strong electron transfer region at least up to 50,000 cm^{-1}. This corresponds to Laporte-allowed transitions $\pi_g \rightarrow 5f$ and perhaps also $\sigma_g \rightarrow 5f$ at higher energy.

By techniques of flash photolysis it has been possible [231] to build up a quasi-stationary concentration of the excited species $(\pi_u)^3 5f^1$ which turns out to be dark blue with a strong absorption band at 17,000 cm^{-1} which can be ascribed to the Laporte-allowed transition $\pi_g \rightarrow \pi_u$. The excited uranyl ion has the two almost contradictory properties of having [4] the conditional oxidation state U[V] by containing one 5f electron, and is *highly oxidizing* by lacking one of the π_u electrons. Actually, the photochemistry of the excited state is quite extraordinary [230, 231] when abstracting hydrogen atoms from organic molecules and other unexpected reactions at room temperature. One of us wanted to prepare uranium(IV) sulphate $U(SO_4)_2$, $4H_2O$ by exposing uranyl nitrate in a mixture of fairly strong sulphuric acid and ethanol in sun-shine for a few days during the German occupation of Denmark in 1944. The supernatant solution smelled exactly like old madeira. This is an interesting analytical problem which might be resolved with modern gas chromatographs. Also the 10^{-8}M uranyl carbonate complexes present in sea water may contribute [234] to the degradation of dissolved organic compounds, using about a-thousandth of the 30 einstein/m^2 day green and blue sunlight. Such uranyl ions in the upper water-layer are excited roughly every 10 minutes.

It is not possible to make a simple decision in the case of $x_{uncorr} = 3.1 - 0.7 = 2.4$ of the uranyl ion because the complexes (normally containing 6, 5 or 4 ligating atoms in the equatorial plane) are highly *anisotropic* [220] exhibiting $x_{uncorr} = 1.8$ toward the equatorial plane (and the neptunyl ion NpO_2^{+2} 2.1) as derived from $UO_2I_4^{-2}$ and

$NpO_2Cl_4^{-2}$ with broad and quite intense electron transfer bands. Entirely reducing ligands produce only fox-red colours of uranyl complexes such as $UO_2(S_2CNR_2)_2$, $UO_2(S_2CNR_2)_3^-$, $K_2[Fe(CN)_6UO_2]$ mentioned above, and the rather amorphous precipitate obtained from uranyl salts with HS^-. A contributing fact is the *much* longer equatorial distances [231] found in crystal structures, but one would wish to have an explanation why the electron transfer bands move so slightly towards lower wavenumbers in the series UO_2^{+2}, NpO_2^{+2}, PuO_2^{+2} and AmO_2^{+2} (though they do not occur in the visible of the corresponding MO_2^+). A related observation [209] is the band in the red of dark green Np(VII) and dark blue Pu(VII) oxo complexes.

A fundamental question [234, 235] is why UO_2^{+2} has so short U–O distances and lacks proton affinity in aqueous solution like $VO(H_2O)_4^{+2}$ when molybdenum(VI) forms *cis*-dioxo complexes (frequently oligomerizing). Only d^2 systems such as $RuO_2Cl_4^{-2}$, $ReO_2(NH_3)_4^+$ and $OsO_2(OH)_4^{-2}$ are *trans*. Much evidence is available that the 5f orbitals are hardly important for the specific stability of the uranyl ion (if 6d and 5f had comparable importance one would expect [6] spontaneous deviations from the centre of inversion). The most plausible diagnosis is strong covalent bonding of π_g with $6d\pi$ (producing the exceedingly broad electron transfer bands above 32,000 cm^{-1} with highly increased internuclear equilibrium distances) and probably even stronger bonding of σ_g with *both* 7s and $6d\sigma$ forming a cigar-shaped linear combination with cylindrical symmetry around the OUO axis. This bonding [235] would be assisted by the relativistic stabilization [6] of the 7s electrons apparent in the marginally higher ionization energies I_1 and I_2 of Ra and Ra^+ compared with Ba and Ba^+. On the other hand, the spectroscopic evidence does not indicate a strong influence of the empty uranium 7p shell. The specific stability of UO_2^{+2} may be compared with the larger stability of CO_2 than of isoelectronic species (NO_2^+ and the unknown BO_2^-) or of N_2 compared with C_2^{-2} and NO^+. Chemistry can be a matter of very delicate balance.

Note added in Press. A careful study of the polarized absorption spectra [236] of $Cs_2UO_2Cl_4$ disclosed 12 electronic origins. The lowest transition at 20,096 cm^{-1} is of the magnetic dipole type, and hence goes to $\Omega = 1$. This would still be compatible with a combination of uranium $^2F_{5/2}$ and the hole ($\omega = 3/2$). The new evidence is strongly against Russell-Saunders coupling. We continue studies of the uranyl ion in strong phosphoric acid and in phosphate glasses [237].

References

1. Moore, C. E.: Atomic Energy Levels. Vol. 1 (H to V), Vol. 2 (Cr to Nb), Vol. 3 (Mo to La and Hf to Ac), Vol. 4 (Ce to Lu). Nat. Bur. Stand. Circular No. 467. Washington, D. C.: 1949, 1952, 1958 and in press
2. Condon, E. U., Shortley, G. H.: Theory of Atomic Spectra (2 Ed.). Cambridge: University Press 1953
3. Jørgensen, C. K.: Orbitals in Atoms and Molecules. London: Academic Press 1952
4. Jørgensen, C. K.: Oxidation Numbers and Oxidation States. Berlin-Heidelberg-New York: Springer 1969
5. Sugar, J.: J. Opt. Soc. Am. *55*, 1058 (1965)
6. Jørgensen, C. K.: Modern Aspects of Ligand Field Theory. Amsterdam: North-Holland 1971
7. Jørgensen, C. K.: Angew. Chem. *85*, 1 (1973); Angew. Chem. Int. Ed. *12*, 12 (1973)
8. Herzberg, G.: Electronic Spectra and Electronic Structure of Polyatomic Molecules. Princeton: Van Nostrand 1966
9. Jørgensen, C. K.: Adv. Chem. Phys. *5*, 33 (1963)
10. Formanek, J.: Die qualitative Spektralanalyse anorganischer und organischer Körper. Berlin: Mückenberger 1905
11. Orgel, L. E.: Introduction to Transition-metal Chemistry. London: Methuen 1960 (2 Ed. 1966)
12. Ephraim, F., Mezener, M.: Helv. Chim. Acta *16*, 1257 (1933)
13. Jørgensen, C. K.: Structure and Bonding *13*, 199 (1973)
14. Hund, F.: Linienspektren und Periodisches System der Elemente. Berlin: Springer 1927
15. Karayianis, N.: J. Math. Phys. *6*, 1204 (1965)
16. Jørgensen, C. K.: Chimia (Aarau) *27*, 203 (1973)
17. Jørgensen, C. K.: J. Inorg. Nucl. Chem. *1*, 301 (1955)
18. Jørgensen, C. K.: Mat. fys. Medd. Danske Vid. Selskab *29*, no. 11 (1955)
19. Racah, G.: Phys. Rev. *76*, 1352 (1949)
20. Wybourne, B. G.: Spectroscopic Properties of Rare Earths. New York: Interscience 1965
21. Judd, B. R.: Operator Techniques in Atomic Spectroscopy. New York: McGraw-Hill 1963
22. Freeman, A. J., Watson, R. E.: Phys. Rev. *127*, 2058 (1962)
23. Johnson, D. A.: J. Chem. Soc. (A) 1525 and 1528 (1969)
24. Judd, B. R.: Proc. Roy. Soc. (London) *A228*, 120 (1955)
25. Elliott, J. P., Judd, B. R., Runciman, W. A.: Proc. Roy. Soc. (London) *A240*, 509 (1957)
26. Trees, R. E.: J. Opt. Soc. Am. *54*, 651 (1964)
27. Trees, R. E., Jørgensen, C. K.: Phys. Rev. *123*, 1278 (1961)
28. Jørgensen, C. K.: Solid State Phys. *13*, 375 (1962)
29. Messmer, P. P., Birss, F. W.: J. Phys. Chem. *73*, 2085 (1969)
30. Katriel, J.: Theoret. Chim. Acta *23*, 309 and *26*, 163 (1972)
31. Colpa, J. P., Islip, M. F. J.: Mol. Phys. *25*, 701 (1973)
32. Colpa, J. P., Brown, R. E.: Mol. Phys. *26*, 1453 (1973)
33. Exman, I., Katriel, J., Pauncz, R.: Chem. Phys. Letters *36*, 161 (1975)
34. Jørgensen, C. K.: Theoret. Chim. Acta *34*, 189 (1974)
35. Ruedenberg, K.: Rev. Mod. Phys. *34*, 326 (1962)
36. Jørgensen, C. K.: Bull. Soc. Chim. France 4745 (1968)
37. Jørgensen, C. K.: J. Inorg. Nucl. Chem. *32*, 3127 (1970)
38. Nugent, L. J.: J. Inorg. Nucl. Chem. *32*, 4385 (1970)
39. Ellis, C. B.: Phys. Rev. *49*, 875 (1936); *55*, 1114 (1939)
40. Gobrecht, H.: Ann. Physik *28*, 673 (1937); *31*, 181 and 755 (1938)
41. Bethe, H., Spedding, F. H.: Phys. Rev. *52*, 454 (1937)
42. Satten, R. A.: J. Chem. Phys. *21*, 637 (1953)
43. Jørgensen, C. K.: Acta Chem. Scand. *9*, 540 (1955)
44. Carnall, W. T., Fields, P. R., Rajnak, K.: J. Chem. Phys. *49*, 4412, 4424, 4443, 4447 and 4450 (1968)

45. Sommerdijk, J. L., Bril, A., De Jager, A. W.: J. Luminescence *8*, 341 and *9*, 288 (1974)
46. Piper, W. W., De Luca, J. A., Ham, F. S.: J. Luminescence *8*, 344 (1974)
47. Crosswhite, H. M., Schwiesow, R. L., Carnall, W. T.: J. Chem. Phys. *50*, 5032 (1969)
48. Gruber, J. B., Cochran, W. R., Conway, J. G., Nicol, A. T.: J. Chem. Phys. *45*, 1423 (1966)
49. Carnall, W. T.: J. Chem. Phys. *47*, 3081 (1967)
50. Carnall, W. T., Rajnak, K.: J. Chem. Phys. *63*, 3510 (1975)
51. Jørgensen, C. K.: Mat. fys. Medd. Danske Vid. Selskab *29*, no. 7 (1955)
52. Johnston, D. R., Satten, R. A., Schreiber, C. L., Wong, E. Y.: J. Chem. Phys. *44*, 3141 (1966)
53. Jørgensen, C. K.: Acta Chem. Sand. *17*, 251 (1963)
54. Bernstein, E. R., Keiderling, T. A.: J. Chem. Phys. *59*, 2105 (1973)
55. Hayes, R. G., Edelstein, N.: J. Am. Chem. Soc. *94*, 8688 (1972)
56. Jørgensen, C. K.: Acta Chem. Scand. *9*, 1362 (1955)
57. Rabinowitch, E., Thilo, E.: Periodisches System, Geschichte und Theorie. Stuttgart: Ferdinand Enke 1930
58. Bethe, H.: Ann. Physik *3*, 133 (1929)
59. Leech, J. W., Newman, D. J.: How to Use Groups. London: Methuen 1969
60. Cotton, F. A.: Chemical Applications of Group Theory. New York: Interscience (John Wiley) 1963
61. Judd, B. R.: Proc. Roy. Soc. (London) *A 241*, 122 (1957)
62. Zalkin, A., Forrester, J. D., Templeton, D. H.: J. Chem. Phys. *39*, 2881 (1963)
63. Jørgensen, C. K.: Chimia (Aarau) *25*, 109 (1971)
64. Jørgensen, C. K., Pappalardo, R., Schmidtke, H. H.: J. Chem. Phys. *39*, 1422 (1963)
65. Griffith, J. S.: The Theory of Transition-metal Ions. Cambridge: University Press 1961
66. Sugano, S., Tanabe, Y., Kamimura, H.: Multiplets of Transition-metal Ions in Crystals. New York: Academic Press 1970
67. Tanabe, Y., Sugano, S.: J. Phys. Soc. Japan *9*, 753 and 766 (1954)
68. Verhaegen, C.: J. Chem. Phys. *49*, 4696 (1968)
69. Wolfsberg, M., Helmholz, L.: J. Chem. Phys. *20*, 837 (1952)
70. Ballhausen, C. J., Gray, H. B.: Inorg. Chem. *1*, 111 (1962)
71. Basch, H., Viste, A., Gray, H. B.: J. Chem. Phys. *44*, 10 (1966)
72. Day, P., Jørgensen, C. K.: J. Chem. Soc. 6226 (1964)
73. Orgel, L. E.: J. Chem. Phys. *23*, 1819 (1955)
74. Jørgensen, C. K.: Chimia (Aarau) *28*, 6 (1974)
75. Kuse, D., Jørgensen, C. K.: Chem. Phys. Letters *1*, 314 (1967)
76. Kibler, M.: Chem. Phys. Letters *7*. 83 (1970) and *8*, 142 (1971)
77. Kibler, M.: Int. J. Quantum Chem. *9*, 403 (1975)
78. Ballhausen, C. J., Dahl, J. P.: Theoret. Chim. Acta *34*, 169 (1974)
79. Schäffer, C. E.: Wave Mechanics – the First Fifty Years (eds.: W. C. Price, S. S. Chissick and T. Ravensdale). London: Butterworths 1973, p. 174
80. Schäffer, C. E., Jørgensen, C. K.: Mol. Phys. *9*, 401(1965)
81. Jørgensen, C. K.: J. Physique *26*, 825 (1965)
82. Yamatera, H.: Bull. Chem. Soc. Japan *31*, 95 (1958)
83. Schäffer, C. E.: Structure and Bonding *5*, 68 (1968); *14*, 69 (1973)
84. Schäffer, C. E.: Pure Appl. Chem. *24*, 361 (1970)
85. Linares, C., Louat, A., Blanchard, M.: Structure and Bonding *33* (1977)
86. Kumar, V., Vishwamittar; Chandra, K.: Chem. Phys. Letters *42*, 561 (1976)
87. Burns, G.: Phys. Letters *25A*, 15 (1967)
88. Smith, D. W.: J. Chem. Phys. *50*, 2784 (1969)
89. Smith, D. W.: Structure and Bonding *12*, 49 (1972)
90. Jørgensen, C. K.: Theoret. Chim. Acta *24*, 241 (1972)
91. Reisfeld, M. J., Crosby, G. A.: Inorg. Chem. *4*, 65 (1965)
92. Edelstein, N., Brown, D., Whittaker, B.: Inorg. Chem. *13*, 563 (1974)
93. Harnung, S. E., Schäffer, C. E.: Structure and Bonding *12*, 201 and 257 (1972)
94. Newman, D. J.: Adv. Phys. *20*, 197 (1971)

95. Schäffer, C. E.: Theoret. Chim. Acta *34*, 237 (1974)
96. Glerup, J., Mønsted, O., Schäffer, C. E.: Inorg. Chem. *15*, 1399 (1976)
97. Dieke, G. H., Crosswhite, H.: Spectra and Energy Levels of Rare Earths Ions in Crystals. New York: Interscience 1968
98. Hellwege, K. H.: Ann. Physik *4*, 95, 127, 136, 143, 150 and 357 (1948)
99. Prather, J. L.: Nat. Bur. Stand. Monograph no. 19. Washington D. C.: 1961
100. Crosswhite, H. M., Crosswhite, H., Kaseta, F. W., Sarup, R.: J. Chem. Phys. *64*, 1981 (1976)
101. Ammeter, J., Schlosnagle, D. C.: J. Chem. Phys. *59*, 4784 (1973)
102. Orgel, L. E.: J. Chem. Phys. *23*, 1824 (1955)
103. Jørgensen, C. K.: Acta Chem. Scand. *11*, 53 (1957)
104. McClure, D. S.: Solid State Phys. *9*, 399 (1959)
105. Solomon, E. I., Ballhausen, C. J.: Mol. Phys. *29*, 279 (1975)
106. Schwochau, K., Jørgensen, C. K.: Z. Naturforsch. *20a*, 65 (1965)
107. Jørgensen, C. K.: Acta Chem. Scand. *16*, 793 (1962)
108. Piper, T. S., Koertge, N.: J. Chem. Phys. *32*, 559 (1960)
109. Ryan, J. L., Jørgensen, C. K.: J. Phys. Chem. *70*, 2845 (1966)
110. Ryan, J. L.: Inorg. Chem. *8*, 2053 (1969)
111. Spedding, F. H.: Phys. Rev. *58*, 255 (1940)
112. Hellwege, K. H., Hüfner, S., Pelzl, J.: Z. Physik 203 (1967) 227
113. Lucken, E. A. C.: Nuclear Quadrupole Coupling Constants. London: Academic Press
114. Bambynek, W., Crasemann, B., Fink, R. W., Freund, H. U., Mark, H., Swift, C. D., Price, R. E., VenugopalaRao, P.: Rev. Mod. Phys. *44*, 716 (1972)
115. Salzmann, J. J., Jørgensen, C. K.: Helv. Chim. Acta *51*, 1276 (1968)
116. Jørgensen, C. K.: Topics in Current Chemistry *56*, 1 (1975)
117. Dieke, G. H., Crosswhite, H. M.: Appl. Optics *2*, 675 (1963)
118. Jørgensen, C. K. in Gmelin: Handbuch der anorganischen Chemie „Seltene Erden", Teil B, Lieferung 1, p. 17 (1976)
119. Jørgensen, C. K.: Halogen Chemistry (ed. V. Gutmann) *1*, 265. London: Academic Press 1967
120. Haszeldine, R. N.: J. Chem. Soc. 1764 (1953)
121. Padrick, T. D., Palmer, R. E.: J. Chem. Phys. *62*, 3350 (1975)
122. Eby, J. E., Teegarden, K. J., Dutton, D. B.: Phys. Rev. *116*, 1099 (1959)
123. Teegarden, K., Baldini, G.: Phys. Rev. *155*, 896 (1967)
124. Balzani, V., Carassiti, V.: Photochemistry of Coordination Compounds. London: Academic Press 1970
125. Jørgensen, C. K.: Structure and Bonding *22*, 49 (1975)
126. Vinogradov, S. N., Gunning, H. E.: J. Phys. Chem. *68*, 1962 (1964)
127. Seitz, F.: J. Chem. Phys. *6*, 150 (1938)
128. Jørgensen, C. K.: Absorption Spectra and Chemical Bonding in Complexes. Oxford: Pergamon Press 1962
129. Reisfeld, R., Boehm, L.: J. Non-crystalline Solids *16*, 83 (1974)
130. Boulon, G., Pedrini, C., Guidoni, M., Pannel, C.: J. Physique (Paris) *36*, 267 (1975)
131. Jørgensen, C. K.: Acta Chem. Scand. *16*, 2406 (1962)
132. Jørgensen, C. K.: Acta Chem. Scand. *11*, 166 (1957)
133. Krumholz, P.: Structure and Bonding *9*, 139 (1971)
134. Isci, H., Mason, W. R.: Inorg. Chem. *14*, 905 and 913 (1974)
135. Bird, B. D., Day, P.: J. Chem. Phys. *49*, 392 (1968)
136. Matthews, R. W., Walton, R. A.: J. Chem. Soc. (A)1639 (1968)
137. Jørgensen, C. K.: Mol. Phys. *2*, 309 (1959)
138. Freed, S.: Phys. Rev. *38*, 2122 (1931)
139. Jørgensen, C. K.: Mat. fys. Medd. Danske Vid. Selskab *30*, no. 22 (1956)
140. Heidt, L. J., Berestecki, J.: J. Amer. Chem. Soc. *77*, 2049 (1955)
141. Jørgensen, C. K., Brinen, J. S.: Mol. Phys. *6*, 629 (1963)
142. Loh, E.: Phys. Rev. *154*, 270 (1967)

143. Blasse, G., Bril, A.: J. Chem. Phys. *47*, 5139 (1967)
144. Reisfeld, R., Hormodaly, J., Barnett, B.: Chem. Phys. Letters *17*, 248 (1972)
145. Stewart, D. C., Kato, D.: Analyt. Chem. *30*, 164 (1958)
146. Jørgensen, C. K.: Acta Chem. Scand. *11*, 981 (1957)
147. Callahan, W. R.: J. Opt. Soc. Am. *53*, 695 (1963)
148. Loh, E.: Phys. Rev. *147*, 332 (1966)
149. Sugar, J., Spector, N.: J. Opt. Soc. Amer. *64*, 1484 (1974)
150. Butement, F. D. S.: Trans. Faraday Soc. *44*, 617 (1948)
151. Reisfeld, R.: Structure and Bonding *30*, 65 (1976)
152. McClure, D. S., Kiss, Z.: J. Chem. Phys. *39*, 3251 (1963)
153. Alig, R. C., Kiss, Z. J., Brown, J. P., McClure, D. S.: Phys. Rev. *186*, 276 (1969)
154. Drotning, W. D., Drickamer, H. G.: J. Chem. Phys. *59*, 3482 (1973)
155. Jørgensen, C. K.: Mol. Phys. *7*, 417 (1964)
156. Hulliger, F.: Helv. Phys. Acta *41*, 945 (1968)
157. Dupont, A.: J. Opt. Soc. Am. *57*, 867 (1967)
158. Hoffmann, M. V.: J. Electrochem. Soc. *118*, 933 (1971); *119*, 905 (1972)
159. Verstegen, J. M. P. J., Sommerdijk, J. L.: J. Luminescence *9*, 297 (1974)
160. Johnson, K. E., Sandoe, J. N.: Mol. Phys. *14*, 595 (1968)
161. Johnson, K. E., Sandoe, J. N.: J. Chem. Soc. (A) 1694 (1969)
162. Catterall, R., Symons M. C. R.: J. Chem. Soc. 3763 (1965)
163. Jortner, J., Kestner, N. R. (editors): Electrons in Fluids. Berlin: Springer-Verlag 1973
164. Jørgensen, C. K.: Acta Chem. Scand. *10*, 1503 (1956)
165. Allen, G. C., Wood, M. B., Dyke, J. M.: J. Inorg. Nucl. Chem. *35*, 2311 (1973)
166. Ryan, J. L., Jørgensen, C. K.: Mol. Phys. *7*, 17 (1963)
167. Fried, S., Hindman, J. C.: J. Am. Chem. Soc. *76*, 4863 (1954)
168. Shiloh, M., Givon, M., Marcus, Y.: J. Inorg. Nucl. Chem. *31*, 1807 (1969)
169. Miles, J. H.; J. Inorg. Nucl. Chem. *27*, 1595 (1965)
170. Nugent, L. J., Baybarz, R. D., Burnett, J. L., Ryan, J. L.: J. Phys. Chem. *77*, 1528 (1973)
171. Kosower, E. M., Skorcz, J. A., Schwarz, W. M., Patton, J. W.: J. Am. Chem. Soc. *82*, 2188 (1960)
172. Pappalardo, R., Jørgensen, C. K.: J. Chem. Phys. *46*, 632 (1967)
173. Birmingham, J. M., Wilkinson, G.: J. Am. Chem. Soc. *78*, 42 (1956)
174. Orgel, L. E.: Quart. Rev. (London) *8*, 422 (1954)
175. Perkampus, H. H.: Wechselwirkung von π-Elektronensystemen mit Metallhalogeniden. Berlin-Heidelberg-New York: Springer 1973
176. Fromherz, H., Menschick, W.: Z. physik. Chem. *B3*, 1 (1929)
177. Katzin, L. E.: J. Chem. Phys. *20*, 1165 (1952)
178. Jørgensen, C. K., Preetz, W.: Z. Naturforsch. *22a*, 945 (1967)
179. Jørgensen, C. K., Preetz, W., Homborg, H.: Inorg. Chim. Acta *5*, 223 (1971)
180. Rabinowitch, E.: Rev. Mod. Phys. *14*, 112 (1942)
181. Linhard, M., Weigel, M.: Z. anorg. Chem. *266*, 49 (1951)
182. Berry, R. S.: Chem. Rev. *69*, 533 (1969)
183. Schmidt-Böcking, H., Bethge, K.: J. Chem. Phys. *58*, 3244 (1973)
184. Jørgensen, C. K.: Chimia (Aarau) *29*, 53 (1975)
185. Jørgensen, C. K.: Structure and Bonding *24*, 1 (1975) and *30*, 141 (1976)
186. Jørgensen, C. K., Berthou, H.: Chem. Phys. Letters *31*, 416 (1975)
187. Wagner, C. D.: Discuss. Faraday Soc. *60*, 291 (1976)
188. Jørgensen, C. K.: Adv. Quantum Chem. *8*, 137 (1974)
189. Jørgensen, C. K.: Progress Inorg. Chem. *12*, 101 (1970)
190. Waysbort, D., Evenor, M., Navon, G.: Inorg. Chem. *14*, 514 (1975)
191. Jørgensen, C. K.: Acta Chem. Scand. *17*, 1034 (1963)
192. Ludi, A., Güdel, H. U.: Structure and Bonding *14*, 1 (1973)
193. Robin, M. B., Day, P.: Adv. Inorg. Chem. Radiochem. *10*, 248 (1967)
194. Emschwiller, G., Jørgensen, C. K.: Chem. Phys. Letters *5*, 561 (1970)

195. Glauser, R., Hauser, U., Herren, F., Ludi, A., Roder, P., Schmidt, E., Siegenthaler, H., Wenk, F.: J. Am. Chem. Soc. *95*, 8457 (1973)
196. Braterman, P. S.: J. Chem. Soc. (A) 1471 (1966)
197. Keller, C., Jørgensen, C. K.: Chem. Phys. Letters *32*, 397 (1975)
198. Medalia, A. I., Byrne, B. J.: Analyt. Chem. *23*, 453 (1951)
199. Banks, C. V., Heusinkveld, M. R., O'Laughlin, J. W.: Analyt. Chem. *33*, 1235 (1961)
200. Jørgensen, C. K.: Mol. Phys. *5*, 271 (1962)
201. Drickamer, H. G.: Solid State Phys. *17*, 1 (1965)
202. Blasse, G.: Structure and Bonding *26*, 43 (1976)
203. Jørgensen, C. K.: J. Inorg. Nucl. Chem. *24*, 1587 (1962)
204. Bird, B. D., Day, P., Grant, E. A.: J. Chem. Soc. (A) 100 (1970)
205. Dickinson, J. R., Piepho, S. B., Spencer, J. A., Schatz, P. N.: J. Chem. Phys. *56*, 2668 (1972)
206. Connick, R. E., Fine, D. A.: J. Am. Chem. Soc. *82*, 4187 (1960) and *83*, 3414 (1961)
207. Jørgensen, C. K.: Acta Chem. Scand. *10*, 1505 (1956)
208. Sugar, J., Reader, J.: J. Chem. Phys. *59*, 2083 (1973)
209. Jørgensen, C. K.: Chem. Phys. Letters *2*, 549 (1968)
210. Jørgensen, C. K.: Chimia (Aarau) *23*, 292 (1969)
211. Vander Sluis, K. L,, Nugent, L. J.: Phys. Rev. *A6*, 86 (1972)
212. Jørgensen, C. K.: J. Inorg. Nucl. Chem. *24*, 1571 (1962)
213. Barnes, J. C.: J. Chem. Soc. 3880 (1964)
214. Barnes, J. C., Day, P.: J. Chem. Soc. 3886 (1964)
215. Barnes, J. C., Pincott, H.: J. Chem. Soc. (A) 842 (1966)
216. Blasse, G., Bril, A., De Poorter, J. A.: J. Chem. Phys. *53*, 4450 (1970)
217. Hoefdraad, H. E.: J. Solid State Chem. *15*, 175 (1975)
218. Reisfeld, R., Boehm, L., Ish-Shalom, M., Fischer, R.: Phys. Chem. Glasses *15*, 76 (1974)
219. Reisfeld, R., Lieblich-Sofer, N., Boehm, L., Barnett, B.: J. Luminescence *12*, 749 (1976)
220. Jørgensen, C. K.: Proceed. Symposium Coordination Chem. Tihany 1964, p. 11. Budapest: Akadémia Kiadó 1965
221. Malkova, T. V., Shutova, G. A., Yatsimirskii, K. B.: Russ. J. Inorg. Chem. *9*, 993 (1964)
222. Carrington, A., Jørgensen, C. K.: Mol. Phys. *4*, 395 (1961)
223. Müller, A., Diemann, E., Jørgensen, C. K.: Structure and Bonding *14*, 23 (1973)
224. Jørgensen, C. K., Rittershaus, E.: Mat. fys. Medd. Danske Vid. Selskab *35*, no. 15 (1967)
225. Hoefdraad, H. E.: J. Inorg. Nucl. Chem. *37*, 1917 (1975)
226. Varga, L. P., Asprey, L. B.: J. Chem. Phys. *48*, 139 and *49*, 4674 (1968)
227. Ivey, H. F.: J. Luminescence *8*, 271 (1974)
228. Jørgensen, C. K.: Chem. Phys. Letters *34*, 14 (1975)
229. Jørgensen, C. K.: Structure and Bonding *25*, 1 (1976)
230. Rabinowitch, E., Belford, R. L.: Spectroscopy and Photochemistry of Uranyl Compounds. Oxford: Pergamon Press 1964
231. Burrows, H. D., Kemp, T. J.: Chem. Soc. Rev. (London) *3*, 139 (1974)
232. Jørgensen, C. K., Reisfeld, R.: Chem. Phys. Letters *35*, 441 (1975)
233. Spector, N., Sugar, J.: J. Opt. Soc. Am. *66*, 436 (1976)
234. Jørgensen, C. K.: Revue chim. min. (Paris) *14*, 127 (1977)
235. Jørgensen, C. K., Penneman, R. A.: Heavy Element Properties (eds.: W. Müller and H. Blank), p. 117. Amsterdam: North-Holland Publishing Co. 1976
236. Denning, R. G., Snellgrove, T. R., Woodwark, D. R.: Mol. Phys. *32*, 419 (1976)
237. Lieblich-Sofer, N., Reisfeld, R., Jørgensen, C. K.: Inorg. Chim. Acta, submitted

2. Rare-Earth Lasers

(References to this chapter are to be found on p. 119, because so many references concern rather inaccessible literature the titles of the publications are also given). In this Chapter we have adopted the nomenclature almost universally prevailing among physicists of Nd^{3+}, Eu^{2+}, . . . rather than Nd(III), Eu(II), . . . used in the other chapters.

A. Introduction

The laser is a source of coherent radiation in the optical region of the spectrum. The common characteristic of all laser types is the population inversion in the active element of the laser. The conditions for laser action were first described by Schawlow and Townes [1] in 1958. The first demonstration of laser action was achieved in ruby by Maiman [2]. In the last 15 years a systematic search was conducted for new laser systems using trivalent and divalent rare earths in a great variety of crystals [3–5], glasses [6–20] and liquids [21–23].

Since these data are available we shall concentrate mostly on the recent development in the rare earth laser art which had a tremendous impact in the first half of the seventies. At the present time of energy crisis it is realized that the most noteworthy property of the laser is its ability to produce an energy concentration in space and time even greater than that found in the heat of nuclear explosion. A huge laser program is now being carried out at the Lawrence Livermore Laboratories, also in Los Alamos Scientific Laboratory, Sandia Laboratories, KMS Fusion Inc. and at Rochester [24] under sponsorship of the United States Energy Research and Development Administration (ERDA).

All these projects move to a final goal to initiate the thermonuclear reaction by irradiating a deuterium-tritium pellet with high energy laser of a few picoseconds duration. The vapor evolved from the pellet surface creates a back pressure that compresses and heats the pellet's center. During the thermonuclear burning the pellet is held together by the inertia of its mass.

The first direct experimental proof that the neutrons produced in laser-driven implosions are of thermonuclear origin, was demonstrated at the end of 1975 at the Lawrence Livermore Laboratory [25]. The other activities of this laboratory are connected with

1) laser isotope separation to demonstrate the scientific possibility of using a laser induced process to alter the isotopic ratios of chemical elements especially uranium,
2) laser research and development to identify fruitful areas of new laser research and provide the necessary scientific data base for evaluating new lasers.

Also recently KMS Fusion Inc. reported a development of "plasma spatial filter" which eliminates hot spots in high-power glass lasers providing an improvement in the quality of the laser beam.

Neodymium doped glass lasers are the best understood high-energy short pulse lasers now available. Because of this understanding we know how to optimize the spectroscopic properties of Nd^{3+} in glass so as to increase the energy storage and energy extraction ability.

In the future laser fusion program, both the Lawrence Livermore Laboratory and KMS Fusion plan to expand the neodymium glass laser to peak power of several terawatts.

The basic parameters of the solid state laser operation are:

1. *The threshold of laser action* is defined as the minimum input power or energy needed to start the laser action.
2. The output power of the laser, P_{out}, for a given peak power pumping: for a pulsed laser, P_{out} is given in joules per pulse and for a continuous operation, CW, given in watts.
3. The spectral distribution of the emitted radiation is defined by a central wavelength λ_e or frequency ν_e and line width $d\lambda$ (or $d\nu$) of the emission.
4. The spatial distribution of energy in the laser beam both in position and direction. The latter is usually specified by a mean angular divergence.

The R.E. laser consists of a resonant cavity, containing the amplifying medium, the excited rare earth ions incorporated in crystals, glasses or liquids. The laser medium (tube or rod) is placed between two parallel mirrors, having reflection coefficients R_1 and R_2. The mirrors may be placed separately or evaporated directly on the rod's ends.

In the standard laser set-up one of the reflecting mirrors is almost totally reflecting, while the other is partially transmitting; therefore its reflection coefficient R (fraction of light intensity reflected) is less than one. The fraction of light remaining after a full round-trip passage through the laser is denoted by $e^{-2\gamma} = R_1R_2$ (assuming that all losses other than reflection may be neglected); γ is positive and is a measure of loss in a single passage, therefore $\gamma = -1/2 \ln R_1R_2$.

Oscillations may be sustained in the laser if the amplification of the radiation through the active material is sufficient to compensate for the fraction of energy lost due to all causes. In other words, in order for a fluorescent material to exhibit laser operation, the round-trip optical gain, resulting from the optical pumping, must exceed round-trip losses within the cavity. In each passage through the laser the intensity of the radiation is increased by a factor $e^{\beta L}$ by virtue of the amplification in the material, where L is the length between the mirrors and β the amplification coefficient, expressed as

$$\beta(\nu) = k(\nu_0)\,\Delta N \qquad (2.1)$$

where $k(\nu_0)$ is the absorption coefficient at maximum wavenumber ν, and ΔN is the population inversion. When $\beta L > \gamma$ is achieved, the intensity of radiation of a proper frequency will build up rapidly until it becomes so large that the stimulated transitions will deplete the upper level and reduce the value of β. This is a dynamic situation that in most lasers gives rise to pulsations. If the level of excitation is such that βL is less than γ for all frequencies, the intensity of radiation does not build up at any frequency.

The threshold of laser oscillation is attained when the peak value β of the amplification curve satisfies the equation

$$\beta L > \gamma \tag{2.2}$$

Equation (2.2) is the simplest formulation of the threshold condition.

Waves reflected several times will reinforce only if the distance between the mirrors is an integral multiple of the half-wavelength within the laser. If the refractive index n does not vary between the mirrors

$$m\lambda = 2Ln$$

(where m is an integer). The condition for constructive interference is:

$$\frac{\nu}{c} = \frac{m}{2Ln} \tag{2.3}$$

If the optical path between the mirrors is inhomogeneous then the distance L should be replaced by the optical distance L′ where

$$L' = \int_0^L n\,dz$$

and

$$\frac{\nu}{c} = \frac{m}{2L'} \tag{2.4}$$

Plane waves of frequency defined by the above equation and directed along the laser axis (perpendicular to the mirrors) are the axial modes of the laser.

Thus, a laser of a given length and mirror reflectivity will operate only if the population inversion is large enough to ensure the amplification per unit length satisfying equation

$$\beta = \Delta N\, k(\nu)_0 \geqslant \frac{\gamma}{L} \tag{2.5}$$

This equation combines the requirement for the qualities of the resonator design (L and γ) and the amplifying medium which is related to the population inversion ΔN and the radiative transition probabilities reflected in absorption coefficient $k(\nu)_0$.

We shall summarize briefly the parameters for the laser design which are extensively discussed elsewhere [26]. The laser medium spontaneously emits light into an extremely large number of oscillator modes. The usual optical resonators [27] provide appreciable feedback for only a limited number of modes which are the resonator modes [28]. From this concentration of radiation in the small number of modes arises the coherence of laser light.

An important parameter of a resonator mode is its decay time t_c. This is the time at which the nonequilibrium energy ΔE decays to 1/e of its initial value, $\Delta E \cong \exp$

$(-t/t_c)$. The value of t_c is related to the fractional round-trip loss of the mode and the round-trip optical length L of the resonator by

$$t_c = -\frac{L}{c}\ln(1-\gamma) \qquad (2.6)$$

where c is the velocity of light. Usually $\gamma \lll 1$ and the equation may be approximated

$$t_c = L/c\gamma \qquad (2.7)$$

The decay time and the quality factor Q of the cavity are related by

$$Q = 2\pi\nu_{12}t_c \qquad (2.8)$$

where Q is the loss rate of the system and depends on the macroscopic properties of the cavity. Any disturbance that decreases Q increases γ and increases ΔN, the population inversion at which the oscillation begins.

Thus far we have discussed the properties dependent on the engineering set-up of a solid state laser. The other half of the game with which we shall be mainly interested in this book depends on the intrinsic properties of amplifying material. These are the radiative and non-radiative transition probabilities between various electronic levels of rare earth ions and the methods by which population inversion can be achieved.

Finally, we shall see how combination of two parts will enable construction of a high-energy laser by use of Q switch and mode locking operation.

B. Spontaneous and Stimulated Emission

Optical transitions between different atomic levels having energies E_2 and E_1 and populations N_2 and N_1 may occur spontaneously or by stimulation of electromagnetic radiation in the appropriate frequency region. Under conditions for which the interaction energy between the radiation field and the atomic system is small enough and can be regarded as a perturbation on the energy of the system, we may define transition probabilities for the process of absorption and emission. The number of absorptions from the ground state per unit time may be written

$$B_{12}N_1\rho(\nu_{12})$$

where $\rho(\nu_{12})$ is the energy density of the corresponding frequency $\nu_{12} = (E_2-E_1)/h$. From the excited state the atoms may return to the ground state by spontaneous emission for which the number per unit time is $A_{21}N_2$, and by stimulated emission the rate of which is given by $B_{21}N_1\rho(\nu_{12})$. Under conditions of equilibrium the number per unit time of atoms absorbing is equal to the number of atoms emitting and we have,

$$B_{12}N_1\rho(\nu_{12}) = A_{21}N_2 + B_{21}N_2\rho(\nu_{12}) \tag{2.9}$$

Using the Boltzmann relation between N_1 and N_2 which is

$$N_2 = N_1 \exp(-h\nu_{12}/kT) \tag{2.10}$$

the radiation density $\rho(\nu_{12})$ may be expressed from (2.9) and (2.10) as

$$\rho(\nu_{12}) = \frac{A_{21}}{B_{12}\exp(h\nu_{12}/kT) - B_{21}} \tag{2.11}$$

When electromagnetic radiation in a cavity is in thermal equilibrium at absolute temperature T the distribution of radiation density follows Planck's Law

$$\rho(\nu) = \frac{8\pi h\nu^3}{c^3} \cdot \frac{1}{\exp(h\nu/kT) - 1} \tag{2.12}$$

Dividing Eq. (2.11) by B_{21} we obtain

$$\rho(\nu) = \frac{A_{21}/B_{21}}{(B_{12}/B_{21})(\exp(h\nu/kT)) - B_{12}/B_{21}} \tag{2.13}$$

Hence, by comparing Eqs. (2.12) and (2.13) we obtain

$$B_{21} = B_{12}$$

and

$$A_{21} = \frac{8\pi h\nu^3}{c^3} \cdot B_{12} \tag{2.14}$$

This equation is valid in vacuum for particles having non-degenerate energy levels. When the energy levels are degenerate the Einstein relation takes the form

$$g_1B_{12} = g_2B_{21}$$

where g_1 and g_2 are the multiplicities of levels 1 and 2, respectively.

In solids, if the index of refraction n differs from unity, the spontaneous emission rate must be replaced by

$$A_{21} = \frac{8\pi h\nu^3 n^3}{c^3} B_{12} \tag{2.15}$$

It should be noted that in optical lasers the population inversion is a typical example of nonequilibrium situation as can be seen from the following. For a system in thermal equilibrium with the radiation field, the ratio of the rate of spontaneous emission to that of stimulated emission is given by Einstein's relation:

$$R = \frac{A_{21}}{B_{12}\rho(\nu)} = [\exp(h\nu/kT) - 1] \tag{2.16}$$

where A_{21} is the spontaneous and B_{12} the stimulated emission coefficients.

Taking $2\pi\nu = 3.4 \times 10^{15}\ \mathrm{sec}^{-1}$ corresponding to a wavelength ~ 550 nm we obtain (for temperature of 300 °K) $R = e^{82} \simeq 10^{35}$. Thus under conditions of thermal equilibrium the proportion of stimulated emission is entirely negligible for transitions in the optical region. It is clear therefore that in order to devise a laser a means must be found of enhancing greatly the number of stimulated emissions relative to the spontaneous ones. The number of stimulated emissions per second is proportional to the upper state population N_2. Since $B_{12} = B_{21}$ we see that the rate of stimulated emission will exceed that of absorption if $N_2 > N_1$. Since the population of the lower state is always greater than that of the upper state this condition is described as population inversion. Under conditions of population inversion it is clear that amplification of wave will occur as it passes through the system of excited atoms.

In a practical situation observations are made on a large number N_0 of atoms the distribution of which among the different states will follow Boltzmann's Law which is the number of atoms in state j

$$N_j' = \frac{N_0 e^{-E_j/kT}}{\sum_i n e^{-E_i/kT}} \tag{2.17}$$

where E_j and E_i are energies of states j and i. If the degeneracy of a level is designated by g, the number of atoms in level n is $N_n = g_n N_n'$ where N_n' refers to the population of each component of state in level n.

In the absorption of light the decrease of light intensity passing through material of thickness dx is given by

$$I(\nu, x) = I_0 e^{-k(\nu)x} \tag{2.18}$$

the frequency ν_0 is the center of the absorption line $k(\nu)$ and is expressed in reciprocal cm.

The width of the curve $k(\nu)$ versus frequency at the place where $k(\nu)$ has fallen to one half of its peak value $k(\nu)_0$ is the half width of the absorption line denoted by $\Delta\nu$.

The relation between the integrated absorption coefficient k and spontaneous emission coefficient A_{21} is given by the Fuchtbauer-Ladenburg formula [29, 30]

$$\int k(\nu)d\nu = \frac{c^2 A_{21} g_2}{8\pi\nu_0^2 n^2 g_1} \quad (N_1 - \frac{g_1}{g_2} N_2) \tag{2.19}$$

where n is the refractive index.

This formula can be rewritten as

$$\int k(\nu)d\nu = \kappa(N_1 - \frac{g_1}{g_2} N_2) \tag{2.20}$$

The constant κ equals

$$\frac{c^2 A_{21}}{8\pi\nu_0^2 n^2} \frac{g_2}{g_1} \quad = \tag{2.20a}$$

$$\frac{\lambda_0^2 A_{21}}{8\pi n^2} \frac{g_2}{g_1} \quad = \tag{2.20b}$$

$$\frac{\lambda_0^2}{8\pi n^2 \tau_{21}} \frac{g_2}{g_1} \tag{2.20c}$$

It should be noted that the radiative probability $A_{21} = \frac{1}{\tau_0} = \frac{1}{\tau_{21}}$ where τ_0 is the radiative lifetime of the emitting level.

These equations are concerned with transition between any lower state 1 and any upper state 2. Finally, oscillator strengths of the transition are given by

$$P_{12}\tau_{21} = \frac{mc}{8\pi^2 e^2 n} \frac{g_2}{g_1} \lambda_0^2 \tag{2.21}$$

The symbols are defined as follows

$k(\nu)$ absorption constant at ν per cm.
N_i population of state i
λ_0 peak wavelength of the transition
n index of refraction
g_i degeneracy of state i
P_{12} oscillator strength of the transition
m, e electronic mass, charge (times -1)
c velocity of light
τ_{21} radiative lifetime of the transition

λ_0 is the wavelength of the fluorescent peak and $\Delta\nu$ is the effective fluorescence linewidth determined from integration of lineshape.

In case of optical transitions, when the density of radiation is not very high, all absorption takes place from the ground state denoting the absorption of the totally unexcited material by $k(\nu)$ we have

$$\int k(\nu)\, d\nu = \kappa N_0 \tag{2.22}$$

where N_0 is the total number of atoms per unit volume.

The relation between the absorption cross-section per atom $\sigma(\nu)$ and the integrated cross-section is

$$\sigma(\nu)_0 = k(\nu)_0 / N_0 \tag{2.23}$$

When population inversion takes place for levels 1 and 2, Eq. (2.5) gives negative values for the absorption coefficients, since amplification can be defined as negative absorption. $\beta = -k(\nu)_0$. The amplification rate β is expressed as:

$$\beta(\nu) = \kappa g(\nu, \nu_0) \left(\frac{g_1}{g_2} N_2 - N_1 \right) \tag{2.24}$$

where g (ν, ν_0) is a line-shape function different from zero only in a small region around ν_0. It is normalized so that

$$\int g(\nu, \nu_0)\, d\nu = 1$$

The peak value g(0) is related to the linewidth for a Lorentz curve per unit volume as: $g(0) = \dfrac{2}{\pi \Delta\nu}$

and for a Gaussian curve: $g(0) = \dfrac{2}{\Delta\nu} \left(\dfrac{\ln 2}{\pi} \right)^{1/2}$

The relative population inversion which is

$$n = \left(\frac{g_1}{g_2} N_2 - N_1 \right) / N_0 \tag{2.25}$$

is -1 for a totally unexcited material and zero for the material for which the absorption and amplification are equal. We then obtain,

$$\beta(\nu) = \kappa N_0 g(\nu, \nu_0) \left(\frac{\frac{g_1}{g_2} N_2 - N_1}{N_0} \right) = \kappa N_0\, g(\nu, \nu_0) n \tag{2.26}$$

Hence, the integrated amplification factor is:

$$\int \beta(\nu)\, d\nu = \kappa N_0 \left\{ \frac{\frac{g_1}{g_2} N_2 - N_1}{N_0} \right\} \tag{2.27}$$

Thus the integrated amplification rates are now expressed in terms of relative population inversion and the measurable absorption properties of the material in the unexcited state.

When population inversion takes place for levels 1 and 2, the basic formula for the integrated absorption is Eq. (2.19) and negative value is obtained for the absorption coefficient which equals the amplification coefficient β.

In principle it is possible to determine experimentally A_{21} by measurements of the intensities of fluorescence spectral lines and decays and β coefficient by absorption. The calculation of β coefficient from the basic principles of quantum mechanics is performed by the means of time-dependent perturbation theory and is developed *in all texts* of quantum chemistry [31–33].

The observed intensity of the spectral line (the mean number of transitions, absorption or luminescence, occurring between states per unit time) is determined by the incidence of transitions between the initial and final states of an atom corresponding to the frequency of the line. The amount of absorption is governed by the intensity of the incident radiation, the path length of the exciting radiation through the sample, the concentration of the absorbing or emitting centers and the probability that a radiative transition will occur between the initial and final state of the absorbing (or emitting) atom. The probability of radiative transition between two electronic states as we have seen is reflected quantum mechanically through the transition moment integral

$$\mu = \int_{-\infty}^{\infty} \Psi_i \, er \, \Psi_f \, dv \tag{2.28}$$

Ψ_i is the wavefunction of the atom in the initial state (before radiative transition) and Ψ_f is the wavefunction of the atom in the state produced by the radiative transition. dv is the volume element in configuration space, e is the electronic charge, and r is the operator corresponding to the displacement of electronic charge in the atom as a result of the transition (er is the dipole moment vector). Under given conditions of illumination intensity, sample size and emitter or absorber concentration, the intensity of the transition is proportional to μ^2. In general the calculation of transition probability is very difficult because the wavefunctions of many-electron atoms are not known.

However, one property of the wavefunctions and the transition moment length r enables us to determine whether a transition is allowed or forbidden, that is whether μ is positive or zero respectively. This property is called the *parity* and is derived from the behaviour of mathematical function when it is inverted through a center of symmetry (*i. e.,* when x, y, z are replaced by $-x$, $-y$, $-z$). Those functions for which

$f(x, y, z) = f(-x, -y, -z)$ are said to have even parity

Those for which

$$f(-x, -y, -z) = -f(x, y, z)$$

are said to have odd parity. The product of two functions having the same parity results in another function having even parity. The product of two functions having different parities results in a function having an odd parity. The integral over the entire configuration space of a function having odd parity *is zero* and the integral of a function having even parity *is non zero,* provided no other compensation occurs. Since the transition moment r for an electric dipole is a vector, it changes sign upon inversion, (*i. e.,* it has odd parity). Therefore, for a given transition moment to be non zero – and for the transition to be allowed – the product of the wavefunctions of the two states involved in the transition must also have odd parity. This will be the case only if the two wavefunctions have opposite parities. Consequently transitions occurring between states of the same parity are forbidden and should have zero probability. Those occurring between states of different parity are allowed and will have probabilities that depend on the magnitude of charge displacement and the states involved in the transitions.

Since the transition rate is proportional to the square of the matrix element,

$$|E|^2 |\mu(n, m)|^2 \cos^2\theta$$

where θ is the angle included between E, electric vector of the electromagnetic wave, and μ. It must be averaged when the atoms are free to arrange themselves with respect to the field.

It can be proven that [34] the emitted radiation is coherent (in phase) with stimulating radiation. In the case of a single plane wave the square of the electric field is proportional to the radiation intensity $\rho(\nu)$. Hence, the rate of stimulated radiation is proportional to the radiation density.

By integrating over entire space

$$B_{21} = \frac{8\pi^3}{3h} |\mu(n, m)|^2 \tag{2.29}$$

The rate of spontaneous emission (in vacuum) may be then calculated from Einstein's relation

$$A_{21} = \frac{64\pi^4\nu^3}{3hc^3} |\mu(n, m)|^2 \tag{2.30}$$

The observed trivalent rare earth laser lines are 4f–4f transitions which for a free ion are electric dipole parity forbidden. Although magnetic-dipole and electric quadrupole transitions are parity allowed, most 4f–4f radiative transitions of rare earths in solid hosts are dominantly of forced electric dipole type. These are electric dipole transitions which proceed through the mixing of levels of opposite parities produced by the crystal field at noncentrosymmetric sites. For example a $4f^q$ level can contain small admixtures of $4f^{q-1}$ 5d levels. The weaker parity component of the wave functions is typically of the order of 10^{-3}. For 4f–4f trivalent rare earth emission lines the oscillator strength P is typically about 10^{-6} [35, 36].

In the case of intraconfigurational transition within the R.E. ions the electric dipole must be replaced by a forced electric dipole.

The laser action in rare earth doped solids results from the fact that reasonably sharp fluorescence bands occur in these systems. Thus sufficient radiant power per band-width can be built up and stimulated emission achieved. Atoms may be excited to a broad band which is either a charge transfer band, f–d transition band or an assembly of closely spaced f levels. Alternatively, a donor ion with a high absorption band can be excited and transfer its energy to the fluorescent acceptor ion (for a detailed discussion of energy transfer see the Chapter 4). Non-radiative rapid relaxation from the broad band populates the lower lying sharp fluorescent levels. Those fluorescent levels may directly radiate to the ground state as it is the case in the three level system, or to a level lying at an energy gap above the ground state as it is in a four level system. The disadvantage of the three level system is that in order to reach the threshold population inversion, 50 percent of the atoms in the ground state must be excited. On the other hand, if the terminal level 2 in a four level system lies sufficiently high with respect to the ground state that $N_2 = N_1 \exp(-\Delta E_g/kT)$ is very small, practically zero, then any atoms which are excited to the fluorescent levels form inverted populations with respect to the terminal level. If the value of ΔE_g is not sufficiently high for level 2 to be depopulated at room temperature, then population may be effected by cooling.

C. Three-level Laser System

Let us first consider a three level system. At a steady state, the number of atoms per unit time arriving to an excited level must be equal to the number of atoms per unit time leaving this level. Assuming that we are dealing with a non-degenerate level, we obtain [37] (Fig. 1).

$$\frac{dN_4}{dt} = \dot{N}_4 = W_{14}N_1 - \alpha N_4 = 0 \tag{2.31}$$

where $\alpha = W_{41} + p_{43} + p_{41}$. Here $W_{ij} = W_{ji}$ are the stimulated transition, and p_{ij} spontaneous transfer probabilities.

$$\frac{dN_3}{dt} = \dot{N}_3 = W_{13}N_1 - \beta N_3 + p_{43}N_4 = 0 \quad \text{and} \quad \beta = W_{31} + p_{31} \tag{2.32}$$

$$N_1 + N_3 + N_4 = N_0 \tag{2.33}$$

The solutions of these equations are

$$N_1 = \frac{\alpha\beta N_0}{\beta(W_{14} + \alpha) + \alpha W_{13} + p_{43}W_{14}} \tag{2.34}$$

$$N_3 = \frac{(\alpha W_{13} + p_{43}W_{14})N_0}{\beta(W_{14} + \alpha) + \alpha W_{13} + p_{43}W_{14}} \tag{2.35}$$

and the population inversion per unit volume is

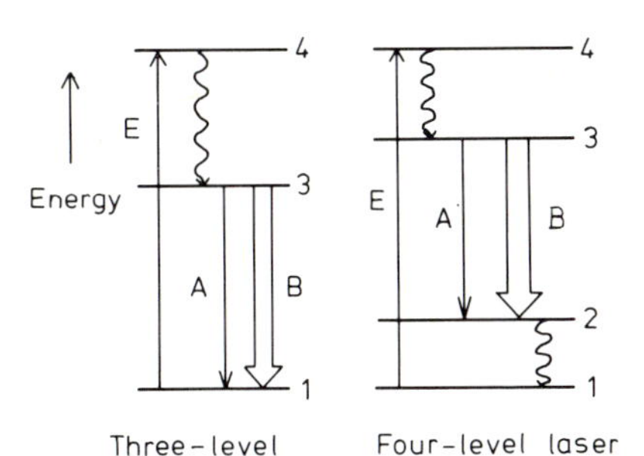

Fig. 1. Energy diagram for 3- and 4-level lasers. Excitation E from the external energy source raises the system to level 4 from which it decays nonradiatively to level 3. In the former case, light emission corresponds to a transition to the groundstate 1, and in the four-level system to a level 2 rapidly decaying nonradiatively to 1. Spontaneous emission (A) preceeds the induced emission (B) responsible for a build-up of light in the laser

$$\frac{\Delta N}{V} = N_3 - N_1 = \frac{\alpha W_{13} + p_{43} W_{14} - \alpha\beta}{\beta(W_{14} + \alpha) + \alpha W_{13} + p_{43} W_{14}} N_0$$

$$= \frac{W_{14}(\epsilon p_4 - p_{31}) - p_{31} p_4}{W_{14} + p_4(2W_{31} + p_{31}) + W_{14}(W_{31} + \epsilon p_4 + p_{31})} N_0 \tag{2.36}$$

where $p_4 = p_{41} + p_{43}$ and $\epsilon = p_{43}/(p_{43} + p_{41})$ = the fraction of atoms which decay from the absorption band to the metastable state. It is well known that the depopulation probability of the ions from the broad absorption band is about three to four orders of magnitude higher than the decay rate from the metastable fluorescence state. Therefore

$$p_4 \gg p_{31}$$

Also in many systems of this type the nonradiative transition probability exceeds any process of level 4. Therefore the population of 4 is small compared with the other levels at any instant. Also, we can assume that the pumping probability from level 1 to level 4 is smaller than the radiationless decay from the absorption band. Therefore:

$$p_{43} \gg W_{14}$$

With this approximation, the population inversion per unit volume may be written as [37]:

$$\Delta N = N_3 - N_1 = \frac{\epsilon W_{14} - p_{31}}{p_{31} + \epsilon W_{14} + 2W_{31}} N_0 \tag{2.37}$$

Before the threshold is reached, W_{31} is very small; therefore putting $W_{31} = 0$ we have for the threshold condition

$$n_{th} = \frac{\Delta N_{th}}{N_0} = \frac{\epsilon W_{14} - p_{31}}{p_{31} + \epsilon W_{14}} \geqslant 0 \tag{2.38}$$

where $\Delta N_{th}/N_0$ is the relative population inversion. By expressing our W and p in terms of Einstein's *B* and *A* coefficients, we can rewrite this equation as

$$N_3 - N_1 = \frac{[B_{14}\rho(\nu_{14}) - \epsilon A_{31}]}{B_{14}\rho(\nu_{14}) + \epsilon A_{31}} \cdot N_0 \tag{2.39}$$

This result enables an estimation of the radiation density (and hence the temperature of the broad band pumping source) in order to obtain the oscillation threshold.

$$(W_{14})_{th} = \frac{1 + n_{th}}{1 - n_{th}} \frac{p_{31}}{\epsilon} = \frac{1 + n_{th}}{1 - n_{th}} \frac{1}{\epsilon\tau_f} \tag{2.40}$$

where $\tau_f = p_{31}^{-1}$ is the measured (fluorescent) lifetime of the fluorescent level. In the case where the quantum efficiency of the fluorescent level differs from unity, $\tau_f = \eta\tau_0$ where η is the quantum efficiency and τ_0 the radiative lifetime of the metastable level 3.

In actual situations only a small excess of population inversion over unity is needed in order to sustain the oscillation. In these conditions

$$(W_{14})_{th}^3 \simeq \frac{1}{\epsilon\eta\tau_0} \tag{2.41}$$

from which we see that long fluorescence lifetimes ensure a low threshold of oscillations. The superscript 3 indicates a three-level system.

D. Four-level Laser System

In a four level system (Fig. 1), an excited atom in level 4 decays rapidly to level 3. From this level it can decay to the ground level 1 or to level 2; if this is the case, the laser transition is from level 3 to level 2.

Similarly to the three level system, the population densities at steady state are determined by the following equations:

$$\dot{N}_4 = (N_1 - N_4)W_{14} - N_4 p_4 \tag{2.42}$$

$$\dot{N}_3 = (N_2 - N_3)W_{32} - N_3 p_3 + N_4 p_{43} \tag{2.43}$$

$$\dot{N}_2 = N_1 p_{12} - (N_2 - N_3)w_{32} - N_2 p_{21} + N_3 p_{32} + N_4 p_{42} \tag{2.44}$$

where

$$p_4 = p_{43} + p_{42} + p_{41}$$

$$p_3 = p_{32} + p_{31}$$

$$N_1 + N_2 + N_3 + N_4 = N_0 \tag{2.45}$$

and assuming that levels 1 and 2 are in thermal equilibrium $p_{21}/p_{12} = \exp(h\nu_{12}/kT)$.

Since the decay probability from level 4 to the laser level 3 is usually a phonon-assisted non-radiative process, it is much higher than other decays so

$$\epsilon = \frac{p_{43}}{p_4} \approx 1$$

and $p_4 \gg W_{14}$ also holds.

Under the above conditions the steady state equations become:

$$N_2W_{32} - N_3\alpha + N_1\epsilon W_{14} = 0 \tag{2.46}$$

$$N_1p_{12} - N_2p_{21} + N_3\beta = 0 \tag{2.47}$$

$$\alpha = p_3 + W_{32}$$

$$\beta = p_{32} + W_{32}$$

and

$$N_1 + N_2 + N_3 = N_0 \tag{2.48}$$

Following DiBartolo [37], the threshold population inversion for a four level laser is given by

$$n_{th} = \frac{W_{14}}{\epsilon W_{14} + p_3 + W_{32}} = \frac{N_3 - N_2}{N_0} \tag{2.49}$$

as

$$W_{32} \ll p_3.$$

We obtain the oscillation threshold for four level system:

$$(W_{14})_{th} = \frac{n_{th}}{\epsilon(1 - n_{th})} \cdot p_3 = \frac{n_{th}}{1 - n_{th}} \cdot \frac{1}{\epsilon\tau_f} \tag{2.50}$$

For $\gamma \ll n_{th} \ll 1$, then

$$W_{14}^4 \approx \frac{n_{th}}{\epsilon\tau_f} = \frac{n_{th}}{\epsilon\eta\tau_0} \tag{2.51}$$

The threshold population inversion n_{th} per unit volume is connected with the amplification coefficient β as explained above by Eq. (2.5)

$$\beta(\nu) = \mathrm{KN_0 g(\nu) n_{th}} = \frac{\gamma}{\mathrm{L}} \tag{2.52}$$

so that

$$\mathrm{n_{th}} = \frac{\beta(\nu)}{\mathrm{kN_0 g(\nu)}}$$

Following Eqs. [2.20 a, 2.51 and 2.52], we obtained conditions for threshold population inversion as:

$$\mathrm{n_{th}} = \frac{\gamma}{\mathrm{LN_0}} \frac{4\pi^2 \Delta\nu\tau_0 \mathrm{n}^2}{\lambda^2} \frac{\mathrm{g_1}}{\mathrm{g_2}} \tag{2.53}$$

(where n is the refractive index) for a Lorentz curve, and for a Gaussian curve

$$\mathrm{n_{th}} = \frac{\gamma}{\mathrm{LN_0}} \frac{4\pi^2 \tau_0 \Delta\nu \mathrm{n}^2}{\lambda^2} \frac{\mathrm{g_1}}{\mathrm{g_2}} \left(\frac{\pi}{\ln 2}\right)^{1/2} \tag{2.54}$$

It should be noted that

$$\mathrm{n_{th}} = \frac{\gamma}{\sigma(0)\mathrm{LN_0}} = \frac{1}{\sigma_0 \mathrm{N_0} \tau_c \mathrm{v}} \tag{2.55}$$

where $\sigma(0)$ is the cross section for absorption, v is the light velocity in the amplifying medium and τ_c is the cavity lifetime. Thus, W_{14}^4 may be written:

$$\mathrm{W_{14}^4} = \frac{\mathrm{n_{th}}}{\eta\tau_0} \approx \frac{4\pi^2 \Delta\nu\gamma}{\mathrm{N_0}\lambda^2 \mathrm{L}\eta} = \frac{4\pi^2 \Delta\nu \nu^2 \mathrm{n}^2}{\mathrm{N_0 v^3} \tau_c \eta} \tag{2.56}$$

It is interesting to compare the threshold values for the pumping probablities for a four level and a three level system (superscripts 4 and 3)

$$\frac{\text{four } \mathrm{W_{14}}}{\text{three } \mathrm{W_{14}}} = \frac{\mathrm{n_{th}}}{1 + \mathrm{n_{th}}} \frac{\tau_f{}^3}{\tau_f{}^4} \approx \mathrm{n_{th}} \frac{\tau_f{}^3}{\tau_f{}^4} = \frac{\mathrm{N_{th}}}{\mathrm{N_0}} \frac{\tau_f{}^3}{\tau_f{}^4}$$

$$\mathrm{N_0} = \mathrm{n_0 V}$$

Comparison of three level to four level system on basis of:

$$(\mathrm{W_{13}})_{th} = \frac{1}{\eta\tau_0} \quad \text{and} \quad (\mathrm{W_{14}})_{th} = \frac{\mathrm{n_{th}}}{\eta\tau_0}$$

three level four level

In a three level system the threshold value of the pumping probability depends essentially on the fluorescent lifetime of the metastable level and is almost entirely insensitive to the linewidth and Q factor of the cavity.

In a four level system the threshold pumping probability does not depend on the radiative lifetime τ_0 but is directly proportional to linewidth and inversely proportional to η and the cavity pumping time τ_c.

E. Modes of Oscillation

Since the threshold condition is met only in a restricted frequency range laser oscillation will occur only for a few discrete frequencies ν satisfying the equation:

$\nu = \dfrac{mc}{2Ln}$ where m is an integer and n the refractive index. Plane waves of a frequency ν satisfying Eq. (2.57) and directed along the laser axis are called the axial modes.

The modes are equally spaced in frequencies. Thus, the output of the laser may consist of several lines separated from each other by the frequency difference

$$\nu_{n+1} - \nu_n = \frac{c}{2Ln} \tag{2.57}$$

It is shown in the theory of electromagnetic radiation [26] that the number of oscillations $p(\nu)$ of a given frequency ν in a cubic enclosure of length a filled with a material of refractive index n is expressed by:

$$p(\nu) = \frac{8\pi\nu^3 n^3 a^3}{3c^3} \tag{2.58}$$

Hence, the mode density per unit volume per unit frequency interval is:

$$p(\nu) = \frac{1}{a^3}\frac{dp}{d\nu} = \frac{8\pi\nu^2 n^3}{c^3} = \frac{8\pi\nu^2}{v^3} \tag{2.59}$$

Thus, the known Einstein relation between spontaneous and stimulated emission can be written as:

$$A_{21} = \frac{8\pi h\nu^3}{c^3} B_{12} = h\nu p(\nu) B_{12} \tag{2.60}$$

from which it can be seen that the probability per unit time of spontaneous emission in a single mode is equal to the probability per unit time of induced emission due to a single photon in the same mode.

For the laser action only those modes are of importance which lie within the spontaneous linewidth $\Delta\nu$ of the laser material. Their number is:

$p = \int g(\nu)p(\nu)d\nu$ where $g(\nu)$ is the lineshape function or approximately

$p = \Delta\nu p(\nu_0)$ with ν_0 line center frequency

For a Lorentzian lineshape

$$g(\nu) = \frac{\Delta\nu}{2\pi[(\nu-\nu_0)^2 + (\Delta\nu/2)^2]} \quad \int_{-\infty}^{\infty} g(\nu)d\nu = 1 \tag{2.61}$$

The number of modes within the Lorentzian linewidth $\Delta\nu$ and in a resonance volume V is:

$$p_L = \frac{\pi V}{2} p(\nu)\Delta\nu = \frac{4\pi^2\nu^2\Delta\nu n^3}{c^3} V \tag{2.62}$$

For a Gaussian line shape

$$g(\nu) = g_{max} \exp\{-(\nu-\nu_0)^2/(\Delta\nu)^2\} \int_{-\infty}^{\infty} g(\nu)d\nu = 1$$

and the number of modes within the Gaussian line width $\Delta\nu$ and the volume v is

$$p_G = \sqrt{\frac{\pi}{4 \ln 2}} V p(\nu)d\nu \tag{2.63}$$

From the equation connecting the stimulated spontaneous emissions and the number of modes we may obtain the connection between the numbers of modes and the threshold population inversion.

The rate of stimulated emission P_{st} due to the number of photons in one mode, is equal to the rate of spontaneous emission $p_{sp} = \frac{1}{\tau_0}$ times the number of photons N_{ph}

$$p_{st} = \frac{N_{ph}}{p(\nu) \cdot \tau_0} \tag{2.64}$$

The loss rate of photons within the p modes in the resonator cavity is given by

$$\frac{dN_{ph}}{dt} = -\frac{N_{ph}}{t_c} \tag{2.65}$$

where $t_c = \frac{Q}{2\pi\nu}$ is the cavity lifetime.

The gain rate of photons is:

$$\frac{dN_{ph}}{dt} = \frac{N_{ph}}{\tau_0 p} \Delta N_{th} = \frac{N_{ph}}{\tau_0 p} (N_3 - N_1) \tag{2.66}$$

So the threshold population inversion is given as:

$$\Delta N_{th} = p \frac{\tau_0}{t_c} \tag{2.67}$$

If the levels between which the laser transition occurs have different degeneracies than Eq. (2.67) should be:

$$\frac{g_2}{g_1} \Delta N_{th} = p \frac{\tau_0}{t_c}$$

So the population inversion of the laser level per unit volume for a Lorentzian is:

$$\frac{\Delta N_{th}}{V} = \frac{4\pi^2 \nu^2 \Delta\nu n^2 \tau_0}{c^3 t_c} \tag{2.68}$$

Here we come again to the Schawlow-Townes threshold condition.

As an example let us consider a neodymium glass rod of 10 cm optical length and a resonator mirror of 1% transmission, *i.e.*, $t_c = 6.7 \times 10^{-8}$ sec.

$$\begin{aligned} \frac{1}{V} N_{th} &= 2 \times 10^6 \Delta\nu\tau_3 && \Delta\nu \text{ in Hz} \\ &= 8 \times 10^{17} \Delta\lambda\tau_3 && \Delta\lambda \text{ in nm} \\ &= 6.5 \times 10^{16} \Delta\bar{\nu}\tau_3 && \Delta\bar{\nu} \text{ in cm}^{-1} \end{aligned}$$

N_{th}/V in cm^{-3}, τ_3 in sec.

For a silicate glass $\Delta\bar{\nu} = 2 \times 10^{13}$ Hz ($\Delta\lambda = 50$ nm, $\Delta\bar{\nu} = 600\ cm^{-1}$ and $\tau_0 =$ 5×10^{-4} sec. so that $N_{th}/V = 10^{16}\ cm^{-3}$. In a four level system $N_2 = 0$ and N_{cr} is directly equal to the density of ions excited in a metastable state (3) $N_{th} = N_3$. As already stated before in a three level system ($N_1 + N_3 = N_0$ = constant) hence, at the verge of inversion $N_3 = 1/2\ N_0$ and at the threshold the population of level 3 is $N_3 = 1/2\ N_0 + N_{cr}$.

On the basis of this discussion we can now list the requirements for laser systems with four levels.

1. The linewidth of the laser transition should be as small as possible
2. The lifetime of the initial level of the laser transition should be purely radiative
3. The terminal level of the laser transition should be connected to the ground level by fast relaxation processes
4. The laser system should have a strong and wide absorption band coupled with the excitation source

5. The ions from the absorption band should decay rapidly to the metastable level (the nonradiative decay process should be the only one).
6. The terminal level should be high in energy with respect to kT, $h\nu_{21} \gg kT$
7. The photon lifetime should be long t_c long, cavity losses small
8. No absorption from the metastable state should take place at either the laser wavelength or pumping wavelength

F. The Threshold of Optical Pumping

There exists a basic difference in the number of ions which must populate the laser level in order to achieve the threshold conditions in three and four level systems.

In a three level system, the number of ions needed is:

$$N_{th} = \frac{N_0}{2} + \Delta N_{th} \tag{2.69}$$

where N_0 is the total number of active ions in an active medium, and $\Delta N_{th} = p\,\dfrac{\tau_0}{t_c}$

Since $\Delta N_{th} \ll N_0$ this equation reduces to

$$N_{th} \approx \frac{N_0}{2} \text{ (3-level)} \tag{2.69a}$$

in a four level system the terminal level is normally unpopulated and the number of ions needed at the threshold is given by

$$N_{th} \approx \Delta N_{th} = p\,\frac{\tau_0}{t_c} \text{ (4-level)} \tag{2.70}$$

The energy input of a laser may be continuous CW (continuous wave) or in the form of short pulses. In the first case we define the minimum power input per unit volume, which is necessary to maintain a threshold population inversion:

$$P_{input} = \frac{a}{\eta}\,\frac{N_{th}}{\eta\tau_0} \cdot h\nu_4 = \frac{a}{\eta^2}\,\frac{N_{th}}{\tau_0}\,k\nu_4 \tag{2.71}$$

where a is the ratio of the average pumping energy to the energy of absorption band of the medium, and is a measure of the spectral coincidence between the exciting light and the absorption medium, where η is the fraction of the absorbed power that is either reradiated at the pump frequency or results in population of the upper laser level, which is by definition the quantum efficiency of the lasing level 3.

By inserting the expression for population inversion N_{th} for three and four level systems we obtain the explicit expressions:

$$P_{input} = \frac{a}{\eta}\,\frac{N_0}{2\eta\tau_0}\,h\nu_4 = \text{(for the 3-level)} = \frac{aN_0}{2\eta^2\tau_0}\,h\nu_4 \qquad (2.71\,a)$$

$$P_{input} = \frac{a}{\eta}\,\frac{p}{\eta t_c}\,h\nu_4 = \text{(for the 4 level system)} = \frac{ap}{\eta^2 t_c}\,h\nu_4 \qquad (2.71\,b)$$

In pulse operation when the laser is excited with a light pulse short in comparison with the fluorescence lifetime of the laser level, the laser of excited ions by spontaneous emission during the pumping pulse is negligible and each absorbed photon will be transformed into p excited ions.

The minimum energy input E_{input} needed to achieve population inversion, is given by:

$$E_{input} = \frac{N_{th} h\nu_4}{\eta} \qquad (2.72)$$

where $h\nu_4$ is the average value of the energy of pump light photon. Inserting the appropriate values of N_{th} we obtain for the input energy the following:

$$E_{input} = \frac{N_0}{2\eta} \cdot h\nu_4 \text{ for the three level system} \qquad (2.72a)$$

and

$$E_{input} = \frac{p\tau_0}{\eta t_c}\,h\nu_4 = \text{for the four level system} = \frac{4\pi^2\nu^2\Delta\nu n^3}{\eta c^3}\,\frac{\tau_0\,h\nu_4}{t_c} \qquad (2.72b)$$

(in the case of a Lorentzian curve).

Let us consider a neodymium glass rod of 10 cm optical length and a 99% reflectivity, assuming $\eta = 1$.

$\nu \;= 9{,}400\ \text{cm}^{-1} = 2.82 \cdot 10^{14}\ \text{sec}^{-1}$
$h\nu_4 = 2.15\,\text{eV} = 3.44 \cdot 10^{-12}\,\text{erg} = 3.44 \cdot 10^{-19}\ \text{J}$
$\Delta\nu \;= 2 \cdot 10^{13}\ \text{Hz}$
$\tau_0 \;= 5 \cdot 10^{-4}\ \text{sec}^{-1}$
$t_c \;= 6.7 \cdot 10^{-8}$
$E_{input} = 7 \cdot 10^{-13} \cdot \Delta\nu \cdot \tau_2 \cdot n^3 = 2.4 \cdot 10^{-2}\ \text{J/cm}^{-3}$

when $\Delta\nu$ is in Hz and τ_0 is in sec. This is the input energy per volume.

G. Laser Output

Similar to the energy input, the output of a laser may be continuous (CW) or pulsed. The magnitude of the power output (CW and pulsed) is governed by various factors including the laser transition and the method and intensity of excitation, the diameter and length of the laser and the rate at which heat can be dissipated in the laser host and pumping source. The power output also depends on the overall gain of the cavity which in turn depends on the intensity of pumping, absorption losses, reflectivity of the output window, and other cavity parameters. The power output of solid state lasers is a function of the volume of the laser material and the illuminated surface area which is usually proportional to length, the diameter being limited due to absorption of the pumping radiation by the host.

In the case of pulsed lasers, the number of ions brought to the lasing level 3, by the pumping photons is $\frac{E_p \eta}{h\nu_4}$ where E_p is the pumping energy, N_{th} photons are required to overcome the losses, thus the number of excited centers which are effective in producing the laser output energy is:

$$E_0 = h\nu_3 \left(\frac{E_p \eta}{h\nu_4} - N_{th} \right) = \eta \frac{\nu_3}{\nu_4} \left(E_p - \frac{N_{th} h\nu_4}{n} \right) = \eta \frac{\nu_3}{\nu_4} (E_p - E_{input}) \qquad (2.73)$$

inserting the expression for population inversion N_{th} for three and four level systems we obtain:

$$E_{output} = \eta \frac{\nu_3}{\nu_4} \left(E_p - \frac{N_0}{2} \frac{h\nu_4}{\eta} \right) \text{ for three level system} \qquad (2.73a)$$

$$E_{output} = \eta \frac{\nu_3}{\nu_4} \left(E_p - p \frac{\tau_0}{t_c} \frac{h\nu_4}{\eta} \right) \text{ for four level system} \qquad (2.73b)$$

The output power for the CW operation is given by:

$$P_{output} = h\nu_3 \left(\frac{\eta P_p}{h\nu_4} - \frac{N_{th}}{\eta \tau_0} \right) \qquad (2.74)$$

P_p is the pumping power and τ_0 is the fluorescence lifetime of the lasing level.

thus,

$$P_{output} = \frac{N_{th} h\nu_3}{\eta \tau_0} \left(\frac{P_p}{P_{input}} - 1 \right).$$ For three level system we obtain:

$$P_{output} = \frac{N_0}{2} \frac{h\nu_3}{\eta \tau_0} \left(\frac{P_p}{P_{input}} - 1 \right) \qquad (2.74a)$$

while

$$P_{output} = \frac{p h\nu_3}{\eta t_c} \left(\frac{P_p}{P_{input}} - 1 \right) \text{ for the four level system} \qquad (2.74b)$$

Optically pumped lasers are operated in the pulsed mode when high peak powers are required. This is due to the difficulties in overcoming the threshold conditions and high thermal dissipation in the case of the CW laser.

Since in pulsed operation of the laser the duration of the pulse may be very short, the output power can be very large with an input of small amounts of energy, *i.e.*, 1×10^{-3} J corresponds to a mean power of one megawatt when the pulse duration is 10^{-9} sec. Since the pulse length can be varied over a wide range, from nsec for Q-switched and mode-locked outputs up to several msec, it is possible to change the power output by several orders of magnitude.

The power output may be increased by Q-switching [40], which is achieved by exciting the laser medium so that a population inversion occurs by delaying the application of feedback from the axial mirrors. This can be achieved by deflecting the beam at the totally reflecting mirror mechanically or by acousto-optic or electro-optic shutters or by using an opaque dye solution that bleaches transparent when the fluorescence light output reaches a given level. The shutter is closed during most of the pumping process so that high population inversion can be built up. When the shutter is opened, laser oscillation is established instantly and most of the excitation energy stored in the laser active species is delivered in one giant pulse.

An additional method for increasing the peak intensity is the mode locking operation.

As mentioned before, the output of a laser consists of a large number of modes which bear no relation to each other.

If several axial modes are locked together in phase, the resultant peak intensity will be correspondingly higher than that obtained from the sum of individual modes randomly related to each other. Mode-locking or phase-locking techniques force the modes to maintain equal frequency spacings and phase relationships to each other. This is used to generate very high peak power pulses of short duration using techniques similar to those used for Q-switching.

H. Examples of Rare Earth Lasers

Rare earths have been shown to exhibit laser fluorescence in liquid, crystal, glass and, lately, even in gaseous media.

Only crystal and glass lasers are of practical use thus far. We shall therefore present a very brief survey of liquid rare earth lasers and enter into greater detail when discussing crystal and glass laser systems with recent references to glass ceramics and vapour lasers.

a) Liquid Rare Earth Lasers

Rare-earth doped liquids have interesting possibilities as laser materials. Their advantages are that the complicated technique of single crystal growing and shaping is avoided and the concentration of the rare earth ions can be varied in a wide range.

The use of liquids enables the cooling of the active medium in an easy way. The disadvantage of liquids is that the spectra of the emitting rare earths are broader in liquids because of the constantly changing environment. In addition the interaction with the solvent increases the nonradiative deactivation of the excited rare earth ion thus decreasing its quantum efficiency.

The potential value of the rare earth chelates as laser material is summarized by Samelson [41]. The fluorescence properties of europium benzoyl-acetonate and europium dibenzoylmethide were especially studied in great detail [42]. The typical range of the overall efficiency of a chelate laser in protic solvents, because of rapid relaxation in which the higher energy vibrations of the OH bond are active, is only 10^{-6}, and therefore is not of practical value. Heller [43] has overcome this difficulty by dissolving neodymium in a liquid system which contains no protons. This system has no vibrations of sufficient energy to carry away the energy corresponding to the gap between the excited ground multiplet levels of neodymium. Using $SeOCl_2$ as a solvent, Heller and coworkers obtained a liquid neodymium laser with 1 joule output from a 15 cm tube of 0.6 mm diameter with an electrical input of 1,000 joules. This efficiency is higher by 3 orders of magnitude than the chelate laser. However, the disadvantage in this case is the high toxicity and corrosivity of the solvent, therefore the liquid laser which is of great academic interest does not seem at the present stage to be of great practical value.

b) Generalities About Solid State Lasers

The fabrication of laser quality crystals is difficult and costly. One method of decreasing the research effort in solid state lasers is to rely on theoretical analysis which provides direction to the experimental work. The main requirements which may be derived from the previous section in order to achieve low threshold values for the onset of laser action at fixed emission frequency or emission wavelength are:

a) Lowest possible halfwidth $\Delta\bar{\nu}$ of the spontaneous fluorescence. In case of a Q switching operation, high $\Delta\bar{\nu}$ is required.

b) Mean lifetime τ_3 of the metastable state, which, if possible, should correspond to the lifetime of the spontaneous fluorescence τ (quantum efficiency $\eta = 1$).

c) Lowest possible absorption coefficient $k(\lambda)$ for the laser wavelength λ, which may be achieved when the intrinsic absorption of the host, the extrinsic absorption due to unwanted impurities and the scattering by inhomogeneities are at a minimum. This is equivalent to the other requirement that t_c should become as large as possible.

d) Sufficiently wide separation between the terminal state 2 of the emission and the ground state 1. $E_2 \gg kT$, so that $N_3 \approx N_{thresh}$.

e) Sufficiently rapid depopulation of the terminal state 2 into the ground state $\tau_2 < \tau_3$.

f) A strong and broad absorption $k(\lambda')$ for the exciting light.

g) The host is required to have an excellent optical homogeneity and a good machinability for producing the close tolerances demanded for resonator geometry.

h) The activator must allow being incorporated into the host in the proper valency and with homogeneous and atomic distribution.
i) The half-width $\Delta\bar{\nu}$ should neither be increased by the heating of the resonator nor by irregularities in coordination. Accordingly, the heat energy must be dissipated, so that a good heat conduction of the host is of great importance.

Fortunately, theories have been developed in the last few years which enable the estimation of empirical parameters needed to describe the laser material from measurements performed on spectroscopic samples which can be done on small crystal samples rather than on laser quality rods.

Several characteristics vital to laser operation can be examined by combining the Judd-Ofelt [44, 45] theory with the theory of multiphonon desexcitation. Good laser transitions are characterized by large cross-sections for stimulated emission. The oscillator strengths connected with these quantities can be calculated by the Judd-Ofelt model. This model may also be helpful in computing absorption cross-sections of levels above the laser level which affect the achievable pump-rates. Obviously, high quantum efficiency is a desirable characteristic for an upper laser level. Laser action is also enhanced by rapid nonradiative decay of the lower level population and by rapid nonradiative relaxation from the upper pump levels to the emitting laser level. These two properties can be predicted by use of the multi-phonon relaxation theory.

The basic measurements for such predictions are the absorption and fluorescence spectra of the rare earth ions in crystals and associated cross-sections and the fluorescence decay rate as a function of time and the rare earth concentration. The absorption spectrum is usually integrated to obtain the radiation intensity parameters. These in turn are used to calculate the radiation decay probability which (when compared to the measured decay rate) yields the quantum efficiency of emission. The fluorescence branching ratios and intensities can be used for prediction of the fluorescence intensity of the laser transitions and thereby the peak cross-section. Such evaluation can be used further to examine the variation in spectroscopic parameters with crystal composition. On the basis of these findings the composition can be tailored to yield the spectroscopic properties favorable to a given laser application.

Relaxation processes are important in several respects for understanding laser performance. Relaxation wherein the energy is taken up by the excitation of multiple vibration is important in determining the fluorescence conversion efficiency, the quantum yield of the upper laser level and the decay of the terminal laser level.

The Judd-Ofelt [44, 45] theory can be applied to laser crystals and glasses and can successfully account for the induced emission cross-sections that are observed. In this theory which is reviewed elsewhere [35, 36] the line-strength of the forced electric-dipole transition between initial J manifold $|(S, L)J>$ and terminal manifold $|(S', L')J'>$ may be written in the form

$$S = \sum_{t=2,4,6} \Omega_t |<(S, L)J|| U^{(t)} ||(S', L')J'>|^2 \qquad (2.75)$$

where the elements $<|| U^{(t)} ||>$ are the doubly reduced unit tensor operators calculated in the intermediate coupling approximation. The three coefficients Ω_2, Ω_4, Ω_6 contain implicitly the odd-symmetry crystal field terms, radial integrals and perturbation denominators.

The matrix elements of the unit tensor operator are taken between the energy eigenstates, and can therefore be wirtten

$$<f^q[\alpha SL]J||U^{(t)}||f^q[\alpha'S'L']J'> = \sum_{\substack{\alpha SL\\ \alpha'S'L'}} C_{\alpha SL}C_{\alpha'S'L'} <f^q\alpha SLJ||U^{(t)}||f^q\alpha'S'L'J'> \tag{2.76}$$

where the $<f^q\alpha SLJ||U^{(t)}||f^q\alpha'S'L'J'>$ are the matrix elements of the unit tensor operators between pure $L-S$ coupled states. These matrix elements can be further reduced by factoring out the J dependence.

The total oscillator strength P of a transition at frequency ν is then given by

$$P(aJ;bJ') = \frac{8\pi^2 m\nu}{3h(2J+1)}\left[\frac{(n^2+2)^2}{9n} S_{ed} + nS_{md}\right] \tag{2.77}$$

The connection between integrated absorbance of an electric dipole with linestrength S is:

$$\int_{band} k(\lambda)d\lambda = \rho \frac{8\pi^3 e^2\bar{\lambda}}{3hc(2J+1)} \frac{1}{n} \frac{(n^2+2)^2}{9} S_{ed} \tag{2.78}$$

where $k(\lambda)$ is the absorption coefficient at wavelength λ, ρ is the Nd^{3+} ion concentration, $\bar{\lambda}$ is the mean wavelength of the absorption band, J is the total angular momentum of the initial level and $n = n(\bar{\lambda})$ is the bulk index of refraction at wavelength $\bar{\lambda}$. The factor $(n^2+2)^2/9$ represents the local field correction for the ion in a dielectric medium.

The spontaneous emission rate is

$$A(aJ;bJ') = \frac{64\pi^4\nu^3}{3hc^3(2J+1)}\left[\frac{n(n^2+2)^2}{9} S_{ed} + n^3 S_{md}\right] \tag{2.79}$$

in case of electric dipole emission this equation becomes

$$A(aJ;bJ') = \frac{64\pi\nu^3}{3hc^3(2J+1)} \frac{n(n^2+2)^2}{9} e^2 \sum_{t=2,4,6} \Omega_t|<(S,L)J||U^{(t)}||(S',L')J'>|^2 \tag{2.80}$$

where n is the index of refraction of the crystal. The integrated absorption cross-section is related to the oscillator strength by:

$$\int \sigma(\bar{\nu})d\bar{\nu} = \frac{\pi e^2}{mc^2} P \tag{2.81}$$

where $\bar{\nu}$ is measured in inverse centimeters. For a Lorentzian line,

$$\sigma(\nu) = \frac{\sigma_0}{1 + 4(\nu - \nu_0)^2/(\Delta\nu)^2} \tag{2.82}$$

where $\Delta\nu$ is the full width at half maximum, ν_0 is the frequency at the peak, and σ_0 is the peak cross-section. Then:

$$\int \sigma(\nu) d\nu = \frac{\pi}{2} \sigma_0 \Delta\nu \tag{2.82a}$$

Which yields

$$\sigma_0 = \frac{2e^2}{mc^2 \Delta\nu} P \tag{2.82b}$$

Thus, for a given linewidth, the peak cross-section for stimulated emission is proportional to the oscillator strength. Eq. (2.82 b) also shows the importance of the line width, $\Delta\nu$, in the determination of the cross-section.

The peak induced emission cross section is related to the radiative probability by transition:

$$\sigma(\lambda_p) = \frac{\lambda_p^4}{8\pi c n^2 \Delta\lambda_{eff}} A(aJ\colon bJ') \tag{2.83}$$

where the effective linewidth $\Delta\lambda_{eff}$ is used, since for glasses the absorption and emission bands are characteristically asymmetrical.

The cross-section may also be written as:

$$\sigma_p \Delta\bar{\nu} = \int \sigma(\bar{\nu}) d\bar{\nu} = \frac{\lambda_p^2}{8\pi c n^2} A(aJ; bJ') \tag{2.83a}$$

The radiative lifetime of level J can be expressed in terms of the spontaneous emission probabilities as

$$\tau_J^{-1} = \sum_{J'} A(J, J'), \tag{2.84}$$

where the summation is over all terminal levels J'. The fluorescence-branching ratio from level J to level J'' is given by

$$\beta_{JJ''} = \frac{A(J, J')}{\sum_{J'} A(J, J')} \tag{2.85}$$

In addition to the spectroscopic properties, the Judd-Ofelt intensity parameters can also be used to estimate excited-state absorption and the probability of ion-ion interactions that enter into energy-transfer and fluorescence-quenching phenomena.

From Eqs. (2.83) and (2.83a) it can be seen that σ_p depends on the intensity parameters Ω and the bandwidth $\Delta\lambda_{eff}$. Both are affected by compositional changes. The bandwidth $\Delta\lambda_{eff}$ is a measure of the overall extent of the Stark splitting of the J-manifold and is inhomogeneous due to the site-to-site variation in the local field seen by the rare-earth ion. In the following we concentrate on the relationship between Ω values and the glass composition. For example the ${}^4F_{3/2} \rightarrow {}^4I_{11/2}$ laser transition of neodymium is dependent only on the Ω_4 and Ω_6 parameters because of the triangle rule $|J - J'| \leqslant t \leqslant |J + J'|$. For a large cross section one wants the Ω values to be as large as possible (assuming constant $\Delta\lambda_{eff}$). In addition, the fluorescence-branching ratio to ${}^4I_{11/2}$ should be as large as possible.

c) Relaxation Processes in Solid State Lasers

Accurate laser design calculation for lasing media require information about the rates of various nonradiative processes. These include ion-ion energy transfer which can give rise to concentration quenching and non-exponential decay and relaxation by multiphonon emission which is usually essential to completing the overall scheme and can effect the quantum efficiency.

For low concentration of rare earth dopant ions the principle nonradiative decay mechanism is multiphonon emission. The results of the theory of multiphonon emission in crystals were recently summarized by Riseberg and Weber [46] and in glasses by Reisfeld [35].

The *ab initio* calculation of the transition rate between two electronic states with the emission of p phonons involves a very complicated sum over phonon modes and intermediate states. Due to this complexity, these sums are extremely difficult to compute; however, it is just this complexity which permits a very simple phenomenological theory to be used. There are an extremely large number of ways in which p phonons can be emitted and the sums over phonon modes and intermediate states are essentially a statistical average of matrix elements. In the phenomenological approach it is assumed that the ratio of the pth and $(p-1)^{th}$ processes will be given by a coupling constant characteristic of the host crystal, but not dependent on the rare-earth electronic states. For a given lattice at low temperature the spontaneous relaxation rate is given by

$$W(0) = Be^{\alpha \Delta E} \tag{2.86}$$

where B and α are characteristic of the host (α is negative). Thus, the graph of the spontaneous rate vs energy gap will be a straight line when this approach is valid. Experimental data has shown that the approach is very good for a large variety of hosts. In this way, all multiphonon rates can be inferred from a few measured rates.

The dominant emission process is the one which requires the least number of phonons to be emitted. The minimum number of phonons required for a transition between states separated by an energy gap ΔE is given by

$$p = \frac{\Delta E}{\hbar\omega_{max}} \tag{2.87}$$

where $\hbar\omega_{max}$ is the maximum energy of optical phonons. With increased temperature, stimulated emission of phonons by thermal phonons increases the relaxation rate W according to

$$W(T) = W(0)\,(1 + \bar{n}\,(\hbar\omega_{max}))^{P} \tag{2.88}$$

where $\bar{n}$ is the average occupation number of phonons at energy $\hbar\omega_{max}$

$$\bar{n}\,(\hbar\omega_{max}) = \frac{1}{\exp\left(\frac{\hbar\omega_{max}}{kT}\right) - 1} \tag{2.89}$$

Combining the last four equations yields

$$W = Be^{\alpha\Delta E}\left\{1 - e^{-\frac{\hbar\omega_{max}}{kT}}\right\}^{-\frac{\Delta E}{\hbar\omega_{max}}} \tag{2.88a}$$

The quantities B, α and $\hbar\omega_{max}$ are regarded as phenomenological parameters describing the process.

Based on the above theories an extensive program was performed at the Center of Laser Studies to find potential new rare earth CW solid state lasers capable of operating at room temperature. The data presented below are based on this report [47].

d) Crystal Rare Earth Lasers

The strong sharp-line fluorescence from rare earth doped crystals is the reason that these materials are used extensively as active media in optically pumped laser devices. The crystalline host material determines the behaviour of impurity ions and laser characteristics in a number of ways. The static crystalline Stark effect removes most of the degeneracy of the free ions state and is a major factor in determining the energy levels of the impurities. The dynamic interaction of the impurity ions with the phonons of the lattice is responsible for the line-width, the nonradiative relaxation process and for the internal efficiency of the system.

All existing trivalent rare earth lasers are 4 – level lasers. The terminal state is either a crystal-field split component of the ground level (Yb^{3+}, Tm^{3+}, Er^{3+}, Ho^{3+}, Pr^{3+}) or is a spin-orbit component above the ground level (Ho^{3+}, Eu^{3+}, Nd^{3+}, Pr^{3+}...).

Crystallographic data on crystal laser hosts can be found in Ref. [48] and values of lattice constants, density and other properties in Refs. [48] and [49]. The properties of a number of neodymium laser materials are reviewed in Ref. [50].

The methods of growth and chemistry of laser crystals may be found in Ref. [51].

Examples of some trivalent and divalent rare earth doped crystal lasers are summarized by Chessler and Geusic [4].

For a CW room temperature, four level laser as designed in Fig. 1, the general requirements are as follows:

a) Level 2 thermally unpopulated at room temperature.
b) A large radiative line strength between levels 2 and 3.
c) Level 3 relaxed predominantly via radiative processes.
d) Lifetime of level 3 longer than level 2.
e) High population of level 4 which may be achieved either by direct absorption to this level or by high energy transfer probability to this level from a strongly absorbing donor system.

The first condition requires the energy of the lower last level to be much greater than $kT = 200\ cm^{-1}$ at room temperature. Figure 2 shows the principal energy levels of all the trivalent rare-earth ions. It is seen that the latter condition is satisfied in all, but one ion by any energy level other than the ground state. In the one exception, Eu^{3+}, consideration must begin with the second excited state. In certain cases, transitions to the upper Stark components of the ground state manifold may be applicable, but the Judd-Ofelt theory does not allow computation of individual Stark component line strengths (it averages over two manifolds) so these possibilities will not be considered for CW operation here.

The theories described above can be used to optimize the combinations of ions and hosts in order to obtain good laser qualities.

Because of the electrostatic screening exhibited by the closed 5p shell electrons, the matrix elements of the unit tensor operator $U^{(t)}$ between two energy manifolds in a given rare earth ion do not vary significantly when it is incorporated in different hosts. Therefore, the matrix elements computed in the free ion approximation may be used in calculation of the transition probabilities in different media.

On the other hand, the intensity parameters Ω_t are sensitive to the host due to their dependence on site symmetry, radial integral and energy difference between the 4f and the next perturbing configuration. These parameters may be varied up to five times the value in different media. It is important to emphasize that a parameter Ω_t having the same value t as the highest $U^{(t)}$ is of dominant importance.

For example, if a particular transition is characterized by a large value of U^4, a host should be chosen with a large value of Ω_4 to optimize the cross-section for stimulated emission. Sufficient data is not yet available to do this with many ions, however, in the case of Nd^{3+}, five sets of intensity parameters are available. This data has been used to predict the optimum host for laser action on the 0.9 μ line in Nd^{3+}. Thus, large radiative line strengths are identified using the Judd-Ofelt theory satisfying conditions b and e. The multiphonon and radiative emission rates from level 3 are calculated to obtain the total level lifetime. If the radiative lifetime is essentially equal to the total lifetime, condition c is met. In most cases the fourth condition will be met if the decay of the lowest laser level is predominantly non-radiative. Since the

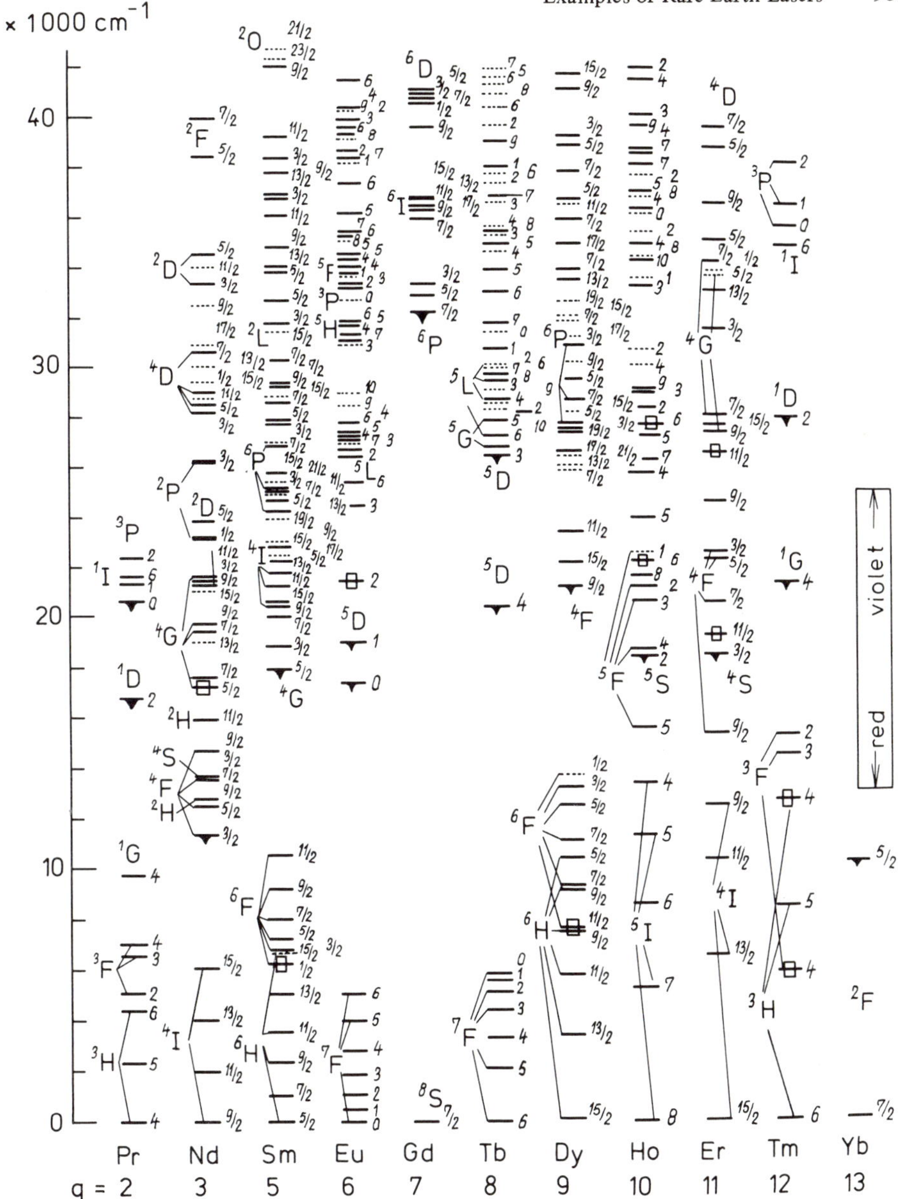

Fig. 2. Energy Levels (J indicated at the right hand) of the trivalent lanthanides (except cerium and promethium) in the unit 1,000 cm^{-1} as a function of the number q of 4f electrons. Excited levels frequently showing luminescence are indicated by a black triangle. The excited levels corresponding to hypersensitive transitions from the groundstate are marked with a square. In cases where the quantum numbers S and L are reasonably well-defined, the Russell-Saunders terms are given at the left. Calculated energy levels are shown as stippled lines. This figure is an extended and modified version of a diagram (known from a book by Dieke) resuming the work of a generation of rare earth spectroscopists. The empty area (shaped like a wine leaf) in the middle of the figure (q = 5 to 9) corresponds to a large gap between the highest level of a term having the same S_{max} as the groundstate, and the lowest (frequently fluorescent) level of the lowest term with $(S_{max} - 1)$

multiphonon rate decreases exponentially with increasing energy gap, this can only occur if the energy gap below level 2 is smaller than the gap below level 3.

Figure 3 indicates most of the known rare earth laser lines. Referring to this figure we can see that many systems of rare earth levels, satisfy the above-mentioned conditions. A complete summary of insulating crystal laser hosts including the space group, and crystallographic data may be found in the Handbook of Lasers [5]. It should be noted that the recently developed ultraphosphate and related crystals which are not included in these Tables will be discussed later.

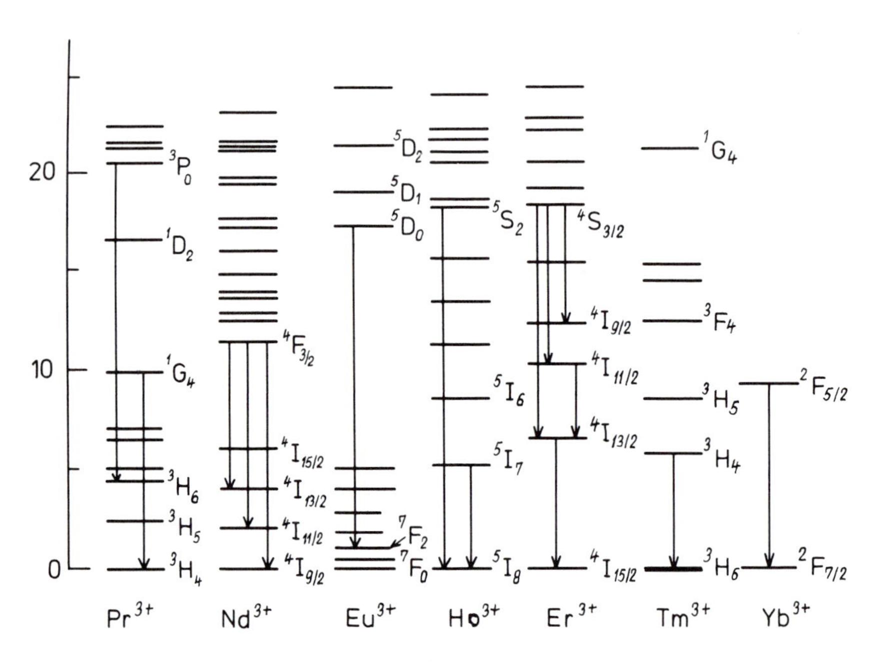

Fig. 3. Energy levels (in the unit 1,000 cm^{-1}) and known laser transitions (Ref. [8]) in trivalent lanthanides. The same levels below 25,000 cm^{-1} are given as in Fig. 2, but only a few important levels are identified with their Russell-Saunders symbols. Most authors call the first excited state of thulium $^3H_4$ in spite of the eigen-vector containing only 7 percent squared amplitude of this component, but 65 percent $^3F_4$ (as we name this level in Fig. 2 and elsewhere in the text

Intensity parameters Ω_t for rare earths in various crystals are presented in Table 6. Those parameters are calculated from the experimentally measured electric dipole oscillator strengths and the calculated matrix elements between various levels of the rare earth ions as explained in detail in the Chapter of Radiative and Non-radiative transitions.

Recently, the Ω_t parameters given in Table 6 were used in a computer program developed by Caird [47] for calculation of the electric dipole transition probabilities of rare earth ions in various crystal lattices. The matrix elements in the intermediate coupling scheme needed for such calculation may be found in references summarized below.

Table 6. Judd-Ofelt intensity parameters (in unit 10^{-20} cm^2) for trivalent lanthanides in various solids

	Host material	Ω_2	Ω_4	Ω_6
Nd	$Y_3Al_5O_{12}$	0.2	2.7	5.0
Pr	Y_2O_3	17.21	19.8	4.88
Nd	Y_2O_3	8.55	5.25	2.89
Eu	Y_2O_3	6.31	0.66	0.48
Er	Y_2O_3	4.59	1.21	0.48
Tm	Y_2O_3	4.07	1.46	0.61
Pr	$YAlO_3$	2.0	6.0	7.0
Nd	$YAlO_3$	1.24	4.68	5.85
Eu	$YAlO_3$	2.66	6.32	0.80
Tb	$YAlO_3$	3.25	7.13	2.00
Ho	$YAlO_3$	1.82	2.38	1.53
Er	$YAlO_3$	1.06	2.63	0.78
Tm	$YAlO_3$	0.67	2.30	0.74
Pr	LaF_3	0.12	1.77	4.78
Nd	LaF_3	0.35	2.57	2.50
Eu	LaF_3	1.19	1.16	0.39
Ho	LaF_3	1.16	1.38	0.88
Er	LaF_3	1.07	0.28	0.63

Pr^{3+}	M. J. Weber [52]
Nd^{3+} and Er^{3+}	K. Rajnak [53]
Sm^{3+} and Dy^{3+}	B. G. Wybourne and J. Conway [54]
Eu^{3+} and Tb^{3+}	G. S. Ofelt [45]
Ho^{3+}	V. L. Donlan [55]
Tm^{3+}	W. F. Krupke [56]

Caird's program also calculates the magnetic dipole probabilities using the equation

$$(aJ'; bJ') = \frac{\hbar^2}{4m^2c^2} \, |< f^q[\alpha SL]J \,||\, \vec{L} + 2\vec{S} \,||\, f^q[\alpha' S' L']J >|^2 \qquad (2.90)$$

From these electric and magnetic dipole line strengths, the total oscillator strengths are obtained. The branching ratio as given by formula (2.85) for the laser transition were also calculated.

The calculation of oscillator strengths and radiative transition rates required knowledge of the index of refraction of the material at the wavelength corresponding to the transition. Index of refraction data has been obtained from the literature for all crystals for which multiphonon parameters were available.

The relevant data for various rare earth ions in $YAlO_3$ are presented in Tables 7 (Pr^{3+}), 8 (Nd^{3+}), 9 (Dy^{3+}), 10 (Ho^{3+}), 11 (Er^{3+}), and 12 (Tm^{3+}) [47].

Table 7. Transitions in praseodymium in $YAlO_3$ with high oscillator strengths *P*

Transition	Wavelength (microns)	Oscillator strength ($P \times 10^6$)	Upper level lifetime (msec)	Lower level decay rate (sec^{-1})	Spontaneous emission rate (sec^{-1})	Total branching ratio
$^3H_5 \rightarrow {}^3H_4$	4.65	2.42	0.003		27.5	0.0001
$^3H_6 \rightarrow {}^3H_5$	4.47	2.27	0.005	3.3×10^5	27.9	0.0001
$^1G_4 \rightarrow {}^3F_4$	3.26	2.03	0.149	8.5×10^8	46.7	0.007
$^1G_4 \rightarrow {}^3H_6$	1.81	3.53	0.149	2.2×10^5	266	0.040
$^1G_4 \rightarrow {}^3H_5$	1.29	4.97	0.149	3.3×10^5	740	0.110
$^1D_2 \rightarrow {}^1G_4$	1.35	5.00	0.151	6.7×10^3	676	0.102
$^1D_2 \rightarrow {}^3F_4$	0.95	4.85	0.151	8.5×10^8	1,320	0.199
$^1D_2 \rightarrow {}^3F_2$	0.81	2.61	0.151	3.9×10^8	988	0.149
$^1D_2 \rightarrow {}^3H_6$	0.77	2.28	0.151	2.2×10^5	951	0.143
$^1D_2 \rightarrow {}^3H_4$	0.58	3.21	0.151		2,430	0.367
$^3P_0 \rightarrow {}^1G_4$	0.87	5.97	0.011	6.7×10^3	1,950	0.021
$^3P_0 \rightarrow {}^3F_4$	0.69	21.7	0.011	8.5×10^8	11,500	0.122
$^3P_0 \rightarrow {}^3F_2$	0.61	19.9	0.011	3.9×10^8	13,500	0.143
$^3P_0 \rightarrow {}^3H_6$	0.59	17.9	0.011	2.2×10^5	13,000	0.138
$^3P_0 \rightarrow {}^3H_4$	0.47	46.1	0.011		54,000	0.575

Table 8. Transitions in neodymium in $YAlO_3$ with high oscillator strength *P*

Transition	Wavelength (microns)	Oscillator strength ($P \times 10^6$)	Upper level lifetime (msec)	Lower level decay rate (sec^{-1})	Spontaneous emission rate (sec^{-1})	Total branching ratio
$^4I_{11/2} \rightarrow {}^4I_{9/2}$	5.37	2.53	0.0008		21.5	$< 10^{-4}$
$^4I_{13/2} \rightarrow {}^4I_{11/2}$	5.04	2.54	0.0014	1.2×10^6	24.6	$< 10^{-4}$
$^4I_{13/2} \rightarrow {}^4I_{9/2}$	2.60	1.52	0.0014		55.2	10^{-4}
$^4I_{15/2} \rightarrow {}^4I_{13/2}$	4.85	2.55	0.0020	7.1×10^5	26.6	10^{-4}
$^4I_{15/2} \rightarrow {}^4I_{11/2}$	2.47	1.28	0.0020	1.2×10^6	51.4	10^{-4}
$^4F_{3/2} \rightarrow {}^4I_{13/2}$	1.35	4.66	0.157	7.1×10^5	629	0.099
$^4F_{3/2} \rightarrow {}^4I_{11/2}$	1.06	14.6	0.157	1.2×10^6	3,170	0.498
$^4F_{3/2} \rightarrow {}^4I_{9/2}$	0.89	8.07	0.157		2,530	0.398
$^2P_{3/2} \rightarrow {}^2G_{9/2}$	1.95	3.25	0.007	4.75×10^9	210	0.0015
$^2P_{3/2} \rightarrow {}^4G_{9/2}$	1.49	2.11	0.007	8.4×10^8	235	0.0017
$^2P_{3/2} \rightarrow {}^2K_{13/2}$	1.39	5.15	0.007	2.0×10^6	653	0.0048
$^2P_{3/2} \rightarrow {}^4F_{9/2}$	0.87	3.51	0.007	2.6×10^7	1,140	0.0083
$^2P_{3/2} \rightarrow {}^2H_{9/2}$	0.74	3.35	0.007	2.4×10^9	1,520	0.0111
$^2P_{3/2} \rightarrow {}^4I_{9/2}$	0.38	1.12	0.007	1.2×10^6	1,700	0.0124

Table 9. Transitions in dysprosium in $YAlO_3$ with high oscillator strength P

Transition	Wavelength (microns)	Oscillator strength ($P \times 10^6$)	Upper level lifetime (msec)	Lower level decay rate (sec^{-1})	Spontaneous emission rate (sec^{-1})	Total branching ratio
$^6H_{13/2} \rightarrow {}^6H_{15/2}$	2.89	2.65	1.12		77.6	0.087
$^6H_{11/2} \rightarrow {}^6H_{13/2}$	4.31	1.92	0.007	9.0×10^2	25.4	2×10^{-4}
$^6H_{11/2} \rightarrow {}^6H_{15/2}$	1.73	1.59	0.007		130	9×10^{-4}
$^4F_{9/2} \rightarrow {}^6F_{11/2}$	0.75	0.59	0.415	4.0×10^8	219	0.091
$^4F_{9/2} \rightarrow {}^6H_{11/2}$	0.65	0.43	0.415	1.5×10^5	219	0.091
$^4F_{9/2} \rightarrow {}^6H_{13/2}$	0.57	1.53	0.415	9.0×10^2	1,070	0.442
$^4F_{9/2} \rightarrow {}^6H_{15/2}$	0.47	0.59	0.415		610	0.253

Table 10. Transitions in holmium in $YAlO_3$ with high oscillator strength P

Transition	Wavelength (microns)	Oscillator strength ($P \times 10^6$)	Upper level lifetime (msec)	Lower level decay rate (sec^{-1})	Spontaneous emission rate (sec^{-1})	Total branching ratio
$^5I_7 \rightarrow {}^5I_8$	2.0	2.48	6.44		155	1.00
$^5I_6 \rightarrow {}^5I_7$	2.86	1.72	1.01	1.6×10^2	51.9	0.052
$^5I_6 \rightarrow {}^5I_8$	1.17	1.56	1.01		280	0.282
$^5I_5 \rightarrow {}^5I_6$	3.91	1.26	0.019	9.9×10^2	20.2	4×10^{-4}
$^5I_5 \rightarrow {}^5I_7$	1.65	1.58	0.019	1.6×10^2	143	0.003
$^5I_4 \rightarrow {}^5I_5$	4.86	1.22	0.002	5.2×10^4	12.7	$< 10^{-4}$
$^5I_4 \rightarrow {}^5I_6$	2.17	1.13	0.002	9.9×10^2	59.0	1×10^{-4}
$^5F_5 \rightarrow {}^5I_6$	1.45	1.39	0.006	9.9×10^2	164	9×10^{-4}
$^5F_5 \rightarrow {}^5I_7$	0.96	2.86	0.006	1.6×10^2	767	0.004
$^5F_5 \rightarrow {}^5I_8$	0.65	5.46	0.006		3,270	0.019
$^5S_2 \rightarrow {}^5I_4$	1.99	1.00	0.038	5.0×10^5	62.5	0.002
$^5S_2 \rightarrow {}^5I_6$	1.04	1.12	0.038	9.9×10^2	258	0.010
$^5S_2 \rightarrow {}^5I_7$	0.76	3.38	0.038	1.6×10^2	1,450	0.055
$^5S_2 \rightarrow {}^5I_8$	0.55	2.62	0.038		2,190	0.084
$^5F_3 \rightarrow {}^5I_4$	1.35	1.81	0.002	5.0×10^5	246	4×10^{-4}
$^5F_3 \rightarrow {}^5I_5$	1.05	1.51	0.002	5.2×10^4	335	5×10^{-4}
$^5F_3 \rightarrow {}^5I_6$	0.83	1.91	0.002	9.9×10^2	689	0.001
$^5F_3 \rightarrow {}^5I_7$	0.64	4.29	0.002	1.6×10^2	2,600	0.004

Table 11. Transitions in erbium in $YAlO_3$ with high oscillator strength P

Transition	Wavelength (microns)	Oscillator strength ($P \times 10^6$)	Upper level lifetime (msec)	Lower level decay rate (sec^{-1})	Spontaneous emission rate (sec^{-1})	Total branching ratio
$^4I_{13/2} \rightarrow ^4I_{15/2}$	1.54	2.04	4.75		211	1.00
$^4I_{11/2} \rightarrow ^4I_{13/2}$	2.78	1.29	1.63	2.1×10^2	41.0	0.22
$^4F_{9/2} \rightarrow ^4I_{11/2}$	1.95	1.32	0.077	6.2×10^2	85.4	0.007
$^4F_{9/2} \rightarrow ^4I_{15/2}$	0.657	5.67	0.077		3,290	0.255
$^4S_{3/2} \rightarrow ^4I_{9/2}$	1.67	1.21	0.14	2.5×10^5	107	0.015
$^4S_{3/2} \rightarrow ^4I_{11/2}$	1.22	0.38	0.14	6.2×10^2	51.2	0.007
$^4S_{3/2} \rightarrow ^4I_{13/2}$	0.85	1.61	0.14	2.1×10^2	557	0.078
$^4S_{3/2} \rightarrow ^4I_{15/2}$	0.55	1.66	0.14		1,410	0.197
$^2P_{3/2} \rightarrow ^2G_{9/2}$	1.42	2.63	0.095	6.5×10^5	320	0.030
$^2P_{3/2} \rightarrow ^4F_{5/2}$	1.07	0.93	0.095	3.6×10^6	201	0.019
$^2P_{3/2} \rightarrow ^4F_{9/2}$	0.62	1.15	0.095	1.3×10^4	759	0.072
$^2P_{3/2} \rightarrow ^4I_{9/2}$	0.52	1.13	0.095	2.5×10^5	1,050	0.100
$^2P_{3/2} \rightarrow ^4I_{11/2}$	0.49	3.37	0.095	6.1×10^2	3,940	0.375
$^2P_{3/2} \rightarrow ^4I_{13/2}$	0.40	1.54	0.095	2.1×10^2	2,490	0.237

Table 12. Transitions in thulium in $YAlO_3$ with high oscillator strength P

Transition[a]	Wavelength (microns)	Oscillator strength ($P \times 10^6$)	Upper level lifetime (msec)	Lower level decay rate (sec^{-1})	Spontaneous emission rate (sec^{-1})	Total branching ratio
$^3F_4 \rightarrow ^3H_6$	1.8	2.72	4.83		207	1.00
$^3H_5 \rightarrow ^3H_6$	1.22	2.24	0.025		370	0.009
$^3H_4 \rightarrow ^3H_5$	2.35	1.29	0.86	4.0×10^4	57.3	0.050
$^3H_4 \rightarrow ^3F_4$	1.46	1.08	0.86	2.1×10^2	126	0.11
$^3H_4 \rightarrow ^3H_6$	0.80	2.51	0.86		964	0.83
$^1G_4 \rightarrow ^3H_4$	1.17	1.06	0.41	1.2×10^3	190	0.08
$^1G_4 \rightarrow ^3H_5$	0.78	1.84	0.41	4.0×10^4	752	0.31
$^1G_4 \rightarrow ^3H_6$	0.48	1.03	0.41		1,150	0.48

[a] It may be noted that we describe the lowest level with $J = 4$ as $^3F_4$ since it has a smaller squared amplitude of $^3H_4$.

The multiphonon relaxation rates may be obtained experimentally from the measured lifetime of a fluorescence state, and the radiative transition probabilities of this state, by use of formula:

$$\tau(aJ) = \frac{1}{A(aJ) + W(aJ)} \tag{2.91}$$

For a given lattice, at low temperature, the spontaneous relaxation rate is given by:

$$W(0) = Be^{\alpha \Delta E} \tag{2.86}$$

Thus, the plot of W(0) versus ΔE will be a straight line, from the intercept and slope of which, the multiphonon parameters B and α may be obtained. As mentioned previously those parameters are independent of the electronic levels or of a specific rare earth ion, and depends solely on the vibration frequency (phonon) of the host lattice. Multiphonon parameters for crystal laser hosts are presented in Table 13.

As seen from these tables, most of the oscillator strengths of the R. E. Ions in $YAlO_3$ have values of 10^{-6} to 10^{-5}, the striking exception being the transition from the $^3P_0$ level to $^3F_{4,2}$ and $^3H_{6,4}$ in the Pr^{3+} system (Table 7). Praseodymium however, cannot be described accurately by the Judd-Ofelt theory without introducing additional corrections to this theory, the main reason for such a discrepancy being the fact that the 5d level is low lying and may contribute considerably to the oscillator strengths of the 4f–4f transition. A similar perturbation on the high lying levels of Sm^{3+} by the charge transfer band was observed in glasses [57].

Recently, Hormadaly and Reisfeld showed that Ω_2 behaves more regularly if the hypersensitive transitions are not included in the evaluation.

The case of $YAlO_3$ in the aforementioned tables may serve as a fairly good prediction for the value of branching ratios in other crystalline hosts. This is based on the fact that when hypersensitive transitions are absent, the branching ratios are similar in different crystalline hosts. Usually, the branching ratios are the highest for a good laser transition. This may be seen in Table 14 which presents an example of several laser transitions in crystals. CW operation is achieved only in cases when in addition to high branching ratios the lifetime of the lasing level is at least 0.26 ms.

Based on this knowledge, Caird [47] predicted new laser transitions, taking also into account a high multiphonon relaxation from the upper and lower laser levels. For example, the transitions $^3H_5 \rightarrow {}^3H_4$ and $^3H_6 \rightarrow {}^3H_5$ of Pr^{3+} show large oscillator strengths in the 4.5 μ region, however, because of the high multiphonon relaxation in $YAlO_3$ the total branching ratio (which consists of the sum of the radiative and non-radiative transitions) will be very small, therefore this medium is not suitable for

Table 13. Multi-phonon parameters of various laser hosts

Host material	B(sec^{-1})	α (cm)	$\hbar\omega_{max}$ (cm^{-1})
$Y_3Al_5O_{12}$	$2.235 \cdot 10^8$	$-3.50 \cdot 10^{-3}$	700
$YAlO_3$	$6.425 \cdot 10^9$	$-4.69 \cdot 10^{-3}$	600
Y_2O_3	$1.204 \cdot 10^8$	$-3.53 \cdot 10^{-3}$	600
LaF_3	$3.966 \cdot 10^9$	$-6.45 \cdot 10^{-3}$	305
$LaCl_3$	$3.008 \cdot 10^{10}$	$-1.37 \cdot 10^{-2}$	240
SrF_2	$3.935 \cdot 10^8$	$-4.60 \cdot 10^{-3}$	350

Table 14. Data for known lasers containing trivalent lanthanides.
"YAG" is yttrium aluminium garnet $Y_3Al_5O_{12}$. Divalent crystal means that a major part of the lanthanide is incorporated in its bivalent form

Ion	Crystal	Transition	Wave-length (microns)	Oscillator strength ($P \times 10^6$)	Oscillator strength in $YAlO_3$ ($P \times 10^6$)	Upper level lifetime (msec)	Spontaneous emission rate (sec^{-1})	Total branching ratio	Comments
Pr	$CaWO_4$	$^1G_4 \rightarrow {}^3H_4$	1.04	–	0.26	0.15	63.3	0.009	Divalent crystal used, Pulsed, 77 °K
Pr	LaF_3	$^3P_0 \rightarrow {}^3H_6$	0.60	9.3	–	0.065	4,620	0.0302	Pulsed, 77 °K
Nd	YAG	$^4F_{3/2} \rightarrow {}^4I_{13/2}$	1.35	3.64	–	0.26	438	0.114	CW, 295 °K
Nd	YAG	$^4F_{3/2} \rightarrow {}^4I_{11/2}$	1.06	10.6	–	0.26	2,060	0.535	CW, 295 °K
Nd	YAG	$^4F_{3/2} \rightarrow {}^4I_{9/2}$	0.9	4.75	–	0.26	1,330	0.346	Pulsed, 295 °K, ground state is terminal level
Eu	Y_2O_3	$^5D_0 \rightarrow {}^7F_2$	0.61	0.73	–	1.60	483	0.772	Pulsed, 220 °K
Ho	$YAlO_3$	$^5I_7 \rightarrow {}^5I_8$	2.0	2.48	2.48	6.44	155	0.996	CW, 77 °K, ground state is terminal level
Ho	CaF_2	$^5S_2 \rightarrow {}^5I_8$	0.55	–	2.62	0.04	2,190	0.083	Divalent crystal used, Pulsed, 77 °K
Er	$YAlO_3$	$^4S_{3/2} \rightarrow {}^4I_{9/2}$	1.66	1.21	1.21	0.14	107	0.015	Pulsed, 295 °K
Er	CaF_2	$^4S_{3/2} \rightarrow {}^4I_{11/2}$	1.2	–	0.31	0.14	51.2	0.007	Divalent crystal used, Pulsed, 77 °K
Er	CaF_2	$^4S_{3/2} \rightarrow {}^4I_{13/2}$	0.85	–	1.61	0.14	557	0.078	Divalent crystal used, Pulsed, 77 °K
Er	YAG	$^4I_{11/2} \rightarrow {}^4I_{13/2}$	2.69	–	1.29	1.63	41.0	0.067	Pulsed, 298 °K
Er	YAG	$^4I_{13/2} \rightarrow {}^4I_{15/2}$	1.6	–	2.04	4.75	211	1.00	Pulsed, 295 °K, ground state is the terminal level
Tm	$YAlO_3$	$^3H_4 \rightarrow {}^3H_5$	2.35	1.29	1.29	0.86	57.3	0.050	Pulsed, 295 °K
Tm	$YAlO_3$	$^3F_4 \rightarrow {}^3H_6$	1.9	2.72	2.72	5.0	207	1.00	Pulsed, 295 °K, ground state is the terminal level

lasers based on these transitions. On the other hand, a crystal having lower phonon energy than $YAlO_3$ (750 cm^{-1}) [58] such as $LaCl_3$ (260 cm^{-1}) or $LaBr_3$ (175 cm^{-1}) may yield the proper conditions for laser action [59].

As another example we can consider the $^5I_6 \rightarrow {}^5I_8$ transition of Ho^{3+} at 1.7 μ which should lase at liquid nitrogen temperature, just as the $^5I_7 \rightarrow {}^5I_8$ transition does at 2 microns. The $^5I_6 \rightarrow {}^5I_7$ line at 2.86 microns is the first member of a series of strong transitions terminating at the $^5I_7$ level. It appears that a number of these lines will lase and that CW action could be obtained at room temperature if the $^5I_7$ level could be effectively depopulated. Unfortunately, the 6.4 msec lifetime of this level is the longest lifetime of all the Ho^{3+} states. However, a number of rare earth ion co-dopants could be used to quench the $^5I_7$ level, specifically, Pr^{3+}, Eu^{3+} and Tb^{3+}. The particular choice would depend on which upper level of holmium is involved. In the case of the $^5I_6 \rightarrow {}^5I_7$ transition, Eu^{3+} is a good quencher because the uppermost level of the ^7F multiplet is further separated from the $^5I_6$ level of Ho^{3+}. With quenching of the $^5I_7$ level the $^5I_6 \rightarrow {}^5I_7$ transition should provide a CW, room temperature laser at 2.86 microns.

Additional examples for future solid lasers as predicted by Caird may be found in Table 15.

Table 15. New predicted lasers of highest potential efficiency

Rare earth ion	Host material	Sensitizer, and/or quenching ions	Transition	Wavelength (microns)	Oscillator strength ($P \times 10^6$)
Pr	Y_2O_3	Yb or Nd	$^1G_4 \rightarrow {}^3F_4$	3.26	3.26
Pr	Y_2O_3	Yb or Nd	$^1G_4 \rightarrow {}^3H_6$	1.81	9.88
Pr	Y_2O_3	Yb or Nd	$^1G_4 \rightarrow {}^3H_5$	1.29	6.29
Pr	$YAlO_3$	Sm	$^1D_2 \rightarrow {}^3F_4$	0.95	4.85
Pr	$YAlO_3$	Sm	$^1D_2 \rightarrow {}^3F_2$	0.81	2.61
Pr	$YAlO_3$	Sm	$^1D_2 \rightarrow {}^3H_6$	0.77	2.28
Pr	$YAlO_3$	Sm	$^1D_2 \rightarrow {}^3H_4$	0.58	3.21
Pr	$YAlO_3$	–	$^3P_0 \rightarrow {}^3F_4$	0.69	21.7
Pr	$YAlO_3$	Dy	$^3P_0 \rightarrow {}^3F_2$	0.61	19.9
Pr	$YAlO_3$	Dy	$^3P_0 \rightarrow {}^3H_4$	0.47	46.1
Nd	Y_2O_3	–	$^4F_{3/2} \rightarrow {}^4I_{9/2}$	0.89	7.79
Pm	$SeOCl_2$ (liquid)	–	$^5F_1 \rightarrow {}^5I_5$	0.92	–
Dy	$YAlO_3$	–	$^6H_{13/2} \rightarrow {}^6H_{15/2}$	2.89	2.65
Dy	$YAlO_3$	Eu	$^4F_{9/2} \rightarrow {}^6H_{13/2}$	0.60	1.53
Ho	$YAlO_3$	–	$^5I_6 \rightarrow {}^5I_8$	1.17	1.56
Ho	$YAlO_3$	Eu	$^5I_6 \rightarrow {}^5I_7$	2.86	1.72
Ho	$YAlO_3$	Eu or Tb	$^5S_2 \rightarrow {}^5I_7$	0.76	3.38
Tm	$YAlO_3$	Cr + Tb	$^3H_4 \rightarrow {}^3F_4$	1.46	1.08
Tm	$YAlO_3$	Cr	$^3H_4 \rightarrow {}^3H_6$	0.80	2.51

e) The Specific Case of Nd^{3+} Crystal Lasers

In view of our latest understanding of these laser requirements it is not surprising that Nd:YAG, $(Nd_xY_{1-x})_3Al_5O_{12}$, is one of the best known crystal lasers. The transition $^4F_{3/2} \rightarrow {}^4I_{11/2}$ (1.06 μ) of Nd^{3+} has an oscillator strength of 10.6×10^{-6}, the upper level lifetime is 0.26 msec and the total branching ratio is high 0.535, finally the transition $^4I_{11/2} \rightarrow {}^4I_{9/2}$ depopulating the terminal level is very fast.

As a result of these favorable properties, continuous oscillations have been obtained from a (YAG) yttrium aluminium garnet Nd^{3+} rod of 3 cm length at room temperature with only 360 W input power to an exciting tungsten lamp [26]. Continuous operation of a $CaWO_4 - Nd^{3+}$ laser under similar conditions requires almost three times as much power. The 1.0641 μ (1 μ = 1,000 nm = 10,000 Å) laser line of Nd:YAG is actually a combination of two lines corresponding to the $^4F_{3/2}$(b) $\rightarrow$ $^4I_{11/2}$(c) and $^4F_{3/2}$(a) $\rightarrow$ $^4I_{11/2}$(b) transitions. Thus the laser operates with a wavelength of 1.0648 μm at room temperature and 1.0612 μm at 77 °K.

In most of the crystal laser hosts such as YAG the Nd^{3+} fluorescence lifetime drops so sharply with increasing Nd^{3+} concentration that laser action is restricted to doping levels of about 5×10^{20} cm^{-3} or less. The effect of high concentration is due to a nonradiative self-quenching process,

$$Nd^{3+}(^4F_{3/2}) + Nd^{3+}(^4I_{9/2}) \rightarrow \text{two } Nd^{3+}(^4I_{15/2}) \text{ or}$$
$$Nd^{3+}(^4F_{3/2}) \rightarrow (^4I_{15/2}) = Nd^{3+}(^4I_{9/2} \rightarrow {}^4I_{15/2})$$

resulting from an electric dipole interaction between Nd^{3+} ion pairs. At ambient temperature, the energy mismatch of the levels corresponding to the above transitions may be overcome by phonon-assisted energy transfer involving the creation or annihilation of one or more crystal phonons (optical or acoustical).

Recently, Singh *et al.* [60], and Weber and Danielmeyer [61] found that the amount of concentration quenching is much lower in NdP_5O_{14} and $La_xNd_{1-x}P_5O_{14}$ crystals (also termed ultraphosphate) making it possible to achieve laser action in these materials even though their Nd^{3+} concentration is almost 4×10^{21} cm^{-3}. An X-ray structure determination showed that the Nd–O polyhedra in those crystals are isolated from each other (*i.e.* do not share O-atoms), and it was proposed that this isolation causes the decrease in concentration quenching. This explanation has been confirmed by the subsequent preparation of a number of additional compounds with isolated Nd–O polyhedra. All of these compounds exhibit strongly reduced concentration quenching and CW laser action has been obtained in three of them: $NdLiP_4O_{12}$ [62], $KNdP_4O_{12}$ [63] and $NdAl_3(BO_3)_4$ [63]. The lifetime τ is increased up to 120 μsec when the crystal is annealed at Nd^{3+} concentration higher by a factor of 30 compared to Nd:YAG [64].

Singh *et al.* [60] measured the stimulated emission cross section and fluorescence lifetimes for the transitions $^4F_{3/2} \rightarrow {}^4I_{9/2}$ and $^4F_{3/2} \rightarrow {}^4I_{11/2}$ of Nd^{3+} in NdP_5O_{14} and $La_{0.75}Nd_{0.25}P_5O_{14}$ and compared their results with those of Nd:YAG. The values of these measurements are presented in Table 16.

It should be noted that if the transition from the $^4F_{3/2}$ level of Nd^{3+} is totally radiative, then the fluorescence lifetime would be longer in ultraphosphate crystals

Table 16. Comparison of laser spectroscopic properties of neodymium in ultraphosphates and in yttrium aluminium garnet

Material	λ(laser) (nm)	λ_p (nm)	α_p (cm^{-1})	σ (10^{-20} cm^2)	τ_f (msec)	N (10^{20} cm^{-3})	F	β_y (10^{-5})	z
$La_{0.75}Nd_{0.25}P_5O_{14}$	1,051	798	8.1	$\sigma_{E\parallel b} = 18$	0.238	9.91	0.63	8.5	2.55
$Y_{2.97}Nd_{0.03}Al_5O_{12}$	1,064.2	808	1.1	46	0.23	1.39	0.4	4	2.15
NdP_5O_{14}	1,051	798	32.4	$\sigma_{E\parallel b} = 18$	0.12	39.6	0.63	8.5	2.55

compared to YAG since it is inversely proportional to the cross section. This can be due to various radiation traps which can be introduced during the growing processes of the ultraphosphate crystals.

This was confirmed by Huber *et al.* in Laser action in pentaphosphate crystals [65] who have added $AuCl_3$ and D_2O to the growing process of pentaphosphate crystals and found that similarly to Dy and Ho these molecules reduce the fluorescence time of Nd^{3+} from 110 to 60 μsec.

Similarly Tofield *et al.* [64] have recently shown that the characteristic lifetime of NdP_5O_{14} is 120 μsec rather than 60 μsec, and that the usual reduction of τ_f results from the trapping of hydrogen in the crystals during growth.

Huber [65] *et al.* report that the NdP_5O_{14} CW lasers have been operated with 600 W pump threshold.

Because of their high concentration (assuming the same absorption cross section and concentration, which is higher by a factor of thirty in comparison to Nd:YAG) it is possible to pump thirty times as much energy into ultraphosphate crystals than into YAG. Cross sections of ultraphosphate as expressed by oscillator strengths have been recently measured by Auzel [66] and were shown to be independent of concentration. However, it should be remembered that this high concentration also leads to deterioration of the crystal and higher resonant losses.

We feel that higher pumping efficiency of the lasers could be achieved by energy transfer from ions which have allowed transitions and therefore can be introduced into the crystal in small amounts. This subject is developed in detail in the chapter on energy transfer.

Very recently lithium neodymium tetraphosphate $LiNdP_4O_{12}$ (LNP) laser characteristics were examined by Otsuka *et al.* [67]. This crystal belongs to the monoclinic system and its Nd content is 4.37×10^{21} cm^{-3}. The lifetime of the $^4F_{3/2}$ is 120 μsec which is one-half of 230 μsec observed in Nd:YAG. The largest effective emission cross section at the line peak (1,047 nm) is 3.2×10^{-19} cm^2 for the electric vector parallel to the c axis and quantum efficiency 0.48. Quasi-continuous laser oscillation was achieved with 1.85 mm long crystal. The peak absorption cross section of the most intense absorption band at 800.9 nm ($^4I_{9/2} \rightarrow {}^4F_{5/2}, {}^2H_{9/2}$) is $\sigma_p = 1.25 \times 10^{-19}$ cm^2. This crystal may be useful as a high gain material for pumping by luminescence diodes, provided that high quality large crystals could be grown.

Potassium Neodymium Orthophosphate $K_3Nd(PO_4)_2$ crystal was selected by Hong and Chinn [68] as another high Nd-concentration laser. The concentration of Nd is 5×10^{21} cm^{-3}. The basic structural units are isolated PO_4 tetrahedra and isolated NdO_7 decahedra. Each decahedron is connected to six PO_4 tetrahedra to form two dimensional $Nd(PO_4)_2^{3-}$ sheets in the a–b plane with K atoms inserted along the c axis between these sheets. The radiative lifetime of Nd^{3+} ions in this structure is 460 μsec for x = 0.005 then decreases monotonically with increasing concentration of Nd to 21 μsec for x = 1.

Due to this increased concentration quenching lasers made from $K_3Nd(PO_4)_2$ have higher threshold and lower efficiencies than those obtained from the other 100% Nd-compounds. Good laser characteristics should be obtained for $K_3Nd_xLa_{1-x}(PO_4)_2$ crystals with x of about 0.2 – 0.4 in which the concentration quenching is reduced but the Nd concentration ($1 - 2 \times 10^{21}$ cm^{-3}) is still high enough for efficient absorption of pump radiation.

The idea underlying the development of ultraphosphate crystals brought the development of a great variety of related crystals. In a recent work [69] Auzel *et al.* have measured the fluorescence intensity and lifetimes of the $^4F_{3/2}$ of Nd^{3+} in highly doped chloroapatites as a function of Nd^{3+} ion concentration, and compared those values with similar measurements on YAG (with 1% Nd^{3+}) and Nd^{3+} ultraphosphate crystals. The results are presented in Figures 4 and 5. It was found that the fluorescence lifetime of the $^4F_{3/2}$ level of Nd^{3+} is slightly shorter in the chloroapatites.

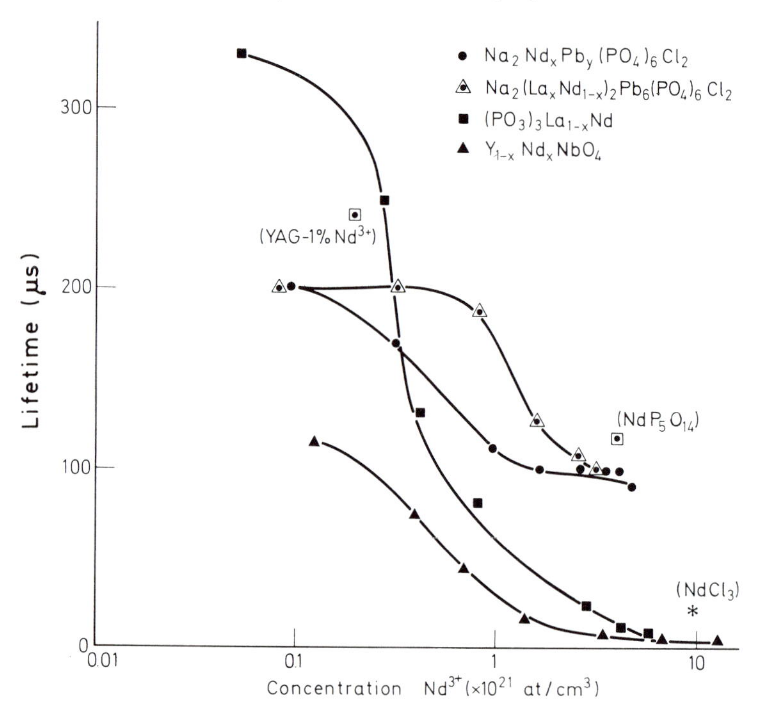

Fig. 4. Lifetime τ (in microseconds) of $^4F_{3/2}$ as a function of the neodymium concentration (in the unit 10^{21} atoms/cm^3) in three materials described by Auzel [69] and compared with the garnet $Y_{2.99}Nd_{0.01}Al_5O_{12}$ and the undiluted compounds $NdCl_3$ and NdP_5O_{14}

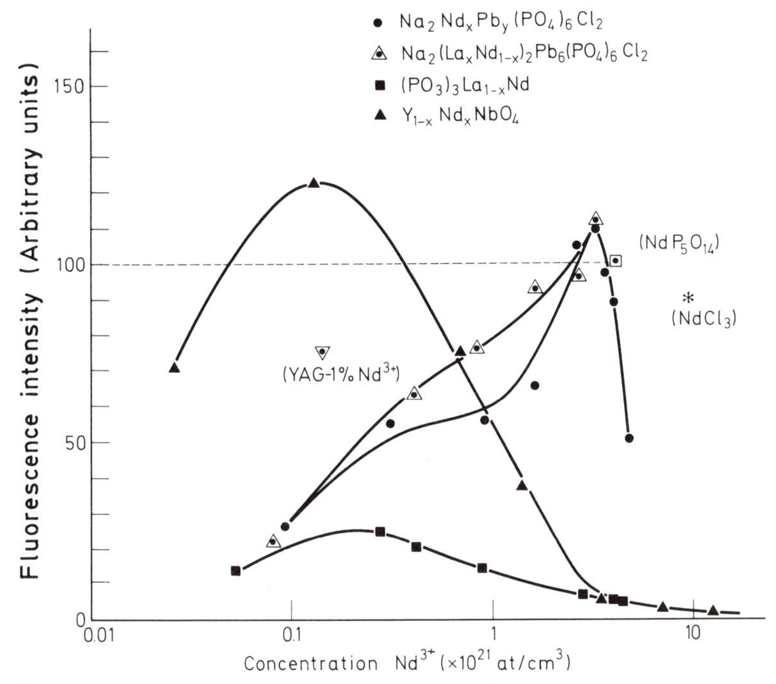

Fig. 5. Fluorescence intensity (in relative units) as a function of the neodymium concentration (note the logarithmic scale) in the six materials shown on Fig. 4

f) Laser Emission of Ho^{3+}, Er^{3+} and Tm^{3+}

Oscillation of Ho^{3+} involving $^5I_7 \to {}^5I_8$ transitions at about 2,000 nm has received considerable attention in the past because efficient optical pumping is possible by codoping with ions such as Er^{3+}, Tm^{3+} and Yb^{3+} [70, 71].

Stimulated emission in Ho^{3+} was first reported in singly doped Ho^{3+}:$CaWO_4$ at 77 K. Observation of energy transfer between Er-Ho [72] led to the first report of efficient CW operation at 77 K in Ho-doped YAG. Ho lasing action assisted by energy transfer at 77 K has also been reported in glass, yttrium iron garnet [73], Er_2O_3 [74], $LiYF_4$ [75] and HoF_3 [76]. Low-threshold room-temperature Ho operation in multiple sensitized hosts has been reported in $Ca_2Er_5F_{19}$ [77] and YAG [78]. "Alphabet materials" is the colloquial name for solids containing several rare earths in variable proportions.

Chicklis *et al.* [79] have observed high-efficiency pulsed laser action in YLF ($LiY_{0.416}Er_{0.5}Tm_{0.067}F_4$) at 2.065 μ at room temperature with a normal-mode threshold of 35 J and efficiency of 1.3%. Q-switched efficiency of ~ 2% and a Q-switched output in excess of 0.5 J has also been obtained. It should be noted that these values of threshold and slope efficiencies are obtained by measurement of the total electrical energy into a 3-in. flash-lamp pumping considerably shorter rods.

The room-temperature excitation spectrum for the 2.06 μ Ho^{3+} fluorescence shows broad excitation through the region of high emission of Xe flashlamps. Room-temperature pulsed fluorescence measurements revealed a 12 msec (±10%) lifetime for the $^5I_7 \rightarrow {}^5I_8$ transition independent of the pumping band.

Lasing action was observed with low threshold and high efficiency with a number of rods with various pumping conditions. Best results to date were obtained with a non-imaging silvered cylinder. The pump was a 3-inch PEK XE-14C-3 xenon lamp with a flash duration of 500 μsec. The rods and lamp were convection cooled with no UV filtering employed. No UV or laser-induced damage has been observed in any of the yttrium lithium fluoride (YLF) rods.

The $^5I_8$ ground state of Ho^{3+} with a multiplicity of 17 is split into 13 levels in S_4 symmetry, four of them being twofold degenerate. Based on this multiplicity and the splitting of the ground-state manifold (~340 cm^{-1}), the terminal state population at room temperature is estimated to be 2% of the total Ho^{3+} population, or $\sim 4.6 \times 10^{19}/cm^3$.

An estimate of the stimulated emission cross section was obtained by direct measurement of the absorption cross section between what is believed to be the initial and final lasing states. The computed value is 1.5×10^{-19} cm^2 compared to 8×10^{-19} cm^2 for Nd^{3+} : YAG and 2×10^{-20} cm^2 for Nd silicate glass [80]. The degeneracy of the upper level state is not unambiguously known, precluding precise determination of the emission cross section. An independent check of the value using a dynamic technique [80] is presently underway.

YLF appears to be a very favorable Ho^{3+} laser material with apparent high resistance to UV and laser-induced damage. The broad excitation spectrum, efficient sensitizer-activator energy transfer, moderately high-emission cross section, and quasi-4 level character result in high overall gain, and low threshold for room temperature pulsed operation. The long Ho^{3+} fluorescent lifetime in this material provides for high energy storage for efficient Q-switched operation. The long lifetime together with projected reduction in threshold with rods of lower Ho^{3+} concentration and improved optical properties point to the possibility of CW operation at room temperature.

Laser action in the green near 550 nm has also been reported on $[{}^5S_2, {}^5F_4] \rightarrow {}^5I_8$ transitions in CaF_2 [81] and BaY_2F_8 [82]. Recently stimulated emission in Ho : BaY_2F_8 near 2,400 nm has been attributed to transition within the $^5F_5 \rightarrow {}^5I_5$ group [83]. Fluorescence originating in the $^5I_6$ state has also been observed in a number of crystals [81–83] but emission from other levels is often quenched by nonradiative processes.

Laser action in $YAlO_3$: Ho^{3+} sensitized with Er^{3+} and Tm^{3+} was studied by Weber *et al.* [84].

The analysis on oscillator strengths and other relevant spectral data associated with known and potential laser transitions in Ho^{3+} are discussed by Caird and DeShazer [85]. It is shown that the $^5F_5 \rightarrow {}^5I_5$ manifolds resulting in stimulated emission in Ho : BaY_2F_8 near 2.4 μ have abnormally low oscillator strengths which is attributed to energy transfer from thulium impurities.

Laser action at 1,663 nm from Er^{3+} in $YAlO_3$ involving a transition from $^4S_{3/2}$ level to a $^4I_{9/2}$ level was described in [86]. Erbium is an attractive element for pulsed solid state lasers pumped by flash tubes, since it generates radiation 1.54 μ, well into the 'eye-safe' region. However, since erbium has only weak pumping bands, it is necessary to assist the pumping in some way, usually by resonant transfer from ytterbium.

Four-level lasing involving $^4S_{3/2} \rightarrow {}^4I_{13/2}$ transitions has been reported for Er^{3+} in several hosts at ambient temperatures [87, 88] and in $YAlO_3$ [84]. This transition is of interest because of the existence of sensitive photo detection at around 850 nm.

Laser action involving transitions between the Stark levels of the $^3F_4$ and $^3H_6$ manifolds of Tm^{3+} has been reported in several host crystals at wavelengths ranging from 1.86 to 2.02 μ [47]. Pulsed oscillation has been achieved at room temperature and CW oscillation at liquid nitrogen temperature.

Laser action of thulium in $YAlO_3$ was observed by Weber *et al.* [86]. Pulsed lasing at 1,861 nm was observed for thulium sensitized with erbium in $YAlO_3$ at 77 K by Weber [86]. The lasing transition terminates in a relatively low-lying level of the $^3H_6$ manifold. Output energies of 150 mJ were obtained but the overall efficiency was low. Table 17 summarizes laser operators of Ho^{3+}, Er^{3+} and Tm^{3+} in $YAlO_3$.

Table 17. Laser characteristics in the perovskite $YAlO_3$

	Transition	Wavelength (nm)	Terminal level (cm^{-1})	Temperature (°K)	Threshold (J)
Ho^{3+}	$^5I_7 \rightarrow {}^5I_8$	2,123	474	77	~ 1
				300	< 300
Er^{3+}	$^4S_{3/2} \rightarrow {}^4I_{13/2}$	851	6,671	300	51
Tm^{3+}	$^3F_4 \rightarrow {}^3H_6$	1,861	~240	77	25

g) High-power Nd^{3+} Glass-Lasers

There are a number of characteristics which distinguish glass from other solid laser host materials. Its properties are isotropic. It can be doped at very high concentrations with excellent uniformity. It is a material which affords considerable flexibility in size and shape, and may be obtained in large homogeneous pieces of diffraction-limited optical quality. It can also be relatively cheap in large volume production and can be fabricated by a number of processes such as drilling, drawing, fusion and cladding, which are generally alien to crystalline materials. Good laser glasses can be chosen with indices of refraction ranging from 1.5 to more than 2.0, or which will set the peak emission wavelength for neodymium at any one of a number of wavelengths between 1.047 and 1.063 microns. Of even greater importance is the flexibility afforded in relation to the physical constants in the ability to adjust the temperature coefficient of the index of refraction and the strain-optic coefficients so as to produce a thermally stable cavity. The major disadvantage of glass is its low thermal conductivity. This imposes limitations on the thickness of pieces which can be used as high average power, necessitating fairly radical configurational modifications for such an operation. These disadvantages may be overcome in future by production of glass ceramics (*vide infra*).

The inherent nature of the glass host produces inhomogeneously broadened lines which are wider than would be found for the same ion in crystals. This raises threshold, but reduces the amplified spontaneous emission losses for the same inversion in amplifier and Q-switched applications. Additionally, shorter pulses can be obtained in mode-locked operation using glass due to its broader fluorescence line.

Because of the possibility of delivering high energy pulses Nd-doped glasses have been of major interest in laser research. Similarly to crystals the Judd-Ofelt theory and the theory of multiphonon relaxation may be applied to neodymium glasses.

Today the most powerful Nd lasers are based on Nd-doped glass. Analysis of the energy levels of the Nd^{3+} ions in laser glass was first performed by Mann and DeShazer [89] who designed energy levels from spectral variations produced by changes in glass temperature from 4K to 300K. The free ion levels of Nd^{3+} are completely split into $(J + 1/2)$ Kramers doublets by the perturbation of the glass environment on the ion. The splitting patterns were similar to those of Nd_2O_3 crystals indicating that the rare earth ion locates in an environment like the sesquioxide crystal environment. Later Ranon *et al.* [90] identified the individual Stark levels of the ground state $^4I_{9/2}$ of the commercial ED-2 glass (Owens Illinois) by a variable temperature absorption from $^4I_{9/2}$ to the single $^2P_{1/2}$ state at 23,213 cm^{-1}.

The essence of the method lies in the fact that at 2.4 °K only the lowest Stark component of the $^4I_{9/2}$ state is thermally populated. Therefore a single absorption line to the $^2P_{1/2}$ state is observed. This line, which is 110 cm^{-1} wide, has clearly a Gaussian shape, confirming the inhomogeneous nature of the broadening of the glass. As the temperature is raised, the first excited Stark level becomes increasingly populated, giving rise to additional absorption which overlaps part of the low temperature line. Raising the temperature further brings into the picture more Stark levels of the ground state; thus, this thermal variation maps out the level structure of the $^4I_{9/2}$ state. Additional absorption measurements at 4.2 °K in the 880 nm band determine the two Stark components of $^4F_{3/2}$ and give the absorption cross-sections for the transitions from the ground level $^4I_{9/2}$ (a) to $^4F_{3/2}$. These cross-sections were used later in emission measurements to determine the cross sections for other transitions from the $^4F_{3/2}$ state.

A check on the assignment of the $^4I_{9/2}$ Stark components is provided by a fluorescence experiment at 4.2K in which the $^4F_{3/2}\,(a) \rightarrow {}^4I_{9/2}$ spectrum is observed. The variable temperature experiments, and particularly the absorption measurements to $^2P_{1/2}$, indicated no change in either the line positions or linewidths with temperature. Thus the knowledge of the form of the emission spectrum for the $^4F_{3/2}\,(a) \rightarrow {}^4I_{9/2}$ transitions at 4.2 °K enables one to compute the form of this spectrum at room temperature and subtract it from the measured spectrum at room temperature. The difference is the $^4F_{3/2}\,(b) \rightarrow {}^4I_{9/2}$ spectrum which can now be decomposed into its five components.

The nature of the concentration fluorescence quenching of Nd^{3+} in silicate glass which is associated with the reaction $Nd^{3+}\,(^4F_{3/2}) + Nd^{3+}\,(^4I_{9/2}) \rightarrow 2Nd^{3+}\,(^4I_{15/2})$ was studied by Chrysochoos [91] who believes that dipole-quadrupole and quadrupole-quadrupole interactions are effective in the cross-relaxation of Nd^{3+} ions.

A method for calculating induced emission cross section of Nd^{3+} laser glasses based the Judd-Ofelt theory described above for crystals was applied by Krupke [92].

The induced-emission cross section σ_p of the ${}^4F_{3/2} \rightarrow {}^4I_{11/2}$ fluorescence transition of Nd^{3+} at $1.06\,\mu$ is one of the most important parameters for laser design. If N is the difference in total population in ${}^4F_{3/2}$ and ${}^4I_{11/2}$, J manifolds, then σ_p gives the maximum spatial growth rate of intensity $I(X) = I_0 \exp\{\sigma_p N(x)\}$. The induced emission cross section σ_p can be related to the radiative transiton probability using a modification of Eq. (2.83a)

$$\sigma_p\, \Delta\bar{\nu}_{eff} = \int_{1.06} \sigma(\bar{\nu})\, d\bar{\nu} = \frac{\lambda_p^2}{8\pi c n^2}\, A({}^4F_{3/2}); ({}^4I_{11/2})$$

where $\lambda = 1/\bar{\nu}$, λ_p is the wavelength of fluorescence peak and $\Delta\bar{\nu}_{eff}$ is the effective fluorescence line shape [25].

The radiative properties of several laser glasses calculated by Krupke are presented in Table 18.

Table 18. Radiative properties of various laser glasses

	3,669A	S33	ED-2	LSG-91H
Calculated radiative life-time τ_{rad}^{c} (msec)	1.0	0.40	0.372	0.384
Measured fluorescence life-time τ_f (msec)	0.52	0.244	0.31	0.30
Calculated radiative quantum efficiency η_c	0.52	0.61	0.83	0.78
Calculated induced-emission cross section $\sigma(10^{-20}\ cm^2)$	1.2	2.8	2.9	2.6
Measured stimulated-emission cross section $\sigma(10^{-20}\ cm^2)$	–	3.2	3.1	2.5
Calculated induced-emission cross section at λ = 1,064 nm, $\sigma(10^{-20}\ cm^2)$	1.05	2.62	2.71	2.45
Measured linewidth $\Delta\bar{\nu}_{eff}$ (cm^{-1})	296	300	300	310
Small-signal gain coefficient at 1,064 nm, $\gamma(cm^2/J)$	0.056	0.139	0.145	0.130
S $({}^4F_{3/2}; {}^2G_{9/2})$ $(10^{-20}\ cm^2)$	0.10	0.22	0.27	0.25
S $({}^4F_{3/2}; {}^4I_{11/2})$ $(10^{-20}\ cm^2)$	0.97	2.6	2.8	2.6

In addition to the spectroscopic properties, the Ω_t parameters can be used to estimate excited state absorption and the probability of ion-ion interactions which are responsible for energy transfer and fluorescence quenching. An extensive discussion on this subject may be found in the chapter on energy transfer.

The induced cross section σ_p depends on Ω_t and the bandwidth $\Delta\lambda_{eff}$. Both values are affected by glass composition. The bandwidths $\Delta\lambda$ is a sum of Stark splitting and inhomogeneous broadening due to site-to-site variation. It should be noted that the spectral lines affected by the inhomogeneous broadening have a Gaussian shape, while the natural spectral lines are more correctly expressed by a Lorentzian shape function.

The relation between Ω_t and glass composition may be understood from the following consideration [25]: the ${}^4F_{3/2} \rightarrow {}^4I_{11/2}$ laser transition of Nd^{3+} is dependent only on Ω_4 and Ω_6, because of the triangle rule $|J - J'| \leqslant \lambda \leqslant (J + J')$, $U^{(2)} = 0$. For a large cross section it is desirable that these Ω_t values be as large as possible. In addition, the fluorescence branching ratio β of ${}^4F_{3/2} \rightarrow {}^4I_{11/2}$ transition should be as large as possible. The effect of glass composition on the Ω_t parameters was investigated by Weber and his group [93] and the result of his measurements are presented in Table 19. The magnitude of σ is directly related to the magnitude of Ω_4 and Ω_6 parameters since these are greater when small alkali ions are incorporated in the glass. The glasses with small alkali ions are better for laser use. This phenomeonon arises from the fact that the Ω_t parameters express the amount of the mixing of the 4f level with the perturbing configuration of Nd^{3+} via the asymmetric terms in the crystal field expansion. Smaller alkaline ions induce lower symmetry around the Nd^{3+} ion thus enhancing the Ω_4 and Ω_6 parameters.

Actual laser glasses are more complicated than the simple examples above. In general, they contain both alkali and alkaline-earth network-modifier ions. The variation of the Ω values with changes in alkali and alkaline-earth content in silicate glasses are summarized in Table 19. In both cases one observes a trend toward larger Ω values for smaller ions. Larger values, in turn, mean shorter lifetime and, for equal bandwidths, higher peak cross sections. It should be noted that in a case of the alkali series, the alkali represents 26% of the modifier ions, whereas in the case of the alkaline-earth series, they total only 17% of the cation concentration. Hence, the somewhat larger effect might be anticipated for the former. Inspection of Table 19 also reveals that the other criterion ($\Omega_6 > \Omega_4$) is satisfied in the lithia-containing glass. High cross-section silicate laser glasses, such as ED-2, indeed have large lithia contents.

The above results show clearly the variation in line strength with changes in the network-modifier ion in silicate base glass. Changes in intensity parameters also occur when the glass network former is varied. This is shown for five oxide glasses in Table 18. Although the network modifier represents two-thirds of the mole percent composition, the glass network itself depends on the glass former ions [94]. In the case of the borates the basic network is made up by BO_3 triangles. Silicates, phosphates, and germanates have tetrahedra (SiO_4, PO_4, GeO_4), and tellurites have octahedral TeO_6.

Another variation with glass composition arises from the dependence of the spontaneous-emission probability and cross sections on the refractive index of the host via the local field correction. This is usually treated in a simple isotropic tight-binding approximation, where the field is given by

$$\frac{E_{eff}}{E_0} \approx 1 + \frac{n^2 - 1}{3} + \ldots$$

This introduces the factor $(n(n^2 + 2)^2)/9$ in the expression for transition probability A. Since n ranges from ~ 1.5 to > 2 for the oxide glasses listed in Table 20, this last factor can range from 3 to 8. The factor for the cross sections is multiplied by $1/n^2$ and, though smaller, is still a non-negligible variation with host material. Because of the strong dependence on refractive index the radiative decay rates for equal Ω

Table 19. Intensity and fluorescence ($^4F_{3/2} \rightarrow {}^4I_{11/2}$) parameters of neodymium in a variety of glasses of compositions given in molar percent

Nd_2O_3	0.5	0.5	0.5	0.4	0.4	0.4	0.4	0.4	0.3	0.3	0.3	0.3	0.3	0.3	0.3	0.4	0.4	0.4	0.4	0.4
SiO_2	66	66	66	65	70	75	80	99	65	65	65	65	65	65	65	0	66	0	0	0
Li_2O	33	0	0	0	0	0	0	0	0	0	0	0	15	0	0	0	0	0	0	0
Na_2O	0	33	0	35	30	25	20	0	0	0	0	0	0	15	0	0	0	0	0	0
K_2O	0	0	33	0	0	0	0	0	15	15	15	15	0	0	15	15	15	15	15	15
MgO	0	0	0	0	0	0	0	0	20	0	0	0	0	0	0	0	0	0	0	0
CaO	0	0	0	0	0	0	0	0	0	20	0	0	0	0	0	0	0	0	0	0
SrO	0	0	0	0	0	0	0	0	0	0	20	0	0	0	0	0	0	0	0	0
BaO	0	0	0	0	0	0	0	0	0	0	0	20	20	20	20	18	18	18	18	18
B_2O_3	0	0	0	0	0	0	0	0	0	0	0	0	0	0	0	66	0	0	0	0
P_2O_5	0	0	0	0	0	0	0	0	0	0	0	0	0	0	0	0	0	66	0	0
GeO_2	0	0	0	0	0	0	0	0	0	0	0	0	0	0	0	0	0	0	66	0
TeO_2	0	0	0	0	0	0	0	0	0	0	0	0	0	0	0	0	0	0	0	66
$\Omega_2 (10^{-20} cm^2)$	3.3	4.2	5.0	4.0	4.6	4.3	4.1	6.1	5.1	4.6	3.8	3.7	2.8	3.6	4.2	3.9	4.4	2.7	6.0	1.6
$\Omega_4 (10^{-20} cm^2)$	4.9	3.2	2.4	3.3	3.4	3.1	2.9	4.5	4.1	3.6	3.3	2.8	4.0	3.4	2.9	3.9	3.0	5.4	3.7	3.2
$\Omega_6 (10^{-20} cm^2)$	4.5	3.1	2.0	3.0	2.9	2.7	2.6	4.3	2.8	2.9	2.7	2.3	3.9	3.5	2.5	4.3	2.6	5.4	2.9	2.9
$\sigma (10^{-20} cm^2)$	2.6	1.8	1.1																	
λ_p (nm)	1.061	1,060	1,059																	
$\Delta\lambda_{eff}$ (nm)	33.8	33.4	35.6																	
τ_R (msec)	0.395	0.60	0.91																	
τ_M (msec)	0.37–	0.45–	0.6–																	
(Measured on	0.46	0.6	1.0																	
samples with 0.02																				
mole % Nd_2O_3)																				

Table 20. Radiative properties of neodymium in glasses as a function of network forming anion. The three last parameters are for $^4F_{3/2} \rightarrow {}^4I_{11/2}$

	Phosphate	Borate	Germanate	Silicate	Tellurite	Aluminate	Titanate	Fluoro-phosphate
Composition (in molar %; all except last sample 0.3 to 0.5 % Nd_2O_3)	66 P_2O_5 18 BaO 15 K_2O	66 B_2O_3 18 BaO 15 K_2O	66 GeO_2 18 BaO 15 K_2O	66 SiO_2 18 BaO 15 K_2O	80 TeO_2 20 BaO –	11 SiO_2 32 Al_2O_3 52 CaO 5 BaO	25 TiO_2 50 SiO_2 25 Na_2O	80 LiF 20 $Al(PO_3)_3$ 0.5 NdF_3
$A_{total}(^4F_{3/2})$, sec^{-1}	2,888	2,386	2,299	1,602	4,193	2,660	3,232	2,695
$\beta(^4F_{3/2}; {}^4I_{9/2})$	0.43	0.42	0.46	0.45	0.45	0.45	0.43	0.42
$\beta(^4F_{3/2}, {}^4I_{11/2})$	0.48	0.49	0.45	0.46	0.46	0.46	0.47	0.48
$\beta(^4F_{3/2}; {}^4I_{13/2})$	0.09	0.09	0.08	0.08	0.08	0.08	0.09	0.09
$\beta(^4F_{3/2}; {}^4I_{15/2})$	0.005	0.005	0.004	0.004	0.004	0.004	0.004	0.005
$\tau_{rad}(^4F_{3/2})$, msec	0.346	0.419	0.435	0.624	0.239	0.392	0.309	0.371
λ_p, nm	1,055	1,061	1,062	1,060	1,063	1,069	1,064	1,054
$\Delta\lambda_{eff}$, nm	25.3	36.8	34.7	34.9	28.9	43.1	38.6	27.2
$\sigma_p(10^{-20}\ cm^2)$	4.1	2.2	1.9	1.5	2.9	1.8	2.5	3.5

values can be significantly different. This is observed, for example, for the tellurite glasses.

As can be seen from the table, the intensity parameters for the phosphates are large. Since it is well known that the emission bandwidths of neodymium-doped phosphate glasses are typically narrower than those for silicates [25], the combination of large Ω values and narrow line widths can yield larger peak cross-sections. As already noted, a larger induced-emission cross section is desirable, provided that parasitic oscillations can be controlled and the pumping efficiency maintained. If the emission bands are narrower, however, the absorption bands may also be narrower, thus resulting in poorer utilization of the broad flashlamp pumping spectrum.

Induced emission cross section for the ${}^4F_{3/2} \rightarrow {}^4I_{13/2}$ transition of Nd^{3+} in several commercial and experimental laser glasses has also been performed by Jacobs and Weber [93] and found that in phosphate glasses the cross section for both ${}^4F_{3/2} \rightarrow {}^4I_{13/2}$ and ${}^4F_{3/2} \rightarrow {}^4I_{11/2}$ can be larger than in the silicates. This is due to a combination of narrower emission bandwidths and larger Ω values in phosphate glasses. For all glasses studied the peak cross section for transmission to the ${}^4I_{13/2}$ state are about a quarter of those to the ${}^4I_{11/2}$ transition.

In addition to the importance of the radiative transition probabilities which are connected with the Ω_t values three important properties of lasers doped with rare earth ions are determined by the nonradiative decay rate of the nonexcited state of these ions. These are:

1. the pump efficiency which is larger when nonradiative decay from the pump band to the laser level is much greater than the radiative decay from the pump band
2. the high quantum efficiency of the fluorescent laser level is determined by small radiative relaxation from this level
3. the high relaxation rate of the terminal laser level assumes high population inversion.

Layne *et al.* [95] have determined the nonradiative decay rates for rare earth ions doped into commercial Owens Illinois glass by directly measuring fluorescent lifetime of these levels and Reisfeld and Eckstein [96] and Reisfeld, Hormadaly and Muranevich [97] have determined multiphonon relaxation rates in phosphate, borate, germanate and tellurite glasses. It has been established that the nonradiative decay of rare earth ions in glasses are dominated by the highest energy lattice vibration. It was also found that the ion-lattice coupling in all oxide glasses is independent of the specific glass or rare earth level in question.

Figure 6 presents multiphonon relaxation rates as a function of energy gap at room temperature. Additional information on multiphonon relaxation in glasses may be found in the chapter on radiative and nonradiative transition probabilities [98].

In Ref. [99] conclusions were drawn about the Nd-doped ED-2 glass in order to estimate the pump conversion efficiency, the radiative quantum efficiency of the ${}^4F_{3/2}$ upper laser level, and the lifetime of the ${}^4I_{11/2}$ terminal laser level. Since the nonradiative rate depends exponentially on the energy gap, the decay time from the optical pump bands to the ${}^4F_{3/2}$ level is determined by the largest energy gap encountered in the cascade. The largest gap, of approximately 2,300 cm^{-1}, occurs below the ${}^2P_{3/2}$ level. The fluorescence lifetime of this level was measured to be 50 nsec. The radiative lifetimes of all the excited states in the pump bands have been calcu-

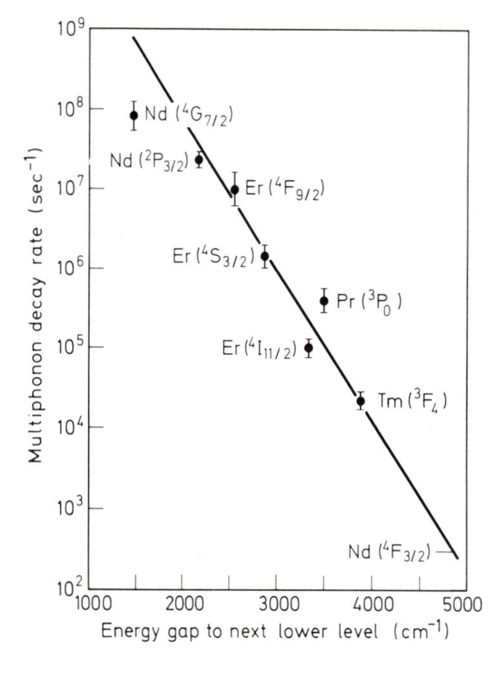

Fig. 6. Multiphonon decay rates for trivalent lanthanides in the silicate glass ED-2 as a function of the energy across which the nonradiative relaxation takes place. The third excited level of thulium $^3F_4$ is elsewhere called $^3H_4$ by us

lated using the Judd-Ofelt intensity parameters for ED-2 glass [25]. These lifetimes are all longer than 10^{-5} sec. Thus, since ion-pair interactions are believed to be negligible, the pump conversion efficiency is found to be greater than 99 percent. The nonradiative decay rate for the $^4F_{3/2}$ level, based on the extrapolation shown in Fig. 6 to an energy gap of 4,800 cm^{-1}, is estimated to be 200 sec^{-1}. The radiative decay rate for this level is about 3×10^3 sec^{-1}, leading to a radiative quantum efficiency of approximately 95 percent in the absence of concentration quenching. An estimate of the radiative quantum efficiency which is in agreement with this value was obtained from measurements of the temperature dependence of the fluorescence lifetime of the $^4F_{3/2}$ level. This temperature dependence arises from thermal population and decay of levels above the $^4F_{3/2}$, in addition to the temperature dependence of W_{NR} shown in [95, 99]. This result is, however, less accurate since the change in the measured rate over the available temperature range is small compared to the uncertainty in the measurement. Finally, the lifetime of the $^4I_{11/2}$ terminal level with a gap of 1,500 cm^{-1} is predicted to be approximately 10 nsec. This value is in good agreement with estimates based on attempts to measure the lifetime directly.

While Nd^{3+} glass laser is by far the most studied, Er^{3+} is an attractive element for solid state lasers, pumped by flash tubes, since it generates radiation at 1,540 nm which is in the eye-safe region. Edwards and Sandoe (100)discuss the prospect of efficient laser generation from Er^{3+} at 1.54 μ using Nd^{3+} and Yb^{3+} in glass as energy donors.

h) Nd-doped Glass Ceramic Laser

The number of host materials for lasers has recently been increased by a new material type – glass ceramics. These are substances in which crystalline areas are embedded in a glass matrix. Such a structural arrangement is obtained by a special thermal treatment of a glass which is prepared by melting and subsequent annealing. The chemical composition of the glass has to be selected in such a way that during the thermal treatment a large number of uniformly dispersed crystal nuclei are formed and increased. The glass ceramic thus formed can have unusual properties. The characteristics of the well known glass ceramics of the basic system $Li_2O-Al_2O_3-SiO_2$ are exceptionally low thermal expansion, and resulting from this, high thermal shock resistance.

In a number of glass ceramics the average dimensions of the crystals are so small (about 500 Å) and the matching of the refractive index of the crystals and residual glass so close that glass ceramics are transparent. This suggested the use of glass ceramics as laser host materials.

However, the neodymium dopant in the first glass ceramics prepared had lower quantum efficiencies of fluorescence than in the laser glasses. Improvement of fluorescence intensity by a factor of 3 to 4 has been obtained in a glass ceramic, the composition of which has been modified by replacing SiO_2 partly by $AlPO_4$ and introducing La_2O_3 as a new component. After crystallization for 10 h at about 800 °C the glass ceramic thus modified contained three crystalline phases, namely, a solid solution with high quartz structure, ensuring the low thermal expansion, a solid solution with fluorite structure containing ZrO_2 and presumably Ta_2O_5 and finally a third crystal type, containing La_2O_3, which is not yet exactly known.

Spectroscopic properties and induced emission of Nd_2O_3-doped glass ceramics of such a type were first presented by Müller and Neuroth [101]. Fluorescence lifetimes of glass ceramics with varied Nd_2O_3 concentration (emission at 1063 nm) below 3 weight percent are all close to 0.20 msec. The half-width of the 1063 nm fluorescence emission band of Nd^{3+} in the glass ceramic is slightly smaller (27 nm) than in the glass (29 nm).

Glass ceramic systems containing neodymium oxide as a dopant are superior to glass lasers in their thermal shock resistance but have somewhat lower laser intensities and the threshold of their operation is much higher than in glass lasers. The first report of a glass ceramic laser of the composition in mol %: 73.14% SiO_2, 13.73% Al_2O_3, 8.69% Li_2O, 1.75% BaO, 1.51% TiO_2, 1.08% ZrO_2 and 0.10% Nd_2O_3 is given by Rapp and Chrysochoos [102]. The glass was melted at 1700 °C and annealed. The annealed glass was heat treated to produce the glass ceramic. Thermal expansion coefficients and laser thresholds for Nd-doped glass and glass ceramic rods were compared [102]. The threshold for lasing was taken as the pump energy at which spiking, a typical characteristic of pulsed solid-state lasers, was detected. The lasing efficiency of the glass ceramic laser rod in this work is lower by one order of magnitude as related to the glass of similar composition.

Improved polycrystalline ceramic laser rods having laser efficiencies of ~0.32% as compared to ~0.1% previously obtained, were produced by controlled powder preparation and processing, composition and cooling rates, and are described in ref.

103. Ceramic laser rods composed of 89–96.5 mole % Y_2O_3, 1.0–2.5% ThO_2 and 1% Nd_2O_3 were synthesized by a conventional sintering process. This material, called Nd-doped yttralox (NDY) ceramic, was produced with laser threshold energies lower than that of the best commercially available Nd: glass rod and with a lasing efficiency ~94%, that of laser glass at 40 J of input energy under pulsed mode conditions. In a similar operating mode an NDY rod, containing 5 mole % ThO_2, and having dimensions 7.6 x 0.46 cm, delivered 0.41 J of optical energy when using an input energy of 162 J, a pump pulse of 150 μsec, and output mirror reflectivity of 70%. The lasing efficiencies depended strongly on the method of powder preparation and processing, composition, and the cooling rate for the sintering temperature. The dependence of the fluorescent linewidth on the NDY composition provides a means of appreciably varying the material gain coefficient. Active attenuation coefficients for AR-coated NDY laser rods were about 2% per cm as compared to 0.76% per cm for an OI ED-2 laser glass rod measured in the same optical cavity. The absorption component of the optical attenuation was measured to be 0.38% per cm at $\lambda = 1.06\ \mu$, indicating that the scattering component is the major contribution to the attenuation coefficient. Considerable evidence is presented which shows that submicroscopic scattering centers exist in the solid-solution matrix and are related to composition fluctuations arising from

(i) chemical segregation in the starting powder which is not entirely eliminated during the high-temperature sintering process and
(ii) the formation of extended defects or ordered zones in the solid-solution phase during specimen cooling from the sintering temperature.

Some of these laser rods can be prepared having lasing efficiencies nearly equal to that of commercially available Nd-Glass laser rods. Chemical segregation of the components during powder preparation and ionic ordering processes occurring during specimen cooling are believed to be the primary sources of composition inhomogeneities and hence optical loss in scattering in these polycrystalline ceramic lasers.

Auzel *et al.* [104] recently proposed vitroceramic materials for infra-red to visible up-conversion, using the energy transfer from Yb^{3+} to Tm^{3+} or Er^{3+} as discussed in Chapter 4. The optimized conditions for the relative concentrations of germanium dioxide, lead difluoride and rare earths were found, and it was shown that the high efficiency is related to a segregation, where the lanthanides are concentrated in the microcrystalline phase rather than in the glassy medium.

i) Lasers from Vapours of Rare-Earth Compounds

Recently, Krupke [105] discussed hot vapours of lanthanide compounds as conceivable materials for high-power lasers with the purpose of initiating thermonuclear reactions (*cf.* Section 5 B). One of the ideas involved was to avoid the second-order effect of variable refractive index at extreme, high light densities, which is known to create considerable problems in solids and probably would do in liquids too. Obviously, it is an advantage to keep the wall temperature needed of the container as low as feasible, and Krupke looked for the somewhat exotic cases of volatile lanthanide compounds. Øye and Gruen [106] demonstrated that aluminium chloride increases the vapour pressure of neodymium-containing species tremendously, compared with pure $NdCl_3$. Though the molecule formed is $NdAl_3Cl_{12}$ it is not known whether it has two chloride bridges $Nd(Cl_2AlCl_2)_3$ with the coordination number $N = 6$ or three, $Nd(Cl_3AlCl)_3$ with $N = 9$. The hypersensitive pseudoquadrupolar transition from $^4I_{9/2}$ to $^4G_{5/2}$ at 17,030 cm^{-1} has $\epsilon = 19$, much lower than $\epsilon = 116$ for gaseous $NdBr_3$ (at 16,470 cm^{-1}) or $\epsilon = 345$ at 16,330 cm^{-1} for NdI_3. Krupke [105] uses the programme of Caird to calculate the radiative transition probabilities in terbium iodide vapour and finds that the transitions $^5D_4 \rightarrow {}^7F_5$ in the green at 18,400 cm^{-1} and $^5G_6 \rightarrow {}^7F_4$ in the violet at 23,100 cm^{-1} shows promising opportunities for work in a four-level laser with radiative life-time slightly below a millisecond.

Actually, anhydrous iodides are frequently more volatile than expected. Some boiling points (°C) given by the "Handbook of Chemistry and Physics" are:

RbI	1,300	TlI	824	SbI_3	401	
NaI	1,300	CaI_2	718	AlI_3	382	
CuI	1,290	CdI_2	713	HgI_2	354	(2.89)
CsI	1,280	ZnI_2	624	SnI_4	341	
LiI	1,190	BeI_2	590	SiI_4	290	
PbI_2	954	AsI_3	403	BI_3	210	

and there is no doubt that mixtures of several iodides frequently produce higher vapour pressure of "double salts" or "molecular adducts" such as $CsNdI_4$, mentioned in Section 3 C.

The classical rare earth laser complexes are the tris(β-diketonates), rarely retaining $N = 6$ (and not at all being octahedral) and Krupke [105] points out that they can be rather volatile. Thus, the dipivaloyl-acetonates (2,2,6,6-tetramethyl-3,5-heptanedionates) $M(thd)_3$ have vapour pressures [107] *increasing* with increasing atomic weight of the lanthanide M, varying at 200 °C from 0.2 torr (mm Hg) for M = La to 5 torr for M = Yb. In solution, these complexes fluoresce in near ultra-violet (366 nm) radiation in the case of M = Sm, Eu, Tb and Dy, and the hopes were expressed [107] that the fluorescence of $Tb(thd)_3$ in the gaseous state might be utilized in a laser if the molecule is sufficiently stable to radiative damage. Since solid $Er(thd)_3$ is monomeric ($N = 6$) with trigonal-prismatic coordination, the dimers in several solvents were ascribed to a complicated structure [108]. It was also shown that partly fluorinated β-diketonates can be used for a gas-chromatographic separation of the various lanthanides. It is known from several areas of organic chemistry that the vapour pressure does not always decrease monotonically as a function of the molecular weight. This is particularly true for fluorine-substituted homologues. A purely inorganic case of this type is the surprising high volatility of $Cr(S_2PF_2)_3$.

We frequently need additional information about the detailed composition of the vapours in equilibrium with a solid at a given temperature, and recent progress in mass spectrometry can help much in this respect. It is interesting how many caesium salts stick together in the gaseous state. A remarkable example [109] is the hexafluoroacetylacetonate $CsY(hfa)_4$ (solid salts of $Y(hfa)_4^-$ have $N = 8$), where the mass spectrum even showed a strong peak, corresponding to $CsY(hfa)_3^+$.

It may be more important that the decomposition products recombine rapidly and reversibly, and that they do not absorb the spectral lines of the luminescence, which may be a difficult requirement for organic compounds, whereas, a regenerating inorganic system (such as a chloride or a bromide) may be less objectionable as far as a small concentration of free halogen goes. As Krupke [105] notes, the pumping of singlet excited states in organic ligands may result in energy transfer from the (otherwise long-lived) lowest triplet state to the lanthanide, and the same type of process might occur in inorganic complexes of ligands having weak bands in the near ultraviolet, such as nitrate or sulphur-containing ligands. Almost resonant energy transfer from the ligand to a lanthanide J-level having an intense (emitting) pseudoquadrupolar transition to another J-level *above* the groundstate may provide unconventional four-level systems. Another aspect of gaseous compounds is the absence of collective vibrational modes (enhancing non-radiative de-excitation) and somewhat modified Boltzmann populations of levels within the first 1000 cm^{-1} above the groundstate. On the other hand thermal excitation of otherwise long-lived excited states may occur.

References

1. Schawlow, A. L., Townes, C. H.: Infrared and optical masers. Phys. Rev. *112,* 1240 (1958)
2. Maiman, T. H.: Stimulated optical radiation in ruby masers. Nature *187,* 493 (1960)
3. Kiss, Z. J., Pressley, R. J.: Crystalline solid lasers. Appl. Optics *5,* 1474 (1966)
4. Chessler, R. B., Geusic, J. E.: Solid-state ionic lasers. In: Laser Handbook (eds. F. T. Arrechi and E. O. Schulz-DuBois). Amsterdam: North-Holland Publ. Co. 1972
5. Weber, M. J.: Insulating crystal lasers. In: Handbook of Lasers with Selected Data on Optical Technology (R. J. Pressley, ed. R. C. West). Cleveland, Ohio: The Chemical Rubber Co. Press 1971
6. Snitzer, E.: Optical laser action of Nd^{3+} in a barium crown glass. Phys. Rev. Lett. *7,* 446 (1961). First demonstration of laser action in glasses
7. Patek, K.: Glass Lasers (ed. J. G. Edwards), London: Butterworth 1970
8. Snitzer, E., Young, C. G.: Glass lasers. In: Advances in Lasers 2 (ed. A. Levine), p. 191–256. New York: Dekker 1968
9. Young, Gilbert: Glass Lasers Proc. IEEE *57,* 1267 (1969)
10. Woodcock, R. W.: Commercial laser glasses. In: Handbook of Lasers (ed. R. J Pressley), p. p. 360. Cleveland: CRC Press 1971
11. Snitzer, E.: Lasers and glass technology. Amer. Chem. Bull. *52* (6), 516 (1973)
12. Deutschbein, O. K., Pautrat, C. C.: CW. laser at room temperature using vitreous substances. IEEE, J. Quantum Electronics *QE-4,* 48 (1968)
13. Sarkies, P. H., Sandoe, J. N., Parke, S.: Variation of Nd^{3+} cross section for stimulated emission with glass composition. Brit. J. Appl. Phys. (J. Phys. D) *4,*1642 (1971)
14. Rapp, C.: The Influence of the Composition on the Gain of Nd-doped Glasses in Physics of Electronic Ceramics, Part B (eds. L. L. Hench and D. B. Dove), pp. 1011–1017. New York: Dekker 1972
15. Hirayama, C., Camp, F. E., Melamed, N. T., Steinbruegge, K. B.: Nd^{3+} in germanate glasses: Spectral and laser properties. J. Non-Cryst. Solids *6,* 324–356 (1971)
16. Lipson, H. G., Buckmelter, J. R., Dugger, C. O.: Neodymium environment in germanate crystals and glasses. J. Non-Cryst. Solids *17,* 1 (1975)
17. Krupke, W. F.: Induced-emission cross section in neodymium laser glasses. IEEE, J. Quantum Electronics *QE-10,* 450 (1974)
18. De Shazer, L. G., Ranon, U., Reed, E. G.: Spectroscopy of neodymium in laser glasses. USCEE Report 479, University of Southern California 1974
19. Mann, M. M., DeShazer, L. G.: Energy levels and spectral broadening of neodymium ions in laser glass. J. Appl. Phys. *41,* 2951 (1970)
20. Reisfeld, R.: Properties of rare earth doped inorganic glasses as related to their lasing abilities. In: Lasers in Phys. Chem. & Biophys. (ed. J. Joussot-Debien), p. 77. Elsevier 1975
21. Samelson, H.: Laser phenomena in europium chelates. J. Chem. Phys. (I) *40,* 2547; (II) *40,* 2553; (III) *42,* 1081 (1965)
22. Heller, A.: Laser action in liquids. Physics Today *20,* 35 (November 1967)
23. Samelson, H., Heller, A., Brecher, C.: Determination of the absorption cross section of the laser transitions of the Nd^{3+} ion in the Nd^{3+} $SeOCl_2$ system. J. Opt. Soc. Am. *58,* 1054 (1968)
24. Krieger, J. H.: Strides made in laser fusion research. Chem. Eng. News *54,* 21 (29. March 1976)
25. Lawrence Livermore Laboratory: Laser Program Annual Report, 1974, sponsored by the U.S. Energy Research & Development Admin. under No. W-7405 Eng. 48
26. Lengyel, B. A.: Lasers. New York: Wiley-Interscience 1971
27. Yariv, A.: Introduction to Optical Electronics. Holt, Reinhart & Winston 1971
28. Kogelnik, H.: In: Lasers, Vol. (ed. A. Levine). New York: M. Dekker, Inc. 1966
29. Mitchel, A. C. G., Zemansky, M. W.: Resonance Radiation and Excited Atoms. New York: Cambridge University Press 1961
30. Lengyel, B. A.: Lasers. New York: John Wiley & Sons, Inc. 1962
31. Pauling, L., Wilson, E. B.: Introduction to Quantum Mechanics. McGraw-Hill Co., Inc. 1935

32. Sargent III, M., O'Scully, M., Lamb Jr., W. E.: Laser Physics. Reading, Mass.: Addison-Wesley Publ. Co. 1974
33. Dieke, R. H., Wiltke, J. P.: Introduction to Quantum Mechanics. Reading, Mass.: Addison-Wesley 1960
34. Heavens, O. S.: Optical Masers. New York: John Wiley & Sons, Inc. 1964
35. Reisfeld, R.: Radiative and nonradiative transitions of rare earth ions in glasses. Structure and Bonding *22*, 129 (1975)
36. Peacock, R. D.: The intensity of lanthanide f-f transitions. Structure and Bonding *22*, 83 (1975)
37. DiBartolo, B.: Optical Interaction in Solids. John Wiley & Sons, Inc.
38. DeMaria, A. J., Ferrar, C. M., Danielson Jr., G. E.: Mode locking of a Nd^{3+}- doped glass laser. Appl. Phys. Letters *8*, 22 (1966)
39. DeMaria, A. J., Glenn, W. H., Brienza, M. J., Mack, M. E.: Picosecond Laser Pulses. Proc. IEEE *57* (1969)
40. Harry, J. E.: Industrial Lasers and Their Applications. London and New York: McGraw-Hill 1974
41. Lempicki, A., Samelson, H.: Liquid lasers. Scientific American *216*, 81 (1967)
42. Samelson, H., Lempicki, A., Brecher, C., Brophy, V.: Room temperature operation of an europium chelate liquid laser. Appl. Phys. Lett. *5*, 173 (1964)
43. Heller, A.: A high-gain, room-temperature liquid laser: trivalent neodymium in selenium oxychloride. Appl. Phys. Letters *9*, 106–108 (1966)
44. Judd, B. R.: Optical absorption intensities of rare earth ions. Phys. Rev. *127*, 750 (1962)
45. Ofelt, G. S.: Intensities of crystal spectra of rare earth ions. J. Chem. Phys. *37*, 511 (1962)
46. Riseberg, L. A., Weber, M. J.: Relaxation phenomena in rare earth luminescence. In: Progress in Optics. Vol. 14 (ed. E. Wolf), in press
47. a) Caird, J. A., Thesis, Ph. D.: On the Evaluation of Rare Earth Laser Materials and the Matrix Elements of Orbital Tensor Operators. Univ. of S. California, also Hughes Aircraft Co. Tech. Rep. (Rep. No. M75-106) 1975
b) Caird, J. A., DeShazer, L. G., Nella, J.: Characteristics of Room-Temperature 2.3 μm Laser Emission from Tm^{3+} in YAG and $YAlO_3$. IEEE J. Quant. Electron. *QE-11*, 874 (1975)
48. Crystal Structures, R. W. G. Wyckoff. Interscience Publishers, N. Y. 1963
49. Handbook of Chemistry and Physics (ed. R. C. Weast). Cleveland, Ohio: The Chemical Rubber Co. 1969
50. Thornton, J. R., Fountain, W. D., Flint, G. W., Crow, T. G.: Properties of neodymium laser materials. Appl. Optics *8*, 1087 (1969)
51. Nassau, K.: Applied Solid State Science, Vol. 2 (ed. R. Wolfe), p. 174. New York: 1971 Academic Press
52. Weber, M. J.: Spontaneous emission probabilities and quantum efficiencies for excited states of Pr^{3+} in LaF_3. J. Chem. Phys. *48*, 4774 (1968)
53. Rajnak, K.: Configuration-interaction effects on the "free-ion" energy levels of $Nd^{3+}$and Er^{3+}. J. Chem. Phys. *43*, 847 (1965)
54. a) Wybourne, B. G.: Structure of f^n configurations. II. f^5 and f^9 configurations. J. Chem. Phys. *36*, 2301 (1962)
b) Conway, J.: Lawrence Radiation Lab., Berkeley, unpublished results.
55. Donlan, V. L.: Wright Patterson Air Force Base, unpublished results.
56. Krupke, W. F., Gruber, J. B.: Optical-absorption intensities of rare-earth ions in crystals: the absorption spectrum of thulium ethyl sulfate. Phys. Rev. *139*, A2008 (1965)
57. Reisfeld, R., Boehm, L., Lieblich, N., Barnett, B.: Ligand field parameters in glasses. Proc. 10th Rare Earth Conf., Arizona, p. 1142 (1973)
58. Weber, M. J.: Probabilities for radiative and nonradiative decay of Er^{3+} in LaF_3. Phys. Rev. *157*, 262 (1967)
59. Thomson, A. J.: Luminescence properties of inorganic compounds. In: Electronic Structure and Magnetism of Inorganic Compounds *4*, 149. Chemical Society Specialist Periodical Report, London 1976

60. Singh, S., Miller, D. C., Potowicz, J. R., Shick, L. H.: Emission cross-section and fluorescence quenching of Nd^{3+} lanthanum pentaphosphate. J. Appl. Phys. *46,* 1191 (1975)
61. Danielmeyer, H. G., Weber, H. P.: Fluorescence in neodymium ultraphosphate. IEEE J. Quantum Electronics *QE-8,* 805 (1972)
62. a) Otsuka, K., Yamada, T.: Transversely pumped LNP laser performance. Appl. Phys. Lett. *26,* 311 (1975)
b) Chinn, S. R., Hong, H. Y.-P.: Low-threshold CW $LiNdP_4O_{12}$ laser. Appl. Phys. Lett. *26,* 649 (1975)
63. Chinn, S. R., Hong, H. Y.-P.: CW laser action in acentric $NdAl_3(BO_3)_4$ and $KNdP_4O_{12}$. Opt. Comm. *15,* 345 (1975)
64. Tofield, B. C., Weber, H. P., Damen, T. C., Pasteur, G. A.: On the growth of neodymium pentaphosphate crystals for laser action. Mater. Res. Bull. *9,* 435 (1974)
65. Huber, G., Jeser, J. P., Kruhler, W. W., Danielmeyer, H. G.: Laser action in pentaphosphate crystals. IEEE J. QE *10,* 766 (1974)
66. Auzel, F.: Oscillator strengths of Nd^{3+} in $Nd_xLa_{1-x}P_5O_{14}$ and concentration quenching in stoichiometric rare earth laser materials. IEEE J. *QE-12,* 258 (1976)
67. Otsuka, K., Yamada, T., Saruwatari, M., Kimura, T.: Spectroscopy and laser oscillation properties of lithium neodymium tetraphosphate. IEEE *QE-11,* 330 (1975)
68. Hong, H. Y.-P., Chinn, S. R.: Crystal structure and fluorescence lifetime of potassium neodymium orthophosphate, $K_3Nd(PO_4)_2$, a new laser material. Mat. Res. Bull. *11,* 421 (1976)
69. Michel, J.-C., Morin, D., Auzel, M. F.: Fluorescence intensity and lifetime . . . . C. R. Acad. Sci. Paris *B281,* 445 (1975)
70. Johnson, L. F., Geusic, J. E., Van Uitert, L. G.: Efficient high power coherent emission from Ho^{3+} ions in yttrium aluminum garnet assisted by energy transfer. Appl. Phys. Lett. *8,* 200 (1966)
71. Devor, D. P., Soffer, B. H.: 2.1 μm laser of 20-W output power and 4% efficiency from Ho^{3+} in sensitized YAG. IEEE J. Quant. Electron. (Part II, Special issue on 1971 IEEE/OSA Conf. Laser Eng. and Applications) *QE-8,* 231 (1972)
72. Johnson, L. F., Van Uitert, L. G., Rubin, J. J., Thomas, R. A.: Energy transfer from Er^{3+} to Tm^{3+} and Ho^{3+} ions in crystals. Phys. Rev. *133,* A494 (1964)
73. Johnson, L. F., Remeika, J. P., Dillon Jr., J. F.: Coherent emission from Ho^{3+} ions in yttrium iron garnet. Phys. Lett. *21,* 37 (1966)
74. Hoskins, R. H., Soffer, B. H.: 8B7-energy transfer and CW laser action in Ho^{3+}: Er_2O_3. IEEE J. Quantum Electron. *QE-2,* 253 (1966)
75. Remski, R. L., James Jr., L. T., Gooen, K. H., DiBartolo, B., Linz, A.: Pulsed laser action in $LiYF_4$:Er^{3+}, Ho^{3+} at 77 °K. IEEE J. Quantum Electron. *QE-5,* 214 (1969)
76. Devor, D. P., Soffer, B. H., Robinson, M.: Stimulated emission from Ho^{3+} at 2 μm in HoF_3. Appl. Phys. Lett. *18,* 122 (1971)
77. Robinson, M., Devor, D. P.: Thermal switching of laser emission of Er^{3+} at 2.69 μ and Tm^{+3} at 1.86 μ in mixed crystals of CaF_2:ErF_3:TmF_3. Appl. Phys. Lett. *10,* 167 (1967)
78. Remski, R. L., Smith, D. J.: Temperature dependence of pulsed laser threshold in YAG:Er^{3+}, Tm^{3+}, Ho^{3+}. IEEE J. Quantum Electron. *QE-6,* 750 (1970)
79. Chickles, E. P., Naiman, C. S., Folweiler, R. C., Gabbe, D. R., Jenssen, H. P., Linz, A.: High efficiency room-temperature 2.06 μm laser using sensitized Ho^{3+}:YLF. J. Appl. Phys. Letts. *19,* 119 (1971)
80. Belan, V. R., Grigoryants, V. V., Zhabotinski, M. E.: Energy transfer between neodymium ions in glass. Opto-Electron. *1,* 33 (1969)
81. Voronko, Yu. K., Kaminskii, A. A., Osiko, V. L., Prokhorov, A. M.: Stimulated emission of Ho^{3+} in CaF_2 at 5,512 Å. Sov. Phys. JETP Lett. *1,* 3 (1965)
82. Johnson, L. F., Guggenheim, H. J.: IR pumped visible laser. Appl. Phys. Letts. *19,* 44 (1971)
83. Johnson, L. F., Guggenheim, H. J.: Electronic and phonon terminated laser emission for Ho^{3+} in BaY_2F_8. IEEE J. Quant. Electron. *QE-10,* 442 (1974)
84. Weber, M. J., Bass, M., Comperchio, E., Riseberg, L. A.: Ho^{3+} laser action in $YAlO_3$ at 2.119 μ. IEEE J. Quant. Electron. *QE-7,* 497 (1971)

85. Caird, J. A., DeShazer, L. G.: Analysis of laser emission in Ho^{3+} doped materials. IEEE J. Quant. Electron. *QE-11*, 97 (1975)
86. Weber, M. J., Bass, M., deMars, G. A.: Laser action and spectroscopic properties of Er^{3+} in $YAlO_3$. J. Appl. Phys. *42*, 301 (1971)
87. Kaminskii, A. A.: Spectroscopic study of stimulated emission of Er^{3+} activated CaF_2-YF_3 crystals. Opt. Spectrosc. *31*, 507 (1971)
88. Chicklis, E. P., Naiman, C. S., Linz, A.: Stimulated emission at 0.85 μm in Er^{3+}:YLF. In: Dig. Tech. Papers, 7th Int. Quantum Electronics Conf., p. 17, May 1972
89. Mann, M. M., DeShazer, L. G.: Energy levels and spectral broadening of Nd^{3+} ions in laser glass. J. Appl. Phys. *41*, 2951 (1970)
90. Ranon, U., DeShazer, L. G., Guha, T. K., Reed, E. D.: Spectroscopy of Nd^{3+} in ED-2 laser glasses and the laser cross section in Nd:YAG. Appendix A, VIII Intl. Quant. Electron. Conf. June 1974 (IEEE, N.Y.), pp. 8-9
91. Chrysochoos, J.: Nature of the interaction forces associated with the concentration fluorescence quenching of Nd^{3+} in silicate glass. J. Chem. Phys. *61*, 4596 (1974)
92. Krupke, W. F.: Induced emission cross sections in neodymium laser glasses. IEEE J. Quant. Electron. *QE-10*, 450 (1974)
93. Jacobs, R. R., Weber, M. J.: Dependence of the $^4F_{3/2} \rightarrow {}^4I_{11/2}$ induced emission cross section for Nd^{3+} on glass composition. IEEE J. Quant. Electron. *QE-12*, 102 (1976)
94. Reisfeld, R., Boehm, L., Ish-Shalom, M., Fischer, R.: 4f → 4f transitions and charge transfer spectra of Eu^{3+}, 4f → 5d spectra of Tb^{3+} and $^1S_0 \rightarrow {}^3P_1$ spectra of Pb^{2+} in alkaline earths metaphosphate glasses. Phys. Chem. Glasses *15*, 76 (1974)
95. Layne, C. B., Lowdermilk, W. H., Weber, M. J.: Measurements of nonradiative relaxation of rare earth in glass using selective laser excitation. IX Intl. Quantum Electronic Conf., Amsterdam, June 1976
96. Reisfeld, R., Eckstein, Y.: Dependence of spontaneous emission and nonradiative relaxation of Tm^{3+} and Er^{3+} on glass host and temperature. J. Chem. Phys. *63*, 4001 (1975)
97. a) Reisfeld, R., Hormadaly, J., Muranevich, A.: Intensity parameters, radiative transitions and nonradiative relaxations of Ho^{3+} in various tellurite glasses. Chem. Phys. Letts. *38*, 188 (1976)
b) Reisfeld, R., Hormadaly, J.: Optical intensities of holmium in tellurite, calibo and phosphate glasses. J. Chem. Phys. *64*, 3207 (1976)
98. Reisfeld, R., Boehm, L., Eckstein, Y., Lieblich, N.: Multiphonon relaxation of rare earth ions in borate, phosphate, germanate and tellurite glasses. J. Luminescence *10*, 193 (1975)
99. Layne, C. B., Lowdermilk, W. H., Weber, M. J.: Nonradiative relaxation of rare earth ions in silicate laser glass. IEEE J. Quant. Electron. *QE-11*, 798 (1975)
100. Edwards, J. G., Sandoe, J. N.: A theoretical study of the Nd:Yb:Er glass laser. J. Phys. D. *7*, 1078 (1974)
101. Müller, G., Neuroth, N.: Glass-ceramic; a laser host material. J. Appl. Phys. *44*, 2315 (1973)
102. Rapp, C. F., Chrysochoos, J.: Neodymium-doped glass ceramic laser material. J. Materials Sci. *7*, 1090 (1972)
103. Greskovich, C., Chernoch, J. P.: Improved polycrystalline ceramic laser. J. Appl. Phys. *45*, 4495 (1974)
104. Auzel, F., Pecile, D., Morin, D.: Rare earth doped vitroceramics: New, efficient, blue and green emitting materials for infrared up-conversion. J. Electrochem. Soc. *122*, 101 (1975)
105. Krupke, W. F.; Prospects for Trivalent Rare Earth Molecular Vapor Lasers for Fusion. Private communication
106. Øye, H. A., Gruen, D. M.: Neodymium Chloride-Aluminium Chloride Vapor Complexes. J. Am. Chem. Soc. *91*, 2229 (1969)
107. Sicre, J. E., Dubois, J. T., Eisentraut, K. J., Sievers, R. E.: Volatile Lanthanide Chelates. J. Am. Chem. Soc. *91*, 3476 (1969)
108. Feibush, B., Richardson, M. F., Sievers, R. E., Springer, C. S.: Gas Chromatographic. Studies of Lanthanide Nuclear Magnetic Resonance Shift Reagents. J. Am. Chem. Soc. *94*, 6717 (1972)
109. Lippard, S. J.: A Volatile Inorganic Salt $CsY(CF_3COCHCOCF_3)_4$. J. Am. Chem. Soc. *88*, 4300 (1966)

3. Chemical Bonding and Lanthanide Spectra

(References to this Chapter are found p. 154).

A major difficulty for evaluating the actual extent of covalent bonding in lanthanide compounds is that it is almost impossible to obtain direct evidence for the major part of the covalent deviations from purely ionic bonding using the empty shells (such as 5d and 6s) whereas, the minor part influencing the partly filled 4f shell has numerically small, but perfectly observable consequences. It is conceivable that the fractional charge on the neodymium atoms has been decreased below + 2 in solid Nd_2S_3 or volatile $Nd(C_5H_5)_3$ by an average contribution of the sum of squared amplitudes of delocalized MO above 0.2 in each of the five 5d-like orbitals in the LCAO model, but at the same time it is beyond doubt that the *J*-level distances are much closer to the configuration $[Xe]4f^3$ in gaseous Nd^{+3} (obtained by a conservative extrapolation from solid NdF_3) than in gaseous Pr^{+2}. These two properties are not incompatible, as we shall see below, but they contribute to a certain atmosphere of "double truth" permeating much theoretical discussion of rare earths. Furthermore, there is a major difference between the 3d, 4d and 5d groups on the one hand, and the 4f group on the other, so that the extent of partly covalent bonding in the former case is mainly determined by the electronegativities of the ligands, whereas, the highly varying distances to the first neighbour atoms add a second dimension to the description of the lanthanides, in particular in mixed oxides.

Since 1968, the *photo-electron spectra* have added valuable and entirely new information about chemical bonding. However, one has to realize [1] that both in the case of ionization of an inner shell (such as 4d, 3d, . . .) or by the removal of one electron $4f^q \rightarrow 4f^{q-1}$ one obtains information about the *ionized state* and only indirectly through a reasonable working-hypotheses, evidence applicable to the ground-state. Seen from the point of view of the solid-state physicist, it is interesting that such evidence can also be obtained for the metallic elements, their alloys and the semi-conducting or metallic stoichiometric compounds, whereas, the optical studies of *J*-levels by necessity are restricted to sufficiently transparent samples. In the metallic or black samples the term distances in M[III] are only marginally smaller than in M(III) compounds and when M[IV] is formed by ionization, the decrease of term distances has the order of magnitude 10 percent, relative to the reliable extrapolated values for gaseous M^{+4}.

A. The Nephelauxetic Effect and the Photo-electron Spectra

The first study of the nephelauxetic effect (this word was proposed [2] by the late Professor Kaj Barr in Copenhagen; it means "cloud-expanding" in Greek) was by Hofmann and Kirmreuther [3], reporting in 1910 that the wave-numbers of the narrow absorption bands of Er_2O_3 are about 1 percent lower than of other erbium(III) salts, such as the hydrated sulphate. These authors suggested that the orbits of the "valence electrons" introduced by Stark are slightly larger in the sesquioxide than in the other compounds. It may be noted, that this comment was made three years before Niels Bohr's model of the hydrogen atom in 1913. Ephraïm and his collaborators [4–7] made a systematic comparison of such minor band shifts in the reflection spectra of praseodymium(III), neodymium(III), samarium(III), holmium(III) and erbium(III) compounds. Comparable work was later done by Boulanger [8] and by one of us [9]. It is not easy to tell exactly when it was realized that the main contribution to the energy differences between the levels of the partly filled shell $4f^q$ is the differing interelectronic repulsion parametrized in the theory of Slater, Condon and Shortley [10] and later refined by Racah [11]. Once Bethe and Spedding [12] applied this theory to the absorption spectra of thulium(III) compounds, there was no longer any legitimate doubt that the small chemical shifts are indirect effects of increased average radii by decreasing $\langle r_{12}^{-1} \rangle$ *both* in the excited and the groundstate and not, as might be true for inter-shell transitions, expanding the radial function of the excited state without modifying the groundstate behaviour. There has always been general agreement that the chemical effect is a "loosening" of the 4f cloud [4] by weak covalent bonding, and that gaseous M^{+3} would show even larger term distances than the fluoride (which was first verified for Pr^{+3} by Sugar [13] in 1965). The variation in the compounds

$$MF_3 > Mg_3M_2(NO_3)_{12}, 24H_2O > M(H_2O)_9X_3 > M_2(SO_4)_3 > MCl_3, 6H_2O > MCl_3 > MBr_3 > MI_3 \gg M_2O_3 > M_2S_3 \quad (3.1)$$

depends to the first approximation on the electronegativity of the ligating atom but the scattering of oxygen is rather enigmatic. It seems today to be a question of *decreasing internuclear distances,* depending among other factors on decreasing *coordination number N.* It was discussed in section 1B that the double nitrate contains the complex $M(O_2NO)_6^{-3}$ with six bidentate nitrate ligands, and the chromophore $M(III)O_{12}$ is almost regular icosahedral.

Suddenly, about 1954, several authors such as Tanabe and Sugano [14] and Orgel [15] became interested in the numerically much more important nephelauxetic effect in the d group compounds. This effect is fairly transparent in the term distance from ^6S to ^4G in $[Ar]3d^5$ of Mn^{++} 26,846 cm^{-1} (to be compared with 20,516 cm^{-1} in Cr^+ and 25,279 cm^{-1} in $[Ar]3d^54s^2$ of the gaseous manganese atom), which should remain invariant (assuming the same radial functions) in octahedral $Mn(II)X_6$ and tetrahedral $Mn(II)X_4$ as far as the two coinciding, excited levels $^4A_1$ and ^4E go. The two lower-lying levels $^4T_1$ and $^4T_2$ have energies also strongly dependent on the sub-shell energy difference Δ. Actually, the term distance is 25,300 cm^{-1} in the rutile-type MnF_2 and perovskite $KMnF_3$ both containing $Mn(II)F_6$, 25,000

cm^{-1} in solution or salts of $Mn(H_2O)_6^{+2}$ and then smoothly decreasing values [16] down to the lowest known, 21,100 cm^{-1} in $Mn(II)Se_4$ occurring in $Mn_xZn_{1-x}Se$. In the general d group chromophore the evaluation of the parameters of interelectronic repulsion needs complicated mathematical manipulation. Schäffer and one of us [2, 17, 18] found that the *nephelauxetic ratio* β between the phenomenological value of Racah's parameter B in the compound and in the corresponding gaseous ion M^{+z} decreases as a regular function of the central atom

$$\begin{aligned} &Mn(II) > Co(II) \sim Ni(II) > Mo(III) > Cr(III) > Fe(III) > Rh(III) \sim Ir(III) \\ &> Tc(IV) \sim Co(III) > Ag(III) > Mn(IV) \sim Cu(III) > Pt(IV) > Pd(IV) > Ni(IV) \end{aligned} \tag{3.2}$$

and as a function of the ligands

$$\begin{aligned} &F^- > H_2O > (NH_2)_2CO > NH_3 > NH_2CH_2CH_2NH_2 > C_2O_4^{-2} > Cl^- \\ &> CN^- > Br^- > N_3^- > I^- > (C_2H_5O)_2PS_2^- > (C_2H_5O)_2PSe_2^- \end{aligned} \tag{3.3}$$

The large majority of d group complexes have β between 0.9 and 0.5, and the two *nephelauxetic series* Eqs. (3.2) and (3.3) can be quantitatively treated with surprising good accuracy [16] by a relation $\beta = 1 - hk$, where h only depends on the ligands and k only on the central atom. Because one sub-shell in octahedral chromophores is approximately non-bonding, and another strongly anti-bonding, a more precise treatment involves three nephelauxetic ratios $1 > \beta_{55} > \beta_{35} > \beta_{33}$ or even 5 ratios [19] if non-diagonal elements involving four orbitals are taken into account.

The model generally invoked to explain the nephelauxetic effect is LCAO (linear combination of atomic orbitals) where the anti-bonding orbital ψ_a of the same symmetry type (in the point-group of the chromophore) as the bonding orbital ψ_b are formed from only *two* atomic orbitals, ψ_M centered on the central atom, and ψ_X on one (or several) of the ligating atoms. As a simple example the dδc orbital of Eq. (1.25) having the angular function A_l proportional to $(x^2 - y^2)$ may combine with a linear combination with the symmetry-determined coefficient 0.5 (then neglecting interligand overlap) multiplied with equal, positive amplitudes of two rotationally symmetric σ orbitals belonging to the two opposite ligating atoms situated on the x axis, and negative amplitudes on the two equivalent ligating atoms on the y axis. The LCAO coefficients

$$\begin{aligned} \psi_b &= k_1\psi_M + k_2\psi_X \\ \psi_a &= k_3\psi_M - k_4\psi_X \end{aligned} \tag{3.4}$$

are restrained by the conditions of *normalization* and *orthogonality*

$$\begin{aligned} &(k_1)^2 + (k_2)^2 + 2k_1k_2S_{MX} = 1 \\ &(k_3)^2 + (k_4)^2 - 2k_3k_4S_{MX} = 1 \\ &k_1k_3 - k_2k_4 + (k_2k_3 - k_1k_4)S_{MX} = 0 \end{aligned} \tag{3.5}$$

where S_{MX} is the *overlap integral* between ψ_M and ψ_X. It is very dangerous to neglect overlap integrals, once one has become interested in covalent bonding [20], but it is always legitimate to conserve one orbital, say ψ_X, and orthogonalize the other $\varphi_M = (\psi_M - S_{MX}\psi_X)/(1 - S_{MX}^2)^{1/2}$, or for that matter conserving ψ_M and construct the orthogonalized linear combination $\varphi_X = (\psi_X - S_{MX}\psi_M)/(1 - S_{MX}^2)^{1/2}$. In the case of two orthogonalized orbitals, the coefficients of Eq. (3.4) can be written in terms of a trigonometric variable ϑ (which is zero when no delocalization occurs)

$$k_1 = k_4 = \sin\vartheta \qquad k_2 = k_3 = \cos\vartheta \tag{3.6}$$

and one can apply the variation principle to minimize the energy of ψ_b having the eigen-values

$$E_b = H_X - (\mathrm{tg}\vartheta)H_{MX} \tag{3.7}$$

$$E_a = H_M + (\mathrm{tg}\vartheta)H_{MX}$$

if one is restricted to the basis set of *two* orbitals in LCAO, having the diagonal elements of energy H_X and H_M and the non-diagonal element H_{MX} and where $(\cot 2\vartheta) = (H_M - H_X)/2\,H_{MX}$ determines ϑ.

To the first approximation the parameters of interelectronic repulsion separating the terms of a partly filled shell in a compound are determined by the square of the squared amplitude of ψ_M. Hence, if LCAO were valid to the extent that the atomic orbital ψ_M had exactly the same radial function as in the corresponding gaseous ion, the nephelauxetic ratio β would be $(k_3)^4$ of Eq. (3.4). However, it was recognized early [21, 22] that one has to take into account both *central-field covalency* where the central field has been modified by the invading electronic density of the ligands, showing the behaviour one would obtain by interpolation of a characteristic *fractional atomic charge,* and *symmetry-restricted covalency* related to $(k_3)^4$. It has been reviewed elsewhere [16] why it is necessary to introduce significant central-field covalency. If this was neglected, the apparent nephelauxetic ratio would be $(k_3)^2$ for the repulsion between a non-bonding and an anti-bonding electron, and 1 for the repulsion between two electrons in the non-bonding sub-shell, in distinct contrast to experimental evidence in $Cr(NH_3)_6^{+3}$. It is possible [16, 23] to obtain consistent estimates of the fractional atomic charge of the central atom (usually between 1 and the oxidation state $+z$, in disagreement with Pauling's electroneutrality principle) by distributing comparable proportions of the nephelauxetic ratio on the effects of the central-field and of symmetry-restricted covalency.

If such arguments were applied to 4f group compounds, fractional atomic charges between 2.8 and 2.9 would be obtained in most cases. These values should not be taken too seriously, because they represent the modification of the central field at positions weighted with the square of the 4f radial function. In other words, it is conceivable that the central field at larger r (the central nucleus at origo) corresponds to more decreased fractional charges, whereas the electronic density supplied by the ligands do not penetrate to so small r values (mainly 0.2 to 0.5 Å) that a large 4f density occurs. Unfortunately, we do not know excited states with two 5d electrons, but there is evidence [24] that the nephelauxetic effect decreases term distances strongly in the excited configuration $4f^{12}5d$ of thulium(II) in $SrCl_2$. Anyhow, we have to accept some *average* nephelauxetic ratio β because of the negligible "ligand field" effects in 4f group chromophores [25]. From a strict group-theoretical point of view the five d-like electrons of octahedral MX_6 present 9 diagonal and one non-diagonal parameter [14, 26, 27] of interelectronic repulsion, and in sufficiently low symmetries, one would have 25 diagonal parameters. It is the assumption of $l = 2$ (i. e. the separability of ψ_M in the product of an angular A_2 and a radial function) which brings these numerous parameters back to definite multiples of three parameters (such as F^0, F^2 and F^4 or Racah's A, B and C). By the same token, if $l = 3$ was not required for seven f-like orbitals, one would need 49 diagonal parameters of interelectronic repulsion in low symmetries.

However, the experimental evidence available from J-level distributions in 4f group compounds produces very valuable *higher* limits on the central-field and the symmetry-restricted covalency, in particular in M(III). For instance, if the average value of $[1 - (k_3)^2]$ in Eq. (3.4) is below a few percent, β is expected to be the value below 1 (determined by the local change of the central field) multiplied by $(k_3)^4 \sim (4\,k_3 - 3)$. The $4f^2$ praseodymium(III) compounds have a particularly pronounced nephelauxetic effect [5, 23] and the absolute values of β are known (since the twelve J-levels have been identified [13] in Pr^{+3}). The $\beta = 0.96$ in PrF_3, 0.95 in $Pr(H_2O)_6^{+3}$ and 0.92 in $BaPr_2S_4$ show that the average $[1 - (k_3)^2]$ cannot be *above* 0.025 in the aqua ion, and only has this limiting value if the central-field covalency can be entirely neglected. These values may be compared with $\beta = 0.94$ in MnF_2, 0.93 in $Mn(H_2O)_6^{+2}$ and 0.81 in MnS. There is little doubt that Nd(III) showing a relative nephelauxetic effect (varying the ligands [28–31]) about 0.6 times the variation in Pr(III), or Er(III) a variation 0.3 times as strong as Pr(III), also have absolute nephelauxetic ratios β closer to 1 under equal circumstances. However, we must realize that the higher limits to the central-field covalency are less significant, since two external 4s electrons in the gaseous manganese atom [15] produces $\beta = 0.94$ compared with Mn^{+2}. Since Cr^+ has the term distance ${}^6S - {}^4G$ 0.76 times as large as Mn^{+2}, it may be argued that the addition of each 4s electron has an effect 0.12 times as pronounced as a unit change of ionic charge interpolated in the isoelectronic series between Cr^+ and Mn^{+2}. By the way, it is an additional complication that we modify *both* the atomic number Z and the ionic charge $+\,z$ by such an interpolation, but it turns out to be possible [16, 23] to write a general expression for the phenomenological value of

$$B = 384 + 58\,q + 124\,(z + 1) - 540/(z + 1) \tag{3.8}$$

(in cm^{-1}) in the 3d group, where q is the integer indicating the number of electrons in the partly filled shell. This expression is hyperbolic in the ionic charge, presenting a singularity for $z = -1$ (this is not physically aberrant because the electron affinity of M^0 forming M^- is very small, in an analogy to Eq. (1.42), and of M^- certainly zero, corresponding to spontaneous loss of an electron from all M^{-2} *in vacuo*), though it cannot be argued that the expression for B in Eq. (3.8) is exactly [23] proportional to $\langle r^{-1} \rangle$ of the 3d shell, since the dielectric type of correlation effects called the *Watson effect* [32, 33] becomes percentage-wise more important for low z. On the other hand, the strictly linear variation with q in Eq. (3.8) is in excellent agreement with the phenomenological parameters derived from atomic spectra. For our purposes it is important to note that electron densities, even *more external* than the 4s orbital in the manganese atom, may influence the parameters of interelectronic repulsion even less in the 4f group compounds.

The reliable comparison of the minute variations of the nephelauxetic effect in the 4f group M(III) cannot be done in the simple statistical study of the ratio of wave-numbers between the baricentre of J-level band intensities in the complex and in the aqua ion (taken as standard reference in the absence of data for M^{+3}) as already seen from the shift [9] toward *higher* wave-numbers observed in the ethylenediamine-tetra-acetate complex $Er(enta)(H_2O)_x^-$ compared with the erbium(III) aqua ion. One of the major problems is that the lowest J-level of M(III) shows a different distribution of sub-levels in the complex and in the aqua ion. This situation introduces a dependence on the absolute temperature T. At very low T (such as crystals studied at liquid helium) all wave-numbers of J-baricenters seen in absorption spectra are *increased* to the extent $d\sigma_{com}$ which is the energy difference (typically 100 to 300 cm^{-1}) between the baricentre and the lowest sub-level of the ground J-level. If we call this quantity $d\sigma_{aqua}$ in a salt of $M(H_2O)_9^{+3}$ the baricentre of each excited J-level is shifted to the extent $(d\sigma_{com} - d\sigma_{aqua})$ which corresponds to *increased* wave-numbers if $d\sigma_{com}$ is larger than $d\sigma_{aqua}$. At room temperature the *Boltzmann population* is proportional to the degeneracy number (the number of mutually orthogonal states in the sub-level at energy E_k) multiplied by $\exp(-E_k/kT)$ where the Boltzmann constant $k = 0.7\ cm^{-1}$. This produces a modification of the relative J-positions in two compounds, and at sufficiently high T, one should observe the unperturbed J-level distances, in sofar they can be reliably identified with the baricentres of band intensity. It was suggested [34, 35] to select the best possible straight lines

$$\sigma_{com} - \sigma_{aqua} = d\sigma - (d\beta)\sigma_{aqua} \qquad (3.9)$$

expressing the difference between the J-level intensity baricentres in the compound and in the aqua ion as a constant $d\sigma$ from which is subtracted the decrease $d\beta$ of the nephelauxetic ratio (compared with the aqua ion) times the wave-number σ_{aqua} of the J-intensity baricentre in the aqua ion. Table 21 gives the parameters $d\sigma$ and $d\beta$ for a large number of solid compounds and complexes of Pr(III), Nd(III), Ho(III), Er(III) and Tm(III). Quite interesting cases are the hexahalide complexes studied by Ryan [36, 37] and the solid double sulphides studied with Flahaut and Pappalardo [38] who also cites results for cyclopentadienides $M(C_5H_5)_3$.

In view of the discussion above, one would have expected the parameter $d\sigma$ of Eq. (3.9) to represent $(d\sigma_{com} - d\sigma_{aqua})$. This is not the case for the thorough studies of Pr(III) [39] in $LaCl_3$, $CeCl_3$, $NdCl_3$, $GdCl_3$ etc. (with decreasing M-Cl distances), Nd(III) [40] in $LaCl_3$ and $LaBr_3$ and Er(III) [41, 42] in $LaCl_3$($N = 9$) and YCl_3($N =$ 6, but not regular octahedral) where all the sub-levels of many of the J-levels (including the ground level) have been localized. It turns out that $d\sigma_{com}$ generally is *smaller* than the sum of the exact $d\sigma_{aqua}$ and the parameter $d\sigma$ determined from Eq. (3.9). The major reason for this discrepancy (besides a considerable experimental scattering, and apparently [36] anomalous low $d\beta$ for a few specific J-levels such as $^2H_{11/2}$ of Nd(III) and Er(III) and $^1I_6$ of Pr(III) falling far away from the optimized straight line) is a fact pointed out by several authors [39, 43, 44] that the *Landé parameter*

Table 21. Nephelauxetic parameters d_β (in percent) and $d\sigma$ (cm^{-1}) relative to the aqua ion according to Eq. (3.9). A, B and C are the sesquioxide types, F fluorite and P pyrochlore [35]

	d_β	$d\sigma$		$d\beta$	$d\sigma$
$Pr_xLa_{1-x}Cl_3$	0.8	− 40	Dy_2O_3(C)	1.8	+ 500
$Pr_xGd_{1-x}Cl_3$	1.2	0	$Dy_{0.5}Zr_{0.5}O_{1.75}$(F)	2.0	+ 500
$PrCl_6^{-3}$	1.9	+ 250	$Dy_{0.2}Zr_{0.8}O_{1.9}$(F)	1.9	+ 500
$Pr_xLa_{1-x}Br_3$	1.3	− 50	$HoCl_6^{-3}$	1.1	+ 150
$PrBr_6^{-3}$	2.3	+ 200	$Ho_2O_3(C)_5$	2.5	+ 450
$Nd_xLa_{1-x}Cl_3$	0.6	− 60	$Ho_{0.5}Zr_{0.5}O_{1.75}$(F)	2.3	+ 500
$NdCl_6^{-3}$	2.2	+ 50	$Ho_{0.2}Zr_{0.8}O_{1.9}$(F)	2.0	+ 450
$Nd_xLa_{1-x}Br_3$	1.2	−	$Er_xLa_{1-x}Cl_3$	0.3	− 30
$NdBr_6^{-3}$	2.3	+ 50	$Er_xY_{1-x}Cl_3$	0.9	0
Nd_2O_3(A)	3.6	+ 200	$ErCl_6^{-3}$	1.2	+ 100
$Nd_{0.1}La_{0.9}O_{1.5}$(A)	3.4	+ 200	Er_2O_3(C)	1.6	+ 200
$NdYO_3$	3.5	+ 250	$Er_2Ti_2O_7$(P)	1.3	+ 250
$Nd_xY_{2-x}O_3$(C)	3.1	−	$Er_{0.8}Zr_{0.2}O_{1.6}$(C)	1.6	+ 200
$Nd_{0.2}Yb_{0.8}O_{1.5}$(C)	3.2	+ 150	$Er_{0.5}Zr_{0.5}O_{1.75}$(F)	1.7	+ 300
$Nd_{0.1}Ce_{0.9}O_{1.95}$(F)	3.7	+ 300	$Er_{0.2}Zr_{0.8}O_{1.9}$(F)	1.1	+250
$Nd_{0.14}Th_{0.86}O_{1.93}$(F)	3.3	+ 150	$LaErO_3$	1.9	+ 250
$Nd_{0.5}Zr_{0.5}O_{1.75}$(F)	2.7	+ 250	$La_{0.2}Er_{0.3}Zr_{0.5}O_{1.75}$(F)	1.6	+ 250
$Nd_{0.2}Zr_{0.8}O_{1.9}$(F)	1.6	+ 100	$Er_xGa_{2-x}CdS_4$	3.6	+ 300
$BaNd_2S_4$	4.2	+ 250	$Er(C_5H_5)_3$	2.8	+ 350
$Nd(C_5H_5)_3$	3.8	+ 400	$TmCl_6^{-3}$	1.3	+ 100
Sm_2O_3(B)	2.3	+ 200	Tm_2O_3(C)	1.5	+ 300
$Sm_{0.5}Zr_{0.5}O_{1.75}$(F)	2.0	+ 350	$Tm_{0.2}Zr_{0.8}O_{1.9}$(F)	1.1	+ 350
$Sm_{0.2}Zr_{0.8}O_{1.9}$(F)	1.5	+ 200	$LaTmO_3$	1.7	+ 350

ζ_{4f} does not seem to vary significantly from one compound to another. The first-order consequence of this invariance is that the last contribution $(d\beta)\sigma_{aqua}$ of Eq. (3.9) really should read $(d\beta)[\sigma_{aqua} - \sigma(S_{max}, L)]$, where the baricentre of the lowest term (situated at $\sigma = (L+1)\zeta_{4f}/2$ for q = 1 to 6, and at $\sigma = L\zeta_{4f}/2$ for q = 8 to 13) is subtracted, since the nephelauxetic effect is a decrease of the *term* distances. Hence, one has to correct the observed shifts of wave-numbers with $d\sigma$ of Eq. (3.9) representing the deviations of the sub-levels from the ground J-baricentre, and with the latter effect representing the almost invariant distance between the baricentre of the lowest term and of the lowest J-level. It must be added in all fairness that ζ_{4f} decreases 2 percent in Pr(III) compounds [71] where E^k decrease 5 to 6 percent, but the variation is much weaker in the subsequent M(III).

Taken at its face value, the much smaller variation of the Landé parameter of spin-orbit coupling ζ_{4f} than of the parameters of interelectronic repulsion suggests the major importance of central-field covalency compared with delocalization effects, such as Eq. (3.4) where one expects the interelectronic repulsion intrinsically proportional to $(k_3)^4$ to vary about twice as much (for a small percentage) as $(k_3)^2$ expected to multiply ζ_{4f} in the LCAO approximation. It is known from $5d^3$ rhenium(IV) hexahalide complexes [45] and the isoelectronic IrF_6 that ζ_{5d} varies to a somewhat smaller extent [16] than parameters of interelectronic repulsion. However, we have to admit that the central field has been modified in a rather special way in order to keep ζ_{4f}

almost constant in the lanthanides. It was shown by Dunn [46] that adding one or two external 4s or 5s electrons to monatomic entities, ζ_{3d} is very close to be invariant in [Ar]$3d^q$ of M^{+2}, [Ar]$3d^q4s$ of M^+ and [Ar]$3d^q4s^2$ of M^0, and ζ_{4d} in the corresponding cases in the palladium group. It may also be noted [47, 48] that all instances of the conditional [16] oxidation state Yb[III] show the same ζ_{4f} within very narrow limits, whether it is the configurations $4f^{13}6s$, $4f^{13}7s$ or $4f^{13}$ 5d of Yb^{+2} or $4f^{13}$ of Yb(III) compounds (it is 10,214/3.5 = 2,918 cm^{-1} for [48] gaseous Yb^{+3}). The most probable conclusion is that the central field U(r) is modified by a constant contribution for quite a long interval of r from zero to a value above the average radius of the 4f shell. For larger r, the neighbour atoms modify U(r) by a contribution varying with r, but here the effective contribution to ζ_{4f} is very small. On the other hand, the parameters of interelectronic repulsion are relatively more sensitive to an expansion of the tails of the 4f radial functions due to the modified central field. It is a fascinating question whether the upper limit (2 percent) to the variation of ζ_{4f} in Pr^{+3} and Pr(III) compounds really represents an upper limit to $[1 - (k_3)^2]$ in the LCAO description Eq. (3.4). Contrary to the case of 5f group chromophores, there is very little evidence available for nuclear hyper-fine structure from the nuclei of ligating atoms influencing electron paramagnetic resonance, and Freeman and Watson [49] pointed out that the density of *uncompensated spin* in the vicinity of the fluorine nuclei in GdF_3 has the *opposite sign* of the gadolinium 4f shell. Such a result cannot be obtained from LCAO models, where the squared amplitude $(k_4)^2$ of Eq. (3.4) has the same (positive sign) as $(k_3)^2$ on the central atom M. The most ready explanation of this unusual situation is a deviation from the behaviour of "restricted" Hartree-Fock wave-functions [32], the radial functions of 5s and 5p orbitals (having larger average radii than 4f) being different for the m_s characterizing the *majority* of the 4f electrons (all seven if we consider states with $M_S = +7/2$ or $-7/2$) and the opposite m_s belonging to the minority. This perturbation is induced by the weaker interelectronic repulsion for parallel m_s (introducing conditions of anti-symmetrization) with the result that the tails of the 5s and 5p electronic density for large r consist mainly of the minority value of m_s.

It is possible to estimate the order of magnitude of the delocalization coefficients k_4 via the second-order perturbation result Eq. (3.7) where the anti-bonding effects, at most between 300 and 500 cm^{-1} in Tables 1 and 2, correspond to 1 percent or less of an effective one-electron energy above 50,000 cm^{-1} for reasons given below. If $(k_4)^2$ is below 0.01, and k_4 below 0.1, it is almost certain that $(k_3)^2$ is above 0.99 and probably above 0.995. Then only a smaller part of the nephelauxetic effect in M(III) is due to symmetry-restricted covalency.

The *photo-electron spectrum* is the probability distribution of kinetic energy E_{kin} of the electrons ejected when a (gaseous or solid) sample is bombarded with monochromatic photons $h\nu$, and ionization energies I are inferred from Einstein's equation

$$I = h\nu - E_{kin} \tag{3.10}$$

for the maxima (or shoulders), *signals* in the probability distribution. Turner [50] started work in 1963 on gaseous molecules ($h\nu$ = 21.2 eV from the resonance line of the helium atom, or 40.8 eV from He^+) where a large amount of interesting information, also about halides and d-group compounds [51], has been discovered. The predictions of the MO theory about *penultimate MO* (having *I* higher than the loosest bound MO) in diatomic and triatomic molecules have been fully confirmed [50]. However, for our purpose, it is more important that one can also bombard solid samples with soft X-rays ($h\nu$ = 1,253.6 eV originating in a magnesium anticathode or 1,486.6 eV from aluminium) and obtain *I* values between 0 and 1,200 eV, though the background (due to inelastically scattered electrons) is very high. In 1968 Fadley, Hagström, Klein and Shirley [52] using a home-made photo-electron spectrometer in Berkeley (comparable to the apparatus originally developed in Uppsala) measured the photo-electron spectra of Eu_2O_3 and of Eu[II] in the alloy $EuAl_2$, showing a large chemical shift 9.6 eV of *I* of all the inner shells. In the following years commercial instruments became available, and a grant from the Swiss National Science Foundation allowed the installation of a Varian IEE-15 photo-electron spectrometer at the University of Geneva 1971, where a comparative survey [53] was made of 600 non-metallic compounds containing 77 elements. The interval of *I* values for an inner shell of a given element is normally 3 to 10 eV wide, and does not depend exclusively on the oxidation state, but also to a large extent on the neighbour atoms. Today, it is perfectly certain [54] that the chemical shifts of *I* do not only represent changes of the Madelung potential Eq. (1.22) superposed the central field U(r) dependent on the fractional atomic charge, but contain equally important contributions from *interatomic relaxation effects* [1, 55].

In the lanthanides, Wertheim, Rosencwaig, Cohen, and Guggenheim [56] and one of us [57, 58] pointed out that the lowest *I*(M4f) is generally higher than that of the loosest bound orbitals of the ligands, say *I*(X2p) of oxide or fluoride, with the exception of $4f^2$Pr(III), $4f^8$Tb(III) and presumably also $4f^1$Ce(III), though the signal is so weak in the latter case that it is difficult to detect. Quite generally, the 4f signals are exceptionally strong [59, 60] helping their identification. For instance, in hafnium(IV) mandelate $Hf(C_6H_5CHOHCO_2)_4$ the fourteen 4f electrons produce two signals (j = 7/2 and 5/2) much stronger than the 232 valence electrons, with the exception of the region containing the signals from the MO, mainly consisting of the 24 oxygen 2s electrons, having an intensity of about a-quarter of the 4f signals. I(M4f) is much higher in M(IV) compounds than in M(III). An instructive example is Tb_4O_7 simultaneously [53] showing I(Tb4f) of the quadrivalent terbium at 24 eV (marginally higher than the value in HfO_2) and the lowest *I* of the terbium(III) corresponding to the ionization process $4f^8(^7F_6) \rightarrow 4f^7(^8S_{7/2})$ at 8 eV. On the other hand, *I*(Eu4f) of europium(II) in $EuSO_4$ is only 6.9 eV, and it has been shown [57] that the variation of *I*(M4f) in $4f^q$ as a function of q (for a definite oxidation state) agrees with the refined spin-pairing energy theory Eq. (1.47). We do not here discuss in detail how the *I* values cited for Tb_4O_7 and $EuSO_4$ really are I' values corrected for the quasi-stationary positive potential prevailing on the non-conducting sample.

By itself it is not too terrifying that *I*(M4f) is larger than *I* of various filled MO because the number q of 4f electrons is kept constant by the *electron affinity* of the 4f shell not being larger than *I* of any other orbital. It is actually from photo-electron spectra that we have the most reliable estimates for the *difference* between *I* and electron affinity of the 4f shell in solids, normally between 7 and 9 eV (the rather unique compound TmTe discussed below suggests 7 eV, and the difference 16 eV between the two observed signals of Tb_4O_7 is cut down to 9 eV, because 7 eV of it consists of differing spin-pairing energy). Nevertheless, we face a major problem in Eq. (3.7) because the chemist generally expects strong covalent bonding when ψ_M and ψ_X have comparable one-electron energies. It may be argued that the non-diagonal element intrinsically [20] is proportional to the small overlap integral S_{MX} and hence, the extent of covalent bonding is moderate even when the diagonal elements H_M and H_X almost coincide. However, from a strict quantum-mechanical point of view, an eigenvalue (here corresponding to the absence of one 4f electron) which can be brought to coincide with another eigenvalue (here corresponding to the absence of an electron of one of the filled MO) of Schrödinger's equation has the specific property that an arbitrary linear combination (with coefficients which can be written cos ϑ and sin ϑ) of the two eigenvalues, *also* is an eigenvalue. In a treatment such as Eq. (3.6) one expects (for reason of symmetry) ϑ to be 45° when the two diagonal elements are identical. There is not an absolute contradiction between the fact that the actual J-levels belonging to $4f^q$ of M(III) are very weakly influenced by covalent bonding, as one sees from nephelauxetic ratios β between 0.95 and 0.99 and from almost invariant ζ_{4f}, and that at the same time the states of the ionized system $4f^{q-1}$ have the conditional oxidation state M[IV] being highly covalent. However, this argument needs careful introduction [61, 62] of interelectronic repulsion before applying the variation principle, and the difficulty remains, as discussed below, that the nephelauxetic effect does not seem excessive in M[IV] produced in photo-electron spectra, though it is more pronounced than in M[III].

In order to discuss the nephelauxetic effect in M[IV] it is necessary to study photo-electron spectra with the highest possible degree of resolution. Whereas, commercial instruments (not using additional devices for ameliorating the monochromacy of the soft X-ray photons, such as the Hewlett-Packard) generally have the one-sided half-width δ of the $Ag3d_{5/2}$ signal, of metallic silver 0.6 eV, (and $Au4f_{7/2}$ of gold 0.65 eV) it is possible to construct home-made apparatus with an instrumental contribution to δ of only 0.3 eV. In the future [63] it is conceivable that one might be able to select an even narrower interval of the continuous spectrum from a synchrotron containing electrons moving with velocities marginally below c. Anyhow, extremely good instrumental resolution does not help much on photo-electron spectra of non-conducting solids tending to a lower limit of δ 0.8 eV, either due to inhomogeneous quasi-stationary positive potential or to the deviating Madelung potential in the outermost atomic layers (it is noted that the photo-electron signals produced by electrons not having suffered inelastic scattering originate in the outermost 20 Å of typical solid samples).

The metallic lanthanides have a tremendous affinity toward oxygen and humidity and it is necessary to work under extremely high vacuo (as contrasted to the pressures 10^{-7} to 10^{-6} torr used for general work) and to monitor the growing-up of the oxygen 1s signal due to the oxidation products (the signal is observed independently of whether they are crystalline or amorphous). Such studies were started by Hagström [64] and continued under high resolution with a home-made instrument by Baer [65]. The spectra of metallic europium and ytterbium clearly show their conditional oxidation state M[II] by having so low I^* values (measured with reference to the *Fermi* *level* of the loosest bound conduction electrons constituting the threshold for Einstein photo-emission; the I values relative to *vacuo* are 2 to 3 eV higher for the metallic lanthanides). The single signal of Eu with $I^* = 2.1$ eV is due to the ionization process $4f^7(^8S) \rightarrow 4f^6(^7F)$ like the isoelectronic gadolinium with $I^* = 8.0$ eV (showing that the difference between I and the electron affinity of the 4f shell is at least 8 eV; otherwise, the conduction electrons would invade the gadolinium cores forming $4f^8$ systems) whereas, the two signals of Yb with $I^* = 1.1$ and 2.4 eV, and of metallic lutetium $I^* = 7.1$ and 8.6 eV are the spin-orbit components $^2F_{7/2}$ and $^2F_{5/2}$ of $4f^{13}$ producing the narrow absorption bands of ytterbium(III) compounds close to 1.3 eV. The I(M4f) are considerably higher [1, 53] in hafnium(IV) and tantalum(V) compounds and increase [57] as a linear function of Z up to gold, 8 eV per element. Thus [66], $I(W4f_{7/2})$ is 37.6 eV in gaseous $W(CO)_6$ and 46.67 eV in gaseous WF_6. The I^* values for metallic Eu and Gd may be compared with $I^* = 1.8$ eV found by Eastman and Kuznietz [67] in EuS and $I^* = 8.9$ eV in the metallic, stoichiometric compound GdS. These authors studied the influence of the photon energy in Eq. (3.10) on the relative intensity of various signals, and $h\nu$ above 40 eV was found to discriminate strongly in favour of the Eu4f signal, whereas the S3p signal at $I^* = 3.9$ eV is stronger for $h\nu$ below 20 eV.

The photo-electron spectra have provided magnificent confirmation of the *Copenhagen principle of final states* [1, 55, 62], that the quantum-mechanical systems are fair gambling machines in the sense that once an atom has absorbed a high-energy photon (which has perhaps traversed thousand other atoms without getting absorbed) the outcome of a variety of alternatives of which (say ejection of a 4p electron, or 4d, or 5p, or 4f) each has a definite probability, but that the distribution of the photon-energy of Eq. (3.10) on the excitation of the system to an ionized state high up in the continuum and on E_{kin} of the ejected electron in an individual process has no sufficient cause in the Medieval sense. Though one may express doubts [68] whether quantum mechanics is valid for systems which are so large that they cannot be reproduced as indiscernible replicas, the whole recent development of photo-electron spectrometry clearly shows that the choice of ionization processes available at a given photon energy obeys quantum mechanics much in the same way as competetive ways of decay of a radioactive nucleus.

If the ionization by soft X-rays of the ground level of $4f^q$ had the probability proportional to the number of states in each level of $4f^{q-1}$, the signals for q = 7 would be dispersed over quite a large interval, in contrast to experiment. A lower limit of the width of the configuration $4f^6$ is $12D$ from Eq. (1.18) or 10 eV. Actually, the ionization process is directed by symmetry-determined selection rules (of the same type as the selection rules for electric dipole transitions in the beginning of section 1C) such as J varying at most 7/2 (this is the higher of the two j levels for an f electron) and in the approximation of Russell-Saunders coupling that S varies (up or down) by half a unit and L at most by 3. Hence, in Russell-Saunders coupling the only term accessible by ionization of $4f^7(^8S)$ is $4f^6(^7F)$. Furthermore many selection rules of great importance in practice are numerical and relate weak intensity of signals allowed for symmetry reasons.

Cox [65, 69] calculated the probabilities (normalized to have the sum q) on the assumption that the squares of the *coefficients of fractional parentage* of the (S, L)-

terms (or J levels) of f^{q-1} in the ground J-level of f^q are proportional to the probabilities of ionization. This hypothesis has been verified to a remarkable extent [65] for the metallic M[III] elements and for the NaCl-type antimonides MSb studied by Campagna, Bucher, Wertheim, Buchanan and Longinotti [70] by cleaving the cubic crystals inside a Hewlett-Packard photo-electron spectrometer. Table 22 gives I^* values for the elements and for the antimonides (all showing a Sb5p signal with $I^* = 2.0$ eV). The photo-electron spectra are particularly rich in details for q between 8 and 13. This is partly due to the distance $(2S + 1) D = (15 - q) D$ between the baricentres [57] of the states characterized by the two values $(S + 1/2)$ and $(S - 1/2)$ accessible to the ionization process, but the terms having $(S - 1/2)$ and strong signal intensities are scattered to a rather large extent around this baricentre. It is possible to find many interesting regularities [51, 65] for the Cox probabilities in Russell-Saunders coupling. Because of the moderate resolution of the photo-electron spectra, spin-orbit coupling plays a minor rôle, except for q = 12 and 13 (like in the absorption spectra for q = 11 and 12) showing strong admixtures of S and $(S - 1)$. The Cox probabilities for forming the individual J-levels of a given term in Russell-Saunders coupling are highly different, and usually strongly concentrated [65, 69] on the J-level having the lowest energy, i.e. $J = |L - S|$ for q below 7 and $J = (L + S)$ for the second half of the f shell.

Table 22. Ionization energies (in eV) I^* relative to the Fermi level (I values relative to vacuo are approximately 3 eV higher) determined from photo-electron spectra of metallic lanthanides [65] and NaCl-type antimonides [70] measured in high vacuo. In addition to the tabulated final states belonging to the configuration $4f^{q-1}$ the elements have weak signals at low I^* due to conduction electrons, and the antimonides a Sb5p signal situated between $I^* = 1.9$ and 2.0 eV. Shoulders are given in parentheses

q				
q = 1:	Ce:	2 (^{1}S)	CeSb:	3 (^{1}S)
q = 2:	Pr:	3.3 ($^2F_{5/2}$)	PrSb:	4.6 ($^2F_{5/2}$)
q = 3:	Nd:	4.8 ($^3H_4$); (5.4)($^3F_2$)	NdSb:	5.8 ($^3H_4$); (6.4)($^3F_2$)
q = 5:	Sm:	5.3 (^{5}I); (6.8)(^{5}F); 7.6(^{5}G); 9.4(^{5}D)		
	SmSb:	6.1(^{5}I); 8.5(^{5}G)		
q = 7:	Gd:	8.0(^{7}F)	GdSb:	9.1(^{7}F)
q = 8:	Tb:	2.3 (^{8}S); 7.4 (^{6}I); 9.3 (^{6}G); 10.3 (^{6}H)		
	TbSb:	3.2 (^{8}S); 8.2 (^{6}I); 10.1 (^{6}G); 11,1 (^{6}H)		
q = 9:	Dy:	3.9 (^{7}F); 6.7 (^{5}D); 7.7 (^{5}L); (8.7)(^{5}H); 9.3 (^{5}I); 10.5 (^{5}K); 12.5 (shake-off to ^{8}S?)		
	DySb:	5.0 (^{7}F); (8.3)(^{5}D); 8.7 (^{5}L); (9.7)(^{5}H); 10.3 (^{5}I); (11.5)(^{5}K); 13.5 (shake-off?)		
q = 10:	Ho:	5.3 ($^6H_{15/2}$); (6)(^{6}F); 8.7 ($^4M_{21/2}$); (9.6)($^4L_{19/2}$)		
	HoSb:	6.0 ($^6H_{15/2}$); 7.0 (^{6}F); 9.4 ($^4M_{21/2}$); (10)(^{4}L)		
q = 11:	Er:	4.8 ($^5I_8$); 5.4 ($^5I_7$); 6.8 (^{5}F); 7.7 ($^5G_6$); 8.7 ($^3L_9$); 9.4 ($^3M_{10}$)		
	ErSb:	5.6 ($^5I_8$); (6.3)($^5I_7$); (8)(^{5}F); 8.7 ($^5G_6$); 9.6 ($^3L_9$); 10.2 ($^3M_{10}$)		
q = 12:	Tm:	4.6 ($^4I_{15/2}$); 5.8 ($^4I_{13/2}$); (7)(^{2}H); 8.3 ($^4G_{11/2}$, $^2K_{15/2}$); 10.1 ($^2L_{17/2}$)		
	TmSb:	5.5 ($^4I_{15/2}$); 6.5 ($^4I_{13/2}$); 8.2 ($^2H_{11/2}$); 9.3 ($^4G_{11/2}$, $^2K_{15/2}$); 11.2 ($^2L_{17/2}$)		
q = 14:	Lu:	7.1 ($^2F_{7/2}$); 8.6 ($^2F_{5/2}$)		

For our purpose it is fascinating that one can determine the nephelauxetic effect in the configuration $4f^{q-1}$ of the ionized states by comparison with the energy levels of M(III), determined by Carnall, Fields and Rajnak [71]. Relative to the lowest signal of metallic terbium, corresponding to formation of ^8S, the following prominent signals have 5.1 (^6I), 7.0 (^6G) and 8.0 eV (^6H) higher I^* (or I for that matter) compared with the positions 4.5, 6.2 and 7.3 eV above the groundstate [71] of Gd(III). It may be argued that this is still an increase of 12 percent of the term distances in Tb[IV] formed by ionization compared with Gd(III). Though the gaseous ions Tb^{+4} and Gd^{+3} have not been sufficiently studied, there is little doubt that parameters of interelectronic repulsion such as E^1 (= 8 $D/9$) and E^3 follow a smooth evolution as function of the ionic charge z like B in Eq. (3.8). Actually, one would expect an increase of the term distances 20 percent from Gd^{+3} to Tb^{+4}. Assuming a nephelauxetic ratio $\beta = 0.98$ for the Gd(III) fluoride and aqua ions, the observed increase 12 percent actually corresponds to a nephelauxetic ratio $\beta = 1.12 \cdot 0.98/1.20 = 0.92$ in metallic terbium. A comparable increase 12 percent of the term distances is seen from Table 22 when comparing the ionized $4f^8$ states of metallic dysprosium with the excited states [71] of terbium(III). On the other hand, the increase of term distances is only 6 percent in ionized $4f^{11}$ states of metallic thulium relative to erbium(III), as seen by the distance 5.5 eV between the first and the last sharp signal, identified by Cox [65] as $^4I_{15/2}$ and $^2L_{17/2}$ with the two highest probabilities of formation among all J-levels of $4f^{11}$ Absorption spectra [71] of Er(III) give this distance 5.2 eV. Hence, the nephelauxetic ratio in metallic thulium is $\beta = 1.06 \cdot 0.99/1.20 = 0.88$. Until recently the only photo-electron spectra of $4f^{13}$ systems reported consisted of fairly broad signals of non-conducting ytterbium(III) compounds [1, 53] but then, a very fortunate example was found [72] in the apparent thulium(II) telluride TmTe showing other, highly unexpected phenomena.

The photo-electron spectrum of this cubic(NaCl-type)crystal [72, 73] shows comparable amounts of $4f^{12}$ Tm[III] and $4f^{13}$ Tm[II] to be present in the groundstate. The former constituent has the same signals as the isotypic [70] TmSb, whereas the other shows a series of signals at lower I^* separated 4.3 eV from $^3H_6$ to $^1I_6$ of the ionized configuration $4f^{12}$. This observation explains why the magnetic susceptibility of TmTe has a value intermediate between the values appropriate for $^3H_6$ of Tm(III) and $^2F_{7/2}$ of Tm(II). In the corresponding selenide TmSe, the photo-electron spectrum [72] shows 80 percent Tm[III] and 20 percent Tm[II].

It is obvious that this question is related to the *time-scale* of the measurements [20], since the crystal structure indicates the time-average content of the unit cell. It is not yet known how much slower the hopping of an electron from Tm[II] to an adjacent Tm[III] is than the almost *instantaneous picture* of the photo-electron spectra, argued [1, 53] to have the time-scale 10^{-17} sec. However, if the hopping of the electron is slower than 10^{-6} sec, one would expect distortions of the crystal structure as in covellite CuS containing at the same time Cu(I), Cu(II), S^{-2} andS_2^{-2} (found also in the iron(II) compounds pyrite and marcasite [16]). In our opinion, the most interesting property of TmTe is that the nephelauxetic ratio β of the Tm[II] sites is, at most, 0.01 below the thulium(III)aqua ions. Hence, the Copenhagen principle of final states has been confirmed to the extent that $\beta = 0.91$ and 0.98 for the two (crystallographically equivalent) thulium sites are greatly different.

Semi-conducting NaCl-type samarium(II)sulphide SmS shows a straight forward photo-electron spectrum [74] due to $4f^6 \rightarrow 4f^5$ ionization. When gadolinium or thorium is introduced in such crystals, for instance $Sm_{0.82}Gd_{0.18}S$ and $Sm_{0.85}Th_{0.15}S$, they remain cubic, but the photo-electron spectra indicate comparable concentrations of Sm[III] and of Sm[II] in agreement with the metallic character showing many conduction electrons.

The major conclusion from the moderate nephelauxetic ratio close to 0.9 in the M[IV] states formed by photo-ionization is that β is definitely above 0.25, the order of magnitude one obtains by roughly equivalent delocalization with $\vartheta = 45°$ in Eq. (3.6). Actually, β close to 0.9 is the same value one finds for the least covalent d group complexes, such as manganese(II) and nickel(II), aqua ions and fluorides. It may be noted that the loosest bound orbitals in the M(III) antimonides, Sb5p have $I^* = 2$ eV several eV below I^* of the M 4f orbitals [70]. It is not known whether it has a physical significance that β is distinctly higher (see Table 22) in TmSb than in elemental Tm, whereas it is marginally the other way round in TbSb and DySb, compared with Tb and Dy.

B. The Intensities and the Hypersensitive Pseudoquadrupolar Transitions

Broer, Gorter and Hoogschagen [75] discussed thoroughly the oscillator strengths P defined Eq. (1.38) for internal $4f^q$ transitions. There is a general feature of the corresponding absorption bands which is better realized today (where the identification of excited J-levels below 30,000 cm^{-1} is almost complete [71]) than in 1945, that not only the parity is apparently conserved, but also that nearly all the spin-allowed transitions (to levels with S_{max}) have roughly the same intensity, without a strong discrimination (both numerically and symmetry-determined, J changing at most one unit for genuine electric dipole transitions) according to the J value, as normal in atomic spectra. However, there is one characteristic common to the spectra of lanthanides in condensed matter and of monatomic entities, that is the multiplicative *factor of spin-forbiddenness,* the ratio between P of spin-forbidden and of typical spin-allowed transitions being a few times $(\zeta_{nl}/d\sigma)^2$ where ζ_{nl} is the Landé parameter and $d\sigma$ the wave-number difference between the spin-forbidden and the closest spin-allowed transition to a level with the same J. In the lanthanides this square is readily 0.1 or 0.2. If it is close to 1, we have a typical situation of intermediate coupling. We may add that there is something wrong with this simple description for most of the spin-forbidden transitions in $4f^7$ Gd(III) and a few in $4f^6$ Eu(III) which have intensities as if the (not directly accessible) spin-allowed transitions were several hundred times stronger than usual. A possible explanation is mixing with parity-allowed ($4f \rightarrow 5d$ or electron transfer) transitions at higher wave-numbers however, they are much less effective in $4f^8$ Tb(III). The percentage ^6P character in the groundstate of Gd(III) is 14 $(\zeta_{4f}/32{,}200 \text{ cm}^{-1})^2 = 2.8$ percent.

The magnetic dipole transitions are only observed in a few cases, such as $^7F_0 \rightarrow {}^5D_1$ in absorption ($P = 1.4 \cdot 10^{-8}$) and $^5D_0 \rightarrow {}^7F_1$ in emission of Eu(III). Theoretically [75], they should only occur in Russell-Saunders coupling within the lowest multiplet from the lowest to the next-lowest *J*-level and are hence expected in the far infra-red, such as $^5I_8 \rightarrow {}^5I_7$ of Ho(III) at 5,200 cm^{-1}, and their *P* is calculated [71] to be close to 2.10^{-7}.

The electric quadrupole transitions are allowed between states of the same parity, but they do not suit the observed spectra of aqua ions very well, since *J* changes at most 2 units, and there is the serious numerical difficulty that their *P* values proportional to $\langle r^2 \rangle^2 / \lambda^2$ may be reasonably large in the X-ray region, but are insufficient, 10^{-9} to 10^{-8} to explain the intensities in the visible with $\langle r^2 \rangle$, now known from Hartree-Fock functions. Hence, Broer, Gorter and Hoogschagen [75] concluded that the mechanism prevailing for the observed *P* between 10^{-7} and 10^{-5} is *forced electric dipole transitions.* Traditionally, two different detailed mechanisms are discussed: the *hemihedric part* of the "ligand field" isolated by the operation Eq. (1.32) and the *odd vibrations* mix excited states with opposite parity to a small extent, either in the excited level of the optical transition considered, or in the groundstate. The main reason why *P* is up to 100 times larger in tetrahedral MX_4 than in comparable octahedral MX_6 in the 3d group is the hemihedric "ligand field" and it seems also to have been established as the major source of intensity in ruby $Al_{2-x}Cr_xO_3$ containing slightly distorted chromophores $Cr(III)O_6$. It is clear that in a solvent or in a vitreous material this distinction may be less significant, because the nuclei of the ligating atoms almost always exhibit much lower symmetry on an instantaneous picture [20] than the site point-group in a crystallographic sense. These deviations may be ascribed to odd vibrations around an equilibrium position possessing a centre of inversion in a crystal, but they may equally well be considered as a fluctuating perturbation of odd parity. We return to this problem below, but note that for both kinds of forced electric dipole transitions, the symmetry-determined selection rules [75] allow *J* to change at most 6 units. Actually, it is exceedingly rare that one has a good prediction [71] of the position of a *J*-level with *J* differing 7 or more units from the groundstate, and then the transition is detected. Pr(III) and Nd(III) do not possess transitions forbidden by this selection rule. Eu(III) have many, $^5G_7$, $^5L_{7,8,9,10}$ (and also $^7F_0 \rightarrow {}^5D_3$, $^5G_3$ and $^5G_5$ forbidden for other reasons) where it may be noted that the transition to $^5L_6$ at 25,400 cm^{-1} is the strongest absorption band ($P = 1.8 \cdot 10^{-6}$) in the whole region from 6,000 to 40,000 cm^{-1}. It is a clear-cut case that Dy(III) is predicted to have the transition $^6H_{15/2} \rightarrow {}^6F_{1/2}$ at 13,700 cm^{-1} in an absolutely empty interval in the absorption spectrum [71]. Another case of spin-allowed transitions changing *J* by more than 6 units are the excited levels of Ho(III) $^5F_1$, $^5D_1$ and $^5D_0$ predicted [71] at 22,300, 43,000 and 42,600 cm^{-1}, respectively. The transition $^4I_{15/2} \rightarrow {}^2P_{1/2}$ in Er(III) is a last case where no absorption has been observed at the predicted position 33,500 cm^{-1} in contrast to $^4I_{9/2} \rightarrow {}^2P_{1/2}$ at 23,250 cm^{-1} in Nd(III) corresponding to one of the best known, sharp bands.

Judd [76] and Ofelt [77] proposed independently in 1962 a quantitative theory for $4f^q$ forced electric dipole transitions between the two levels J_1 and J_2

$$\{\tau_2[U^{(2)}]^2 + \tau_4[U^{(4)}]^2 + \tau_6[U^{(6)}]^2\}(E_2 - E_1) \tag{3.11}$$

where the matrix elements $U^{(2)}$, $U^{(4)}$ and $U^{(6)}$ can be evaluated once and for all in Russell-Saunders coupling (in the actual cases of intermediate coupling [71] one has to evaluate the numerical linear combinations of Russell-Saunders values for a given transition) and where τ_2, τ_4 and τ_6 are parameters characterizing the chromophore studied. In the Judd-Ofelt theory, the physical origin of $U^{(t)}$ is an external perturbation of odd parity, mixing the 4f orbitals (to a very small extent) with the two adjacent *l*-values 2 and 4. In most text-books, these admixed orbitals are called 5d and 5g, but it is quite conceivable that they may derive from electron transfer states. Table 23 gives the positions and the three squared matrix elements $[U^{(t)}]^2$ for the strongest transitions from the groundstate. We follow most of the recent authors by using Ω_t from Eq. (2.75) rather than τ_t from Eq. (3.11). The ratio (τ_t/Ω_t) is $1.086 \cdot 10^{11}$ cm^{-1} times the Lorentz factor $(n^2+2)^2/9n$, depending on the refractive index n. In most cases this ratio is close to $1.3 \cdot 10^{11}$ cm^{-1}. Table 24 gives Ω_t for selected aqua ions and complexes, to be compared with Tables 6 and 19 in Chapter 2.

Table 23. Squared matrix elements of the Judd-Ofelt tensor for transitions from M(III) groundstates (and europium $^7F_1$) with energies (mostly within 50 cm^{-1}) E in cm^{-1} and a million times the oscillator strength *P* in the specific case of the aqua ions

	E	$[U^{(2)}]^2$	$[U^{(4)}]^2$	$[U^{(6)}]^2$	$10^6 P$
Pr(III) $^3H_4$					
$^3H_5$	2,300	0.110	0.202	0.611	–
$^3H_6$	4,500	0.000	0.033	0.140	–
$^3F_2$	5,200	0.509	0.403	0.118	–
$^3F_3$	6,500	0.065	0.347	0.698 }	13
$^3F_4$	6,950	0.019	0.050	0.485 }	
$^1G_4$	9,900	0.001	0.007	0.027	0.32
$^1D_2$	16,850	0.003	0.017	0.052	3.1
$^3P_0$	20,800	0	0.173	0	2.5
$^3P_1$	21,400	0	0.171	0	7.6
$^1I_6$	21,700	0.009	0.052	0.024	–
$^3P_2$	22,600	0	0.036	0.136	15
$^1S_0$	47,000	0	0.007	0	–
Nd(III) $^4I_{9/2}$					
$^4I_{11/2}$	2,000	0.019	0.107	1.165	–
$^4I_{13/2}$	4,000	0	0.014	0.456	–
$^4I_{15/2}$	6,100	0	0	0.045	0.5
$^4F_{3/2}$	11,450	0	0.229	0.055	2.5
$^4F_{5/2}$	12,500	0.001	0.237	0.397 }	8.8
$^2H_{9/2}$	12,600	0.009	0.008	0.115 }	
$^4F_{7/2}$	13,500	0	0.003	0.235 }	8.9
$^4S_{3/2}$	13,500	0.001	0.042	0.425 }	
$^4F_{9/2}$	14,800	0.001	0.009	0.042	0.7
$^2H_{11/2}$	15,900	0	0.003	0.010	0.15
$^4G_{5/2}$	17,300	0.898	0.409	0.036	9.8
$^2G_{7/2}$	17,400	0.076	0.185	0.031	–

Table 23 (continued)

	E	$[U^{(2)}]^2$	$[U^{(4)}]^2$	$[U^{(6)}]^2$	10^6P
$^2K_{13/2}$	19,000	0.007	0	0.031	
$^4G_{7/2}$	19,100	0.055	0.157	0.055	7.0
$^4G_{9/2}$	19,500	0.005	0.061	0.041	
$^2K_{15/2}$	21,000	0	0.005	0.014	
$^2G_{9/2}$	21,200	0.001	0.015	0.014	2.3
$^2D_{3/2}$	21,300	0	0.019	0	
$^4G_{11/2}$	21,600	0	0.005	0.008	
$^2P_{1/2}$	23,250	0	0.037	0	0.4
$^2D_{5/2}$	23,900	0	0	0.002	0.09
$^2P_{3/2}$	26,300	0	0.001	0.001	0.03
$^4D_{3/2}$	28,300	0	0.196	0.017	
$^4D_{5/2}$	28,500	0	0.057	0.028	9.4
$^2I_{11/2}$	28,600	0.005	0.015	0.003	
$^4D_{1/2}$	28,850	0	0.258	0	
$^2L_{15/2}$	29,200	0	0.025	0.010	
$^2I_{13/2}$	30,000	0	0.001	0.002	2.7
$^4D_{7/2}$	30,500	0	0.004	0.008	
Sm(III) $^6H_{5/2}$					
$^6H_{7/2}$	1,100	0.206	0.196	0.095	–
$^6H_{9/2}$	2,300	0.026	0.140	0.327	–
$^6H_{11/2}$	3,600	0	0.024	0.265	–
$^6H_{13/2}$	5,000	0	0.001	0.066	–
$^6F_{1/2}$	6,400	0.194	0	0	0.26
$^6H_{15/2}$	6,500	0	0	0.004	
$^6F_{3/2}$	6,600	0.144	0.136	0	1.0
$^6F_{5/2}$	7,100	0.033	0.284	0	1.7
$^6F_{7/2}$	8,000	0.002	0.143	0.430	2.7
$^6F_{9/2}$	9,200	0	0.021	0.341	2.0
$^6F_{11/2}$	10,500	0	0.001	0.052	0.3
$^4G_{5/2}$	17,900	0	0.001	0	0.03
$^4M_{15/2}$	20,800	0	0	0.031	0.06
$^4I_{11/2}$	21,100	0	0	0.011	
$^4I_{13/2}$	21,600	0	0.003	0.023	0.7
$^6P_{5/2}$	24,000	0	0.026	0	–
$^4L_{13/2}$	24,550	0	0.008	0.010	0.4
$^6P_{3/2}$	24,950	0	0.168	0	3.8
$^6P_{7/2}$	26,750	0	0.002	0.075	1.0
$^4D_{3/2}$	27,700	0	0.025	0	1.2
$^4D_{7/2}$	29,100	0	0.001	0.038	0.8
Eu(III) $^7F_0$					
$^7F_2$	1,000	0.137	0	0	–
$^7F_4$	2,850	0	0.140	0	–
$^7F_6$	5,000	0	0	0.145	–
$^5D_2$	21,500	0.001	0	0	0.01
$^5L_6$	25,400	0	0	0.016	1.8
$^5G_6$	26,700	0	0	0.004	0.2
$^5H_6$	31,500	0	0	0.006	0.7

Table 23 (continued)

	E	$[U^{(2)}]^2$	$[U^{(4)}]^2$	$[U^{(6)}]^2$	10^6P
Eu(III) $^7F_1$					
$^7F_3$	1,500	0.209	0.128	0	–
$^7F_4$	2,500	0	0.174	0	–
$^7F_5$	3,550	0	0.119	0.054	–
$^7F_6$	4,650	0	0	0.377	–
$^5D_1$	18,700	0.003	0	0	0.01
$^5L_6$	25,050	0	0	0.009	–
$^5L_7$	26,000	0	0	0.018	0.3
$^5G_5$	26,350	0	0.001	0.010	
$^5H_7$	30,700	0	0	0.007	–
$^5H_5$	31,200	0	0.002	0.007	–
Gd(III) $^8S_{7/2}$					
$^6P_{7/2}$	32,200	0.001	0	0	0.07
$^6I_{7/2}$	35,900	0	0	0.004	0.12
$^6I_{9/2}$	36,250	0	0	0.010	0.85
$^6I_{17/2}$	36,350	0	0	0.021	
$^6I_{11/2}$	36,550	0	0	0.018	
$^6I_{13/2}$	36,650	0	0	0.024	1.9
$^6I_{15/2}$	36,700	0	0	0.027	
$^6D_{9/2}$	39,700	0.006	0	0	0.08
$^6D_{7/2}$	40,700	0.004	0	0	0.08
Tb(III) $^7F_6$					
$^7F_5$	2,100	0.538	0.642	0.118	–
$^7F_4$	3,350	0.089	0.516	0.265	–
$^7F_3$	4,350	0	0.232	0.413	–
$^7F_2$	5,000	0	0.048	0.470	–
$^7F_1$	5,450	0	0	0.376	–
$^7F_0$	5,700	0	0	0.144	–
$^5D_4$	20,500	0.001	0.001	0.001	0.05
$^5D_3$	26,300	0	0	0.001	–
$^5G_6$	26,500	0.002	0.005	0.012	0.2
$^5L_{10}$	27,100	0	0	0.059	0.7
$^5G_5$	27,800	0.001	0.002	0.014	0.7
$^5L_9$	28,400	0	0.002	0.047	
$^5L_8$	29,300	0	0	0.024	0.3
$^5L_7$	29,450	0.001	0	0.012	
Dy(III) $^6H_{15/2}$					
$^6H_{13/2}$	3,500	0.246	0.414	0.682	–
$^6H_{11/2}$	5,850	0.092	0.037	0.641	1.1
$^6H_{9/2}$	7,700	0	0.018	0.199	1.1
$^6F_{11/2}$	7,700	0.939	0.829	0.205	
$^6F_{9/2}$	9,100	0	0.574	0.721	3.0
$^6H_{7/2}$	9,100	0	0.001	0.039	
$^6H_{5/2}$	10,200	0	0	0.003	0.1

Table 23 (continued)

	E	$[U^{(2)}]^2$	$[U^{(4)}]^2$	$[U^{(6)}]^2$	$10^6 P$
$^6F_{7/2}$	11,000	0	0.136	0.715	2.7
$^6F_{5/2}$	12,400	0	0	0.345	1.6
$^6F_{3/2}$	13,200	0	0	0.061	0.3
$^4F_{9/2}$	21,100	0	0.005	0.030	0.2
$^4I_{15/2}$	22,100	0.007	0	0.065	0.4
$^4G_{11/2}$	23,400	0	0.015	0	0.16
$^4F_{7/2}$	25,800	0	0.077	0.026	2.4
$^4K_{17/2}$	26,400	0.011	0.005	0.094	
$^4M_{19/2}$	27,400	0	0.017	0.102	
$^6P_{5/2}$	27,500	0	0	0.070	–
$^6P_{7/2}$	28,500	0	0.522	0.013	2.8
$^6P_{3/2}$	30,800	0	0	0.110	–
Ho(III) $^5I_8$					
$^5I_7$	5,150	0.025	0.134	1.522	–
$^5I_6$	8,600	0.008	0.039	0.692	1.1
$^5I_5$	11,100	0	0.010	0.094	0.2
$^5I_4$	13,300	0	0	0.008	0.02
$^5F_5$	15,500	0	0.425	0.569	3.8
$^5S_2$	18,400	0	0	0.227	5.2
$^5F_4$	18,600	0	0.239	0.707	
$^5F_3$	20,600	0	0	0.346	1.8
$^5F_2$	21,100	0	0	0.192	1.4
$^3K_8$	21,400	0.021	0.033	0.158	
$^5G_6$	22,100	1.520	0.841	0.141	6.0
$^5G_5$	23,950	0	0.534	0	3.1
$^5G_4$	25,800	0	0.031	0.036	1.0
$^3K_7$	26,200	0.006	0.005	0.034	
$^3H_5$	27,650	0	0.079	0.161	3.2
$^3H_6$	27,700	0.215	0.118	0.003	
$^3L_9$	29,000	0.019	0.005	0.154	0.8
$^3F_4$	30,000	0	0.126	0.005	–
$^3M_{10}$	34,200	0	0.070	0.081	1.0
$^5D_4$	34,800	0	0.304	0.049	2.7
$^3H_4$	36,000	0	0.263	0.004	2.6
Er(III) $^4I_{15/2}$					
$^4I_{13/2}$	6,600	0.020	0.117	1.432	2.2
$^4I_{11/2}$	10,250	0.028	0	0.395	0.8
$^4I_{9/2}$	12,500	0	0.173	0.010	0.3
$^4F_{9/2}$	15,300	0	0.535	0.462	2.3
$^4S_{3/2}$	18,400	0	0	0.221	0.7
$^2H_{11/2}$	19,200	0.713	0.413	0.093	2.9
$^4F_{7/2}$	20,500	0	0.147	0.627	2.3
$^4F_{5/2}$	22,200	0	0	0.223	1.3
$^4F_{3/2}$	22,600	0	0	0.127	
$^2G_{9/2}$	24,600	0	0.019	0.226	0.8
$^4G_{11/2}$	26,400	0.918	0.526	0.117	5.9

Table 23 (continued)

	E	$[U^{(2)}]^2$	$[U^{(4)}]^2$	$[U^{(6)}]^2$	$10^6 P$
$^4G_{9/2}$	27,400	0	0.242	0.124	1.7
$^2K_{15/2}$	27,600	0.022	0.004	0.076	0.9
$^2G_{7/2}$	28,000	0	0.017	0.116	
$^2P_{3/2}$	31,600	0	0	0.017	0.1
$^2K_{13/2}$	33,200	0.003	0.003	0.015	0.1
$^4G_{5/2}$	33,400	0	0	0.003	
$^4G_{7/2}$	34,000	0	0.033	0.003	0.2
$^4D_{7/2}$	39,200	0	0.892	0.029	10
$^2L_{17/2}$	41,700	0.005	0.066	0.033	–
Tm(III) $^3H_6$					
$^3F_4$	5,850	0.537	0.726	0.238	1.1
$^3H_5$	8,300	0.107	0.231	0.638	1.7
$^3H_4$	12,700	0.237	0.109	0.595	2.1
$^3F_3$	14,500	0	0.316	0.841	3.9
$^3F_2$	15,100	0	0	0.258	
$^1G_4$	21,300	0.048	0.075	0.013	0.7
$^1D_2$	27,900	0	0.316	0.093	2.4
$^1I_6$	34,900	0.011	0.039	0.013	0.8
$^3P_0$	35,500	0	0	0.076	
$^3P_1$	36,500	0	0	0.124	0.6
$^3P_2$	38,250	0	0.265	0.022	3

One of the debatable assumptions of the Judd-Ofelt-theory (recently reviewed by Peacock [78]) is that in a given denominator used in the perturbation treatment, the excitation energy of *all* the states of opposite parity is a constant. It has been argued [71, 78] that the reason why the parametrization does not work well for Pr(III) is that this hypothesis of identical excitation energies is less suitable in this particular case having both $4f^2$ and the odd configuration wide. Comparable discrepancies have been noted [85] for Sm(III) in glasses, when comparing transitions in the blue and in the infra-red. By the way, the Judd-Ofelt theory does not work too well either in the 5f group, though the identifications of the individual *J*-levels and the composition of eigen-vectors in intermediate coupling may be less certain than in the 4f group. On the other hand, the M(III) between $4f^3$ Nd(III) and $4f^{12}$ Tm(III) are remarkably well described by smoothly varying parameters of Eq. (3.11). This produces the rather unexpected result that the intensity distribution of *many* absorption bands depends on only the three Ω_t parameters, whereas the positions of the excited *J*-levels depend on ζ_{4f}, E^1, E^2 and E^3 (though it is true that the three latter parameters of interelectronic repulsion are almost proportional in actual cases, and increase as a linear function of q). The Judd-Ofelt theory also explains the curious fact that the transitions to the highest-lying *J*-level in terms with S_{max} are comparatively weak (such as $^4I_{15/2}$, $^4F_{9/2}$ and $^4G_{11/2}$ of Nd(III), $^6H_{15/2}$ and $^6F_{11/2}$ of Sm(III), $^5I_4$ and $^5G_2$ of Ho(III), $^4I_{9/2}$ of Er(III) *etc.*).

Table 24. Judd-Ofelt parameters (in the unit 10^{-20} cm^2) of complexes of trivalent lanthanides

	Ω_2	Ω_4	Ω_6
Nd(III) aqua ion	0.93 ± 0.3	5.0 ± 0.3	7.9 ± 0.4
Nd(III) in molten $LiNO_3 + KNO_3$	11.2 ± 0.3	2.6 ± 0.3	4.1 ± 0.4
$NdW_{10}O_{35}^{-7}$	0.8 ± 0.5	5.6 ± 0.6	7.0 ± 0.3
Nd(α-picolinate)$_3$	3.2 ± 0.8	10.4 ± 1.2	10.3 ± 0.6
Nd(acetylacetonate)$_3$ in alcohol	15.7	0.73	7.4
Nd(acetylacetonate)$_3$ in dimethylformamide	24.5	0.71	9.1
Nd(dibenzoylacetonate)$_3$ in alcohol	34.1	2.5	9.1
$NdCl_6^{-3}$	0.7	0.5	0.8
$NdBr_3$, gaseous	180	9	9
NdI_3, gaseous	275	9	9
Sm(III) aqua ion	0.9 ± 0.8	4.1 ± 0.3	2.7 ± 0.3
Gd(III) aqua ion	2.5 ± 0.5	4.7 ± 0.4	4.7 ± 0.1
Tb(III) aqua ion	0.004	7.2 ± 2.3	3.4 ± 0.3
Dy(III) aqua ion	1.5 ± 4	3.4 ± 0.2	3.5 ± 0.2
Ho(III) aqua ion	0.4 ± 0.2	3.1 ± 0.2	3.1 ± 0.2
$HoW_{10}O_{35}^{-7}$	3.5 ± 0.2	2.3 ± 0.2	2.6 ± 0.2
Er(III) aqua ion	1.6 ± 0.1	2.0 ± 0.2	1.9 ± 0.1
Er(III) in molten $LiNO_3 + KNO_3$	15.5 ± 1.1	1.8 ± 2	1.4 ± 0.8
$ErW_{10}O_{35}^{-7}$	6.7 ± 0.5	2.3 ± 0.3	1.4 ± 0.2
Tm(III) aqua ion	0.8 ± 0.6	2.1 ± 0.3	1.9 ± 0.2
Tm(III) in molten $LiNO_3 + KNO_3$	10.3 ± 1.8	2.7 ±0.7	1.7 ± 0.4
$TmW_{10}O_{35}^{-7}$	1.6	5.8	0.8

With the exception of the hexahalide complexes MX_6^{-3} studied by Ryan [36, 37], there are no clear-cut cases of lanthanide complexes having much *lower* band intensities than the aqua ions. Whatever the mechanism behind the forced electric dipole transitions may be, it seems to be generally present, though Kisliuk, Krupke and Gruber [79] could not detect any absorption lines from that quarter of the erbium(III) ions substituted in cubic C-type Y_2O_3 on C_{3i} sites with a centre of inversion, whereas 75 percent of the M are on C_2 sites.

It may be worthwhile to compare with the behaviour of octahedral chromium(III) and cobalt(III) complexes, where Yamada and Tsuchida [80] established the *hyperchromic series* of ligands. According to the kind information from Claus Schäffer, only fluoro and heteromolybdate complexes (of the type $MO_6Mo_6O_{15}^{-3}$) and the solutions of M(III) in concentrated sulphuric acid (of unknown constitution) have less intense absorption bands than the aqua ions, whereas the oscillator strength P_f of the Laporte-forbidden transitions tends to increase strongly, when the electronegativity of the ligating atoms decreases. Actually, it is a good empirical rule [17, 81, 82] for octahedral d group complexes

$$P_f/P_a = (H_{af})^2/(\sigma_a - \sigma_f)^2 \tag{3.12}$$

that P_f of the Laporte-forbidden transition with the wave-number σ_f is related to the reciprocal square of the distance to the closest Laporte-allowed transition with wave-number σ_a and oscillator strength P_a (typically 0.1). It is beyond doubt today that these Laporte-allowed transitions nearly always are electron transfer bands, explaining not only the hyperchromic series at least to a first approximation, but also the strong increase of d^q intensity M(II) < M(III) < M(IV).It is not known whether it is an accident that the effective non-diagonal element H_{af} in Eq. (3.12) usually is between 1,500 and 2,500 cm^{-1}, about twice the one-sided half-width δ of absorption bands due to the transition from one sub-shell to another. The general consensus is that the d^q transitions are assisted by the co-excitation of one *odd* vibrational mode, but at the same time, δ is the blurred-out version of a larger number of quanta of even vibrational modes (mainly the "breathing" totally symmetric stretching) being co-excited, following Franck and Condon's principle. If the same H_{af} was applicable in Eq. (3.12) to the 4f group M(III), P_f would be close to 10^{-4}, definitely too large, and the observed P values would need H_{af} to be only 200 cm^{-1} or so. The values of Ω_t in Tables 6, 19 and 24 do not vary in any obvious way as the function of the positions of the electron transfer bands, and the mechanism seems rather different from $3d^q$ transitions.

There is one category of transitions very sensitive to the choice of ligands. This behaviour was discovered by Moeller and Ulrich [83] for $^4I_{9/2} \rightarrow {}^4G_{5/2}$ in the yellow of Nd(III) and $^4I_{15/2} \rightarrow {}^2H_{11/2}$ in the green of Er(III) in various tris(β-diketonates). It might have been expected that this was an intensification of the type Eq. (3.12) originating in the very strong transitions (in the ligands) in the near ultra-violet, which had been concentrated for some accidental reason on one of the Laporte-forbidden transitions. Judd and one of us [84] pointed out that these *hypersensitive transitions* are exactly those intensified by an unusually large parameter τ_2 in Eq. (3.11). Since the oscillator strengths of *bone fide* electric quadrupole transitions also are proportional to the squared matrix-element $[U^{(2)}]^2$ multiplied by $\sim 10^{-8}$, such hypersensitive transitions are now called *pseudo-quadrupolar* in view of the fact that the experimental intensities are higher by a factor such as 1,000 or 10^4. The selection rules are either symmetry-determined (J changes at most 2 units; in Russell-Saunders coupling, S is invariant and L changes at most 2 units) or numerical. It turns out [71] that $U^{(2)}$ only has a significant size when J *decreases* 2 units. The combination of these two results in the corollary of each lanthanide (with q from 2 to 12) having *one* pronounced hypersensitive pseudoquadrupolar transition in Russell-Saunders coupling, that is frequently situated in the near infra-red. Thus, Pr(III) has the excited level $^3F_2$ at 5,000 cm^{-1}, Nd(III) $^4G_{5/2}$ at 17,300 cm^{-1}, Pm(III) $^5G_2$ at 17,700 cm^{-1}, Sm(III) $^6F_{1/2}$ at 6,400 cm^{-1}, Dy(III) $^6F_{11/2}$ at 7,700 cm^{-1}, Ho(III) $^5G_6$ at 22,100 cm^{-1}, Er(III) $^4G_{11/2}$ at 26,400 cm^{-1} and Tm(III) $^3F_4$ (according to convention) either at 5,800 or 12,700 cm^{-1}. Because of strong effects of intermediate coupling, Ho(III) has a second such transition to $^3H_6$ at 27,700 cm^{-1} and Er(III) to $^2H_{11/2}$ at 19,200 cm^{-1}. The mixture of the three Russell-Saunders diagonal elements $^3H_4$, $^3F_4$ and $^1G_4$ in the three levels of $4f^{12}$ having $J = 4$ is so extensive in thulium(III) that the two transitions mentioned, and the third at 21,300 cm^{-1} all show fairly large $U^{(2)}$. Though Ω_4 and Ω_6 have the same order of magnitude [71] in M(III) *aqua ions,* the value of Ω_2 seems small and experimentally rather indeterminate.

It is possible [85, 86] to evaluate the three matrix elements $U^{(t)}$, both for transitions in emission and in absorption. For instance, the absorption $^7F_0 \rightarrow {}^5D_2$ in the blue at 21,500 cm^{-1} and the emission $^5D_0 \rightarrow {}^7F_2$ at 16,250 cm^{-1} in Eu(III), both are hypersensitive. Whereas the first use of the red cathodoluminescence of europium(III) in colour television involved $Y_{1-x}Eu_xVO_4$ this lattice has now been replaced [87] by $Y_{2-x}Eu_xO_2S$ giving a deeper red colour because of the pseudoquadrupolar emission. It is not easy to give general rules for the size of $U^{(2)}$ for spin-forbidden transitions. A few such transitions can be perceptibly hypersensitive [78], such as $^8S_{7/2} \rightarrow {}^6P_{7/2}$ and $^6D_{9/2}$ in Gd(III) and $^7F_6 \rightarrow {}^5H_7$ in Tb(III).

We discuss below the relations between chemical bonding and hypersensitivity, but we would like to add that the additivity of ϵ values in Eq. (1.36) would also apply to a mixture of two or more non-equivalent sites for M(III) in a given sample. Since the three Ω parameters do not occur in interfering cross-terms, the observed Ω_2 of a mixture is the average value weighted by the relative percentage concentrations.

C. The Chemistry Behind the Judd-Ofelt Parametrization

Soon after it was established [84] that the hypersensitive pseudoquadrupolar transitions are those having a large $\Omega_2[U^{(2)}]^2$ the ideas about the physical mechanism divided according to two main lines. Judd [88] emphasized the influence of the *point-group symmetry* arguing that the odd (hemihedric) part of the non-spherical part $V_{res}(x, y, z)$ in Eq. (1.22) of the Madelung potential in certain point-groups contain a contribution proportional to the first power of one of the Cartesian coordinates. This is equivalent [20] to at least one of three p angular functions $p\sigma$, $p\pi c$ or $p\pi s$ of Eq. (1.25) possessing total symmetry, as is the case for the point-groups C_1, C_s, C_n (n = 2, 3, 4, . . .), C_{nv} (n = 2, 3, 4, . . . and ∞) all lacking a centre of inversion (this is by no means a sufficient reason; D_{3h} and all other D_{nh} with odd n, or D_{nd} with even n, lack centres of inversion but do not permit totally symmetric A_1 functions). By the way, the same group-theoretical condition is needed for permanent dipole moments. Judd [88] then argued that the parameter Ω_2 only can be strongly positive in these low-symmetry chromophores. It is difficult to discuss this question based on an instantaneous picture almost always [20] lacking all elements of symmetry (the point-group is C_1) when the chromophore contains four or more nuclei, considered as points. However, it is rather striking that the hexa-halide complexes [36, 37] which are so close to the point-group O_h that many electronic transitions are seen only accompanied by one quantum of an odd vibrational mode (such behaviour is known from the intra-sub-shell transitions in $5d^3$ rhenium(IV) hexahalide complexes [45] and 5d group gaseous hexafluorides [89]) still show strong hypersensitive transitions. An extreme case is the vapours of MCl_3, MBr_3 and MI_3 at high temperature, where the molecules seem to be planar [90] like boron(III) halides, with the point-group D_{3h} not permitting Judd's A_1 mechanism. Actually, $^4I_{9/2} \rightarrow {}^4G_{5/2}$ of gaseous NdI_3 has $\epsilon = 345$ and $P = 5.35 \cdot 10^{-4}$ and is the strongest $4f^q$ transition known [91] though $NdBr_3$, ErI_3 and TmI_3 also [90] show fairly high Ω_2 values. It is interesting that Liu and Zollweg [92] found that the volatile adduct with caesium iodide NdI_4Cs have the hypersensitive transitions half as strong as NdI_3 alone. Similar observations were previously reported [93] for the adduct between $NdCl_3$ and aluminium chloride. Besides finding counter-examples where a chromophore belongs to a point-group, not permitting totally symmetric A_1 functions, there is clear-cut chemical evidence that definite types of ligands (either conjugated or highly polarizable) induce high Ω_2. In crystalline or vitreous mixed oxides, it is true that Ω_2 can vary within wide limits, but it does not seem to depend predominantly on the local symmetry.

Extending measurements of aqua ions as perchlorates in deuterium oxide solution, Carnall, Fields and Wybourne [94] studied M(III) in molten nitrates such as the eutectic of $LiNO_3$ and KNO_3 in order to avoid interference in the infra-red by O–H stretching frequencies, and these authors detected very strong hypersensitive transitions. It is probable that NO_3^- functions as a bidentate ligand for lanthanides, like acetate $CH_3CO_2^-$ and other carboxylates RCO_2^-. Actually, a large number of carboxylates [95–98] in addition to the classical case of β-diketonates [83, 99] have high Ω_2 given in Table 24. It is not quite certain that the coordination of nitrogen atoms in complexes [100, 101] of naturally occurring and synthetic amino-acids has additional effect beyond the chelating carboxylate groups. The intensification of the few hypersensitive transitions can be very helpful, *e.g.* establishing [102] the stoichiometry of $Nd(NO_3)^{+2}$ and they have also been detected [103, 104] in biochemical systems. Bidentate carbonate may also occur in $Er(O_2CO)_5^{-7}$.

An alternative to the symmetry-determined occurrence of hypersensitive pseudo-quadrupolar transitions is [84] the *inhomogeneous local dielectric*. The classical description of the interaction between electromagnetic waves and matter (elaborated by Maxwell) is pro-Heraclitean and anti-Democritean in the sense that the dielectric constant is assumed to be indeed constant at least in regions with a diameter comparable to the wave-length λ of the light. It turns out that inhomogeneities in the dielectric constant at the several thousand times shorter distances corresponding to diameters of adjacent atoms may actually explain the observed intensities, and it can be shown [78, 84] that their contribution is almost exclusively concentrated on Ω_2.

There has recently been much discussion [20, 105] about the physical mechanism behind the concept of *hard and soft anti-bases* (Lewis acids) *and bases* proposed by Pearson [106, 107], pointing out a trend toward enhanced affinity between hard anti-bases and hard bases (essentially electrostatic attraction) and between soft anti-bases and soft bases (having a much more sophisticated origin, we may vaguely call it *"chemical polarizability"*). One aspect is the *inorganic symbiosis* [108] that soft central atoms such as Cu(I), Pd(II), Ag(I), Au(III), Au(I), Hg(II) and Tl(III), and in general low, zero and negative oxidation states, prefer to be surrounded by ligands selected from the soft class H^-, CH_3^- and other carbanions, CN^-, CO, phosphines R_3P, sulphur-containing ligands such as RS^-, and I^-. Actually, $Co(NH_3)_5X^{++}$ still shows hard behaviour, preferring $X^- = F^-$ rather than I^-, whereas $Co(CN)_5X^{-3}$ isoelectronic with $Mn(CO)_5X$ both show soft behaviour, preferring H^-, CH_3^- and I^-. It has been shown convincingly [105, 109] that the approximately additive *electric* (dipolar) *polarizabilities* derived from the Lorentz formula applied to refractive indices have very little to do with "chemical polarizabilities", though the two properties vary in a parallel way in the halides.

Though all the rare-earth M(III) definitely are hard central atoms like Li(I), Be(II) and all the alkaline-earths, B(III), Al(III), Si(IV) and Th(IV) preferring F^-

and ligating oxygen atoms, there is the interesting connection with our subject, i.e. the large Judd-Ofelt parameter Ω_2 occurs by a mechanism closely resembling "chemical polarizability".

Another alternative which may not be exceedingly different from the inhomogeneous dielectric constant on an instantaneous picture, is the hypersensitivity through *dynamic vibrational coupling* which was shown by Peacock [78, 110] to be able to explain the very large Ω_2 of gaseous NdI_3, even if the equilibrium positions of the nuclei are coplanar. This mechanism is strongly decreasing with increasing internuclear distances, showing less oscillator strength in NdI_6^{-3} and then generally decreasing with increasing coordination number N.

However much the pseudoquadrupolar transitions between two different *J*-levels of a given chromophore have been shown [78] to have intensities proportional to $[U^{(2)}]^2$, it had not previously been established that the transitions between the individual sub-levels of each level follow selection rules for electric quadrupole transitions. Lagerwey and Blasse [111] reported selection rules for polarized light for the emission $^5D_0 \rightarrow {}^7F_2$ in the hexagonal crystal $GdAl_3B_4O_{12}$ containing a little europium(III), definitely showing the transition not to be of the conventional forced electric dipole type. Unfortunately, this point is not fully clarified because Peacock [112] claims to have found other results.

Recently there has been some question as to whether d^q transitions may be pseudoquadrupolar. Gale, Godfrey, Mason and Peacock [113] argue that the unusually high $P \sim 10^{-2}$ observed for the third spin-allowed transition $^4A_2 \rightarrow {}^4T_1$ in the red (close to 15,000 cm^{-1}) for the three tetrahedral CoX_4^{-2} (X = Cl, Br, I) is due to polarization of the ligand electronic density by the quadrupole moment of the *transition* in the partly filled shell $3d^7$. The dynamic vibronic coupling [78, 110] in NdX_3 is similarly related to the quadrupole moment of the transition $^4I_{9/2} \rightarrow {}^4G_{5/2}$.

It is remarkable [105] that the oscillator strengths of spin-allowed transitions in octahedral nickel(II) complexes are close to 10^{-4} but increase to 10^{-3} in copper(II) complexes of the same or comparable ligands (as pointed out by Jannik Bjerrum [114]) and rather unpredictably are scattered between 10^{-3} and 10^{-2} in apparently quadratic palladium(II) complexes. It cannot be explained by Eq. (3.12) alone; it is as if the complexes of sufficiently Pearson-soft central atoms spontaneously deviate [20] on an instantaneous picture from the highest symmetry available to the nuclei. Such systematic deviations are frequent in post-transition group compounds of Sn(II), Sb(III), Te(IV), I(V), Xe(VI), Pb(II) and Bi(III) as known from pyramidal $SnCl_3^-$, $SbCl_3$, $TeCl_3^+$, IO_3^-, XeO_3 and $Pb(OH)_3^-$ and square-pyramidal TeF_5^-, IF_5 and XeF_5^+ looking like an octahedral complex lacking one ligand.

There is no doubt that systematic deviations from the highest conceivable symmetry of the nuclear skeleton in lanthanide chromophores cannot be ascribed to the partly filled shell. From a purely formal point of view, the *Jahn-Teller theorem* demands spontaneous deviations [20] when the lowest sub-level contains more than one state (excepting one Kramers doublet) but such effects have very rarely been observed in the 4f group. Anyhow, the Jahn-Teller distortions do not remove a centre of symmetry, if present, and their influence on copper(II)complexes [17, 114] is independent of the weak *l*-mixing producing the high intensities [105]. On the other hand, empty orbitals (such as 5d) very well may contribute to deviations from the highest possible symmetry. It cannot be argued that closed-shell systems do not behave this way. As reviewed elsewhere [115] gaseous alkaline-earth dihalides XMX are bent when the ratio of the radii (r_M/r_X) is sufficiently large (BaF_2, $BaCl_2$, $BaBr_2$, BaI_2, SrF_2, $SrCl_2$ and CaF_2) which almost suggests a weak binding between the two X atoms, contrary to all electrostatic arguments. M.O. theory readily explains the rules of Walsh regarding the bending of triatomic molecules such as NO_2, NO_2^- and O_3 (in contrast to linear CO_2) but a very simple case without evident reasons is the triplet ($S = 1$) groundstate [116] of the reactive molecule methylene CH_2 also being bent.

Obviously, the weak spontaneous distortion of 4f group chromophores may occur in several equivalent ways contributing the same amount to the Judd-Ofelt parameters and giving exactly the same sub-level energies. Nevertheless, it is a serious problem that even the sharpest absorption lines of cooled crystals have a few cm^{-1} as one-sided half-width δ, and aqua ions in solution [117] at room temperature typically have δ between 30 and 100 cm^{-1}. Stepanov [118] suggested that the life-time of the excited state was the time acoustic waves propagate away to the surrounding medium. If we consider a sound velocity 5 km/sec through 5 Å, the time 10^{-13} sec corresponds to δ close to 20 cm^{-1} according to Heisenberg's uncertainty principle. At first, this seems to contradict the half-lifes of the order of magnitude 10^{-3} sec found for luminescence of $^5D_0$ of Eu(III), $^6P_{7/2}$ of Gd(III), $^5D_4$ of Tb(III), and the electron transfer state (discussed in section 1 D) of UO_2^{++} having $\Omega = 4$ or 1 and even parity. However, it is quite conceivable that the de-excitation by "radiationless transitions" of the vibrational states of the uranyl ion (which was the example studied by Stepanov [118]) is indeed 10^{10} times more rapid than the luminescence of the lowest vibrational state of the excited electronic state.

There has perhaps been too great a tendency in the "ligand field" theory [20] to ascribe a passive rôle to the partly filled shell being influenced by an external perturbation. Various stereochemical arguments [105] in the d-group and post-transition group compounds, and the almost invariant Ω_4 and Ω_6 in most lanthanide compounds [71, 78] suggest the displacement of the equilibrium positions of the ligating atoms determined by the specific electronic structure of the central atom.

It is frequent that crystals (seen from the point of view of X-ray diffraction) show a high symmetry (e.g. cubic) at high temperature but go through a phase-transition to a lower symmetry by cooling. An extreme case is the ferroelectricity (dielectric constants above 10^3) in the *perovskite* $BaTiO_3$ in the low-symmetry modifications, which was ascribed by Orgel [119] to the "rattling" of the small Ti(IV) in the fairly large holes in the octahedra formed by the six adjacent oxide ligands, a random mobility (with very small amplitude) which is readily collectively aligned by an external electric field creating co-planar carpets of electric dipoles (perpendicular on the carpet). Faucher and Caro [120] recently studied the luminescence of Eu(III) incorporated in the rhombohedric perovskite $LaAlO_3$ at 77,300 and 500 °K. Though this crystal does not become a cubic perovskite before 700 °K, the T-values approaching the transition temperature show a closer spacing of the sub-levels coinciding in the perfectly cubic chromophore with $N = 12$. These authors also calculated the effect of next-nearest neighbour atoms on the parameters of the electrostatic model of the "ligand field". Relative to the twelve first oxide ligands at 2.68 Å the 8 aluminium, 6 lanthanum and 24 oxygen atoms at increasing distances (3.28, 3.79 and 4.64 Å) modify the parameters proportional to $\langle r^4 \rangle / R^5$ by large amounts and those dependent on $\langle r^6 \rangle / R^7$ by about 25 percent. On the other hand, the subsequent 122 atoms with R below 7.58 Å hardly modify the asymptotic values. However, in our opinion, these values only translate the seven one-electron energies one might treat in the angular overlap model, and agreement is only obtained with experiment [120] if $\langle r^2 \rangle / R^3$ is multiplied by 0.26 and $\langle r^6 \rangle / R^7$ by 7.5.

Faucher and Caro [121] also studied the absorption and emission spectra of the *pyrochlores* $Eu_2Ti_2O_7$, $Eu_2Sn_2O_7$ and $Eu_2Zr_2O_7$ where the sharp lines (excepting the magnetic dipole transition $^7F_0 \rightarrow {}^5D_1$) are weak in the two former crystals in agreement with the centre of inversion in the point-group D_{3d} of the M(III) site in the cubic pyrochlore, which is a super-structure of fluorite with the oxygen atoms slightly displaced and lacking an eighth. The larger M(III) has $N = 8$ (a cube compressed along a body-diagonal) and the smaller M(IV) has $N = 6$ in a roughly regular octahedron. $Eu_2Zr_2O_7$ seems to consist of a mixture of the cubic pyrochlore and another local symmetry without the centre of inversion, possibly C_{3v}. It is possible [35] to prepare many mixed oxides of rare earths and quadrivalent metals by rapid precipitation of the mixed solutions with aqueous ammonia, and subsequent calcination of the filtered and washed, co-precipitated hydroxides. Whereas, all the $M_2Ti_2O_7$ prepared this way by calcination at 850 °C an hour, have Debye *powder-diagrams* as pyrochlores, most $M_xZr_{1-x}O_{2-0.5x}$ with x exactly 0.5 or values in the interval 0.2 to 0.7 are *statistically disordered* fluorites (lacking the characteristic super-structure lines of pyrochlore) like all $M_xTh_{1-x}O_{2-0.5x}$ (which can frequently be prepared with x up to 0.6). By cautious calcination of the coprecipitated hydroxides at 800 °C, it is also possible to prepare thermodynamically unstable perovskites [122] containing equi-molar amounts of two rare earths, such as $LaErO_3$, $LaTmO_3$ and $LaYbO_3$ having reflection spectra [34] suggesting the presence of a centre of inversion for the heavy lanthanide with $N = 6$. By further heating, these samples decompose to a mixture of cubic C-oxide (such as $Er_{1-x}La_xO_{1.5}$) and a new crystal type (first found [35] for $NdYO_3$) which has only been characterized by its powder diagram, but found in 40 different mixed oxides. Faucher and Caro [121] suggest that the indeterminate structure of $Eu_2Zr_2O_7$ corresponding to a mosaic of (undetectably) small domains of the disordered phase and pyrochlore, produces spectral properties very similar to glasses [123].

Caro *et al.* [124, 125] have studied the lamellary structure of many mixed oxides (where electron micrographs [125] indicate twin-formation in B-type Sm_2O_3 and Gd_2O_3 and epitaxy on extremely thin layers of $Nd_{0.9}Y_{0.1}O_{1.5}$ and related samples) and "basic salts" such as the thermally very stable $La_2O_2(CO_3)$ and the bastnaesite syncrystallized mixtures of $MF(CO_3)$ and $CaCO_3$. Caro [124] cites the dialogue "Timaios" by Plato who (in polemic with Demokritos) considered solids to consist of parallel planes, and the two most attractive constitutions of a plane being the hexagonal division in equilateral triangles and the division in triangles having the angles 90°, 45° and 45°. Though one can make the prosaic comment that an arbitrary triangle can be used to cover a plane by repetition, (two and two combined to quadrangles with two sets of parallel sides, and then forming an oblique coordinate system with two variables) Plato then argued that four of the regular polyhedra are formed from such triangles (having angles which are 360° divided by an integer), whereas the fifth, the regular dodecahedron (having the point-group K_h like a regular icosahedron) involves regular pentagons (which are not compatible with a repeated unit cell in a crystal) used by Plato as the exception confirming the rule. Caro [124, 126] points out that the extended layer structures $(MO)_n^{+n}$ in many oxy-salts have either a C_3 axis at each M (such as A-type M_2O_3, the hexagonal modification of $M_2O_2(CO_3)$ *etc.*), or a C_4 axis at each M (such as MOCl and the tetragonal modification of $M_2O_2(CO_3)$ *etc.*). Caro also points out that the seemingly low symmetry of cubic C-type M_2O_3 seen from the point of view of M (having highly distorted $N = 6$) is actually better understood as regular tetrahedral coordination of the oxygen atoms by four M. The same is true for two-thirds of the oxygen atoms in A-type M_2O_3 where one-third have $N = 6$ and 0.3 Å longer M-O distances. The sulphur atoms in M_2O_2S replace the latter type of oxygen atoms. It is sometimes slightly dangerous to divide a three-dimensional lattice in layers. A classical example is the cubic and hexagonal close-packed metals (having the same number of atoms per unit volume) which can be described by a different pattern of repetition of the Platonic hexagonal layers. If the equatorial plane contains a hexagon of 6 neighbour atoms, the 6 other neighbour atoms (at the same distance) form a trigonal anti-prism (point-group D_{3d}) in the cubic type (like the chromophore $Sr(II)O_{12}$ in the cubic perovskite $SrTiO_3$) but a trigonal prism (point-group D_{3h}) in the hexagonal type. Anyhow, Caro [126] also discusses the phases intermediate between C-type M_2O_3 and fluorite-type MO_2. Some of these, strictly stoichiometric super-structures are M_nO_{2n-2} (n = 9, 10, 11, 12) or perhaps rather, $M_{6n+3}O_{11n+5}$ (n = 1, 2, 5, ∞) and another type M_7O_{12}. Certain analogies can be drawn between Pauling's classical studies of bridged silicate structures and bridging of OM_4 groups. The complicated crystal chemistry of lanthanide silicates has recently been reviewed [127].

Anybody who has dissolved rare earths in acidic solution has remarked on the high reactivity of A-type oxides comparable to MgO, if not CaO. The dissolution of calcined (carbonate-free) C-type oxides has a long inhibition period, even at 100 °C, and one has the impression that the final stage of dissolution is an auto-catalyzed process, finishing often in less than one minute. Calcined fluorite-type CeO_2 and ThO_2 are even less reactive. Generally this lack of reactivity indicates considerable covalent bonding, though it is not completely excluded that the isotropic three-dimensional lattice with unusually strong Madelung potential might be unreactive by electrostatic attractions. The short internuclear distances known from crystal structures also correspond to exceptionally strong nephelauxetic effect [34, 35, 128]. A related reaction is the slow hydrolysis of oxy-chlorides MOCl and oxy-nitrates $MO(NO_3)$ to one-third aqua ions in solution and two-thirds suspended $M(OH)_3$.

The non-stoichiometric fluorites and their super-structures have the general property that the M atoms remain immobile at their lattice points and diffuse extremely slowly. This is also why one can obtain C-type Nd_2O_3 (and in reducing atmosphere even Pr_2O_3) by cautious calcination of the hydroxides at low temperatures, since the phase transformation to the thermodynamically stable, hexagonal A-type is slow, needing extensive displacements of the M atoms. On the other hand, the oxygen atoms are rather mobile. The *Nernst lamp* is a mixed oxide $Zr_{1-x}Y_xO_{2-0.5x}$ (crystallizing as statistically disordered fluorite) which starts to conduct strongly at

600° to 700 °C (with a rather high Arrhenius activation energy) not by the motion of electrons (as in metals and semi-conductors) but by transport of oxygen atoms (strictly speaking oxide ions) through the vacant positions. Möbius [129] has written an excellent review of such oxide conductors. They also work with certain bivalent cations such as $Zr_{1-x}Mg_xO_{2-x}$ ($x \sim 0.1$), showing a phenomenon normally occurring in solid-state chemistry that the ionic radii can be more important than the oxidation state in systems easily obtaining electroneutrality by charge-compensating substitution or by accommodating vacancies or excess anions (as known from the statistically disordered fluorites UO_{2+x} and yttrofluorite $Ca_{1-x}Y_xF_{2+x}$). Comparable behaviour is known [104] from biochemical molecules, where trivalent lanthanides frequently are able to replace calcium(II).

Gutmann [130] has written a biography of Auer von Welsbach, who developed the *Auer mantle* $Th_{0.99}Ce_{0.01}O_2$ between 1884 and 1890. We have found [131, 132] that it is possible to make mantles of nearly all the mixed oxides previously made [35] by calcination of the co-precipitated hydroxides by mixing strong aqueous nitrate solutions (chlorides are prone to cause trouble with MCl_4 volatility and MOCl formation; perchlorates make the textile burn like gun-powder) in the desired proportions, imbibing the textile, and pyrolyzing it gently in a small flame. Both fluorites, C-type oxides and perovskites can be made into coherent inorganic textiles this way. 110 years ago, Bunsen [133] and Bahr [134] found that oxides such as Er_2O_3 emit narrow bands at the same positions as the absorption bands at room temperature. This phenomenon was very carefully studied by Mallory [135] showing that the green emission at 19,000 cm^{-1} (now known to be ${}^2H_{11/2} \rightarrow {}^4I_{15/2}$) is much stronger than the other emission bands at 15,300 cm^{-1} (due to the excited level ${}^4F_{9/2}$), 18,300 cm^{-1} (${}^4S_{3/2}$ which is known to luminesce [85, 86] in certain compounds) and bands in the blue due to excited levels such as ${}^4F_{5/2}$.

The major question when discussing such narrow emission bands of trivalent lanthanides in incandescent oxides is whether the emission spectrum is selective thermal radiation agreeing with Kirchhoff's theorem in the narrow absorption bands only, or it is due to chemoluminescence with the energetic species in the gas or hydrogen flame (as distinctly true for the violet luminescence of a few parts per million of bismuth(III) in CaO heated in a flame to 600 °C). Recently, Ivey [136] has reviewed *candoluminescence* (the more appropriate word "thermoluminescence" is used for the release of stored energy, due to previous irradiation, as light when heating a solid) and did not find conclusive evidence in favour of one of the two hypotheses. The theorem of Kirchhoff applies to opaque *objects* absorbing all impinging radiation at a given wave-length λ. This is not so much a question of the molar extinction coefficient ϵ in Eq. (1.36) but rather vanishing (I/I_0) due to an appropriate combination of high ϵ, high layer thickness l and high concentration. A body which is opaque for all λ is traditionally called a "black body" but it may be better to speak of an *opaque body* since we normally discuss in the case of incandescency. The opaque body has a standard continuous emission spectrum as a function of λ:

$$E = c_1/\lambda^5 \{\exp(c_2/\lambda T) - 1\} \quad (3.13)$$

where $c_2 = 1.4388$ cm · degree is the reciprocal value of Boltzmann's constant k. The energy distribution (3.13) has a maximum at the wave-length in Å

$$\lambda_{max} = 2.897 \cdot 10^7/T \tag{3.14}$$

whose constant close to 29 million is c_2/ξ and ξ is the root 4.965114 . . . of the transcendent equation

$$5(e^{-x} - 1) + x = 0 \tag{3.15}$$

with the result that the corresponding photon energy is ξkT. However, it might be more appropriate to consider the energy distribution E as a function of the wave-number $\sigma = (\nu/c)$ or frequency ν in which case the maximum occurs at $h\nu = \upsilon kT$ where υ is the root 2.8215 . . . of the transcendent equation

$$3(e^{-x} - 1) + x = 0 \tag{3.16}$$

Finally, it might be argued that one should consider the photon (rather than the energy) distribution already peaking at $h\nu = \omega kT$ where ω is the root 1.5936 . . . of

$$2(e^{-x} - 1) + x = 0 \tag{3.17}$$

The wave-number corresponding to ξkT is 3.45 cm^{-1} times the absolute temperature T. Correspondingly, the photon energies are much higher than the average energy per degree of freedom in statistical mechanics.

Today the technology of illumination is mainly based on transparent lamps such as electric discharges in atomic vapours of sodium or mercury. It is generally believed [136] that the Auer mantle circumvent the unfavourable conditions of Eq. (3.13) where a tungsten filament operated at 3,000 °K still has the maximum in the near infra-red, and where the yield of visible light decreases exponentially almost as $\lambda^{-5}\exp(-c_2/\lambda T)$ (though the opaque gases of the Sun just below the photosphere showing the Fraunhofer lines in absorption, as first explained by Bunsen and Kirchhoff in 1860, have the maximum in the green with T close to 5,800 °K) by being almost transparent in the infra-red and having an absorption band in the visible. The optimum condition [137] is then to add 1 percent cerium to ThO_2. It is known that silica crucibles heated to 1,000 °C or more hardly emit any radiation (because they are transparent) whereas, coloured or black materials on the silica are intensely incandescent. Mantles of very pure Al_2O_3 can be seen to emit a pale blue light in a flame.

We do not believe that the emission of narrow bands from M(III) in mixed-oxide mantles can be entirely explained [131,132] by the standard opaque-body spectrum being emitted in the middle of the narrow absorption bands. Thus, $Er_{0.02}Th_{0.98}O_{1.99}$ still emits the three emission bands also seen in $Er_{0.1}Th_{0.9}O_{1.95}$, $Er_{0.2}Zr_{0.8}O_{1.9}$, $ErAlO_3$ and in pure Er_2O_3 and the bands are not perceptible *narrower.* There is, of course, a lower limit for the erbium concentration which can be detected, but the three bands of the perovskite $La_{0.98}Er_{0.02}GaO_3$ can still be seen as in undiluted $ErGaO_3$, and the green edge toward higher wave-numbers (this asymmetry is a specific characteristic of $^2H_{11/2}$) can still be perceived in $La_{0.993}Er_{0.007}GaO_3$. There does not seem to be any competition with narrow-band emission in the near infra-red, since the emerald-green candoluminescence of $Er_{0.2}Yb_{0.8}O_{1.5}$ is at least as bright as that of $Er_{0.2}Y_{0.8}O_{1.5}$. We would like to emphasize that mantles containing europium(III) do not show narrow-band emission in the red and in the orange. Hence, the mechanism of the candoluminescence does not seem to build up a quasi-stationary concentration of the excited state $^5D_0$. Compatible (but not compelling evidence) with the idea of selective thermal radiation is the general observation that the emission bands occur only at positions where M(III) is known to have absorption bands at room temperature. However, it is strange that the hypersensitive pseudo-quadrupolar transitions are by far the most prominent in candoluminescence (exceptions like $^4F_{9/2}$ of Er(III) in the red and $^4G_{7/2}$ of Nd(III) in the green, are quite weak) unless it is due to very high oscillator strengths at T above 1000 °K.

The incandescence of Nd(III) in mantles [131, 132] has the same yellow-orange colour as light from sodium atoms and is indeed concentrated in a band around 17,000 cm^{-1} corresponding to the hypersensitive transition to the excited level $^4G_{5/2}$. Holmium(III) shows a mauve light due to emission bands, due to $^5F_5$ at 15,500 cm^{-1} in the red, to $^5F_4$ (and $^5S_2$) at 18,500 cm^{-1} in the green, and the most intense (corresponding to a hypersensitive transition) to $^5G_6$ at 22,000 cm^{-1} in the blue. Thulium(III) has a very characteristic, deep purple colour of the incandescence, with a weak band ($^1G_4$) at 21,000 cm^{-1} and several features in the red due to the ^3F levels. It may be noted that the colours of candoluminescence: yellow-orange Nd(III), mauve Ho(III), emerald-green Er(III) and purple Tm(III) are indeed the complementary reflected colours at room-temperature: sky-blue Nd_2O_3, orange Ho_2O_3, pink Er_2O_3 and pale green Tm_2O_3.

Already Haitinger [130] discovered the broad emission band in the orange light from a mantle of $Al_{2-x}Cr_xO_3$, undoubtedly due to the first spin-allowed transition $^4A_2 \rightarrow {}^4T_2$ in the ruby. Mantles [132] of the spinels $CoAl_2O_4$ (Thenard's blue) and $Mg_{0.9}Co_{0.1}Al_2O_4$ emit a reddish orange candoluminescence probably due to the third spin-allowed transition in the tetrahedral chromophore $Co(II)O_4$.

It is conceivable that in the future, with easy access to transportable hydrogen one would find it an advantage to obtain visible light by candoluminescence of Auer mantles consisting of selected mixed oxides, rather than to obtain a higher yield of light by electric discharges using more expensive current. From the general point of view of this book, it is important to note the high photon energy inherent in the constant ξ of Eq. (3.15) having unexpected consequences for certain types of light emission.

References

1. Jørgensen, C. K.: Adv. Quantum Chem. *8,* 137 (1974)
2. Schäffer, C. E., Jørgensen, C. K.: J. Inorg. Nucl. Chem. *8,* 143 (1958)
3. Hofmann, K. A., Kirmreuther, H.: Z. Physik. Chem. *71,* 312 (1910)
4. Ephraim, F., Bloch, K.: Ber. deutsch. chem. Ges. *59,* 2692 (1926); *61,* 65 and 72 (1928)
5. Ephraim, F.: Ber. deutsch. chem. Ges. *61,* 80 (1928)
6. Ephraim, F., Rây, P.: Ber. deutsch. chem. Ges. *62,* 1509, 1520 and 1639 (1929)
7. Ephraim, F., Jantsch, G., Zapata, C.: Helv. Chim. Acta *16,* 261 (1933)
8. Boulanger, F.: Ann. Chim. (Paris) *7,* 732 (1952)
9. Jørgensen, C. K.: Mat. fys. Medd. Danske Vid Selskab *30,* no. 22 (1956)
10. Condon, E. U., Shortley, G. H.: Theory of Atomic Spectra (2. Ed.). Cambridge: University Press 1953
11. Racah, G.: Phys. Rev. *76,* 1352 (1949)
12. Bethe, H., Spedding, F. H.: Phys. Rev. *52,* 454 (1937)
13. Sugar, J.: J. Opt. Soc. Am. *55,* 1058 (1965)
14. Tanabe, Y., Sugano, S.: J. Phys. Soc. Japan *9,* 753 and 766 (1954)
15. Orgel, L. E.: J. Chem. Phys. *23,* 1824 (1955)
16. Jørgensen, C. K.: Oxidation Numbers and Oxidation States. Berlin–Heidelberg–New York: Springer 1969
17. Jørgensen, C. K.: Absorption Spectra and Chemical Bonding in Complexes. Oxford: Pergamon Press 1962
18. Jørgensen, C. K.: Progress Inorg. Chem. *4,* 73 (1962)
19. Lohr, L. L.: J. Chem. Phys. *45,* 3611 (1966)
20. Jørgensen, C. K.: Modern Aspects of Ligand Field Theory. Amsterdam: North-Holland 1971
21. Jørgensen, C. K.: Discuss. Faraday Soc. *26,* 210 (1958)
22. Jørgensen, C. K.: Structure and Bonding *1,* 3 (1966)
23. Jørgensen, C. K.: Helv. Chim. Acta Fascisculus extraordinarius Alfred Werner, p. 131 (1967)
24. Alig, R. C., Duncan, R. C., Mokross, B. J.: J. Chem. Phys. *59,* 5837 (1973)
25. Jørgensen, C. K., Pappalardo, R., Schmidtke, H. H.: J. Chem. Phys. *39,* 1422 (1963)
26. Griffith, J. S.: The Theory of Transition-metal Ions. Cambridge: University Press 1961
27. Jørgensen, C. K.: Acta Chem. Scand. *12,* 903 (1958)
28. Richman, I., Wong, E. Y.: J. Chem. Phys. *37,* 2270 (1962)
29. Henderson, J. R., Muramoto, M., Gruber, J. B.: J. Chem. Phys. *46,* 2515 (1967)
30. Tandon, S. P., Mehta, P. C.: J. Chem. Phys. *52,* 4896 (1970)
31. Toledano, J. C.: J. Chem. Phys. *57,* 4468 (1972)
32. Jørgensen, C. K.: Orbitals in Atoms and Molecules. London: Academic Press 1962
33. Jørgensen, C. K.: Solid State Phys. *13,* 375 (1962)
34. Jørgensen, C. K., Pappalardo, R., Rittershaus, E.: Z. Naturforsch. *19a,* 424 (1964); *20a,* 54 (1965)
35. Jørgensen, C. K., Rittershaus, E.: Mat. fys. Medd. Danske Vid. Selskab *35,* no. 15 (1967)
36. Ryan, J. L., Jørgensen, C. K.: J. Phys. Chem. *70,* 2845 (1966)
37. Ryan, J. L.; Inorg. Chem. *8,* 2053 (1969)
38. Jørgensen, C. K., Pappalardo, R., Flahaut, J.: J. chim. physique *62,* 444 (1965)
39. McLaughlin, R. D., Conway, J. G.: J. Chem. Phys. *38,* 1037 (1963)
40. Carlson, E. H., Dieke, G. H.: J. Chem. Phys. *34,* 1602 (1961)
41. Varsanyi, F., Dieke, G. H.: J. Chem. Phys. *36,* 2951 (1962)
42. Rakestraw, J. W., Dieke, G. H.: J. Chem. Phys. *42,* 873 (1965)
43. Crosswhite, H. M., Dieke, G. H., Carter, W. J.: J. Chem. Phys. *43,* 2047 (1965)
44. Sinha, S. P., Schmidtke, H. H.: Mol. Phys. *10,* 7 (1965)
45. Schwochau, K., Jørgensen, C. K.: Z. Naturforsch. *20a,* 65 (1965)
46. Dunn, T. M.: Trans. Faraday Soc. *57,* 1441 (1961)
47. Bryant, B. W.: J. Opt. Soc. Am. *55,* 771 (1965)
48. Kaufman, V., Sugar, J.: J. Opt. Soc. Am. *66,* 439 (1976)

49. Freeman, A. J., Watson, R. E.: Phys. Rev. *127,* 2058 (1962)
50. Turner, D. W., Baker, C., Baker, A. D., Brundle, C. R.: Molecular Photoelectron Spectroscopy. London: Wiley-Interscience 1970
51. Jφrgensen, C. K.: Structure and Bonding *24,* 1 (1975)
52. Fadley, C. S., Hagström, S. B. M., Klein, M. P., Shirley, D. A.: J. Chem. Phys. *48,* 3779 (1968)
53. Jφrgensen, C. K., Berthou, H.: Mat. fys. Medd. Danske Vid. Selskab *38,* no. 15 (1972)
54. Jφrgensen, C. K.: Chimia (Aarau) *29,* 53 (1975)
55. Wagner, C. D.: Discuss. Faraday Soc. *60,* 291 (1976)
56. Wertheim, G. K., Rosencwaig, A., Cohen, R. L., Guggenheim, H. J.: Phys. Rev. Letters *27,* 505 (1971)
57. Jφrgensen, C. K.: Structure and Bonding *13,* 199 (1973)
58. Jφrgensen, C. K.: Chimia (Aarau) *27,* 203 (1973)
59. Jφrgensen, C. K., Berthou, H.: Discuss. Faraday Soc. *54,* 269 (1973)
60. Berthou, H., Jφrgensen, C. K.: Analyt. Chem. *47,* 482 (1975)
61. Jφrgensen, C. K.: Chimia (Aarau) *28,* 6 (1974)
62. Jφrgensen, C. K.: Structure and Bonding *22,* 49 (1975)
63. Watson, R. E., Perlman, M. E. (editors): Research Applications of Synchrotron Radiation (BNL 50381) Springfield, Virginia: National Technical Information Service. U.S. Department of Commerce 1973
64. Hedén, P. O., Löfgren, H., Hagström, S. B. M.: Phys. Rev. Letters *26,* 432 (1971)
65. Cox, P. A., Baer, Y., Jφrgensen, C. K.: Chem. Phys. Letters *22,* 433 (1973)
66. Jolly, W. L., Perry, W. B.: Inorg. Chem. *13,* 2686 (1974)
67. Eastman, D. E., Kuznietz, M.: J. Appl. Phys. *42,* 1396 (1971)
68. Jφrgensen, C. K.: Theoret. Chim. Acta *34,* 189 (1974)
69. Cox, P. A.: Structure and Bonding *24,* 59 (1975)
70. Campagna, M., Bucher, E., Wertheim, G. K., Buchanan, D. N. E., Longinotti, L. D.: Proceed. 11. Rare Earth Conference, Traverse City, Michigan October 1974 (ed. H. A. Eick). East Lansing 1974
71. Carnall, W. T., Fields, P. R., Rajnak, K.: J. Chem. Phys. *49,* 4412, 4424, 4443, 4447 and 4450 (1968)
72. Campagna, M., Bucher, E., Wertheim, G. K., Buchanan, D. N. E., Longinotti, L. D.: Phys. Rev. Letters *32,* 885 (1974)
73. Campagna, M., Wertheim, G. K., Bucher, E.: Structure and Bonding *30,* 99 (1976)
74. Campagna, M., Bucher, E., Wertheim, G. K., Longinotti, L. D.: Phys. Rev. Letters *33,* 165 (1974)
75. Broer, L. J. F., Gorter, C. J., Hoogschagen, J.: Physica *11,* 231 (1945)
76. Judd, B. R.: Phys. Rev. *127,* 750 (1962)
77. Ofelt, G. S.: J. Chem. Phys. *37,* 511 (1962)
78. Peacock, R. D.: Structure and Bonding *22,* 83 (1975)
79. Kisliuk, P., Krupke, W. F., Gruber, J. B.: J. Chem. Phys. *40,* 3606 (1964)
80. Yamada, S., Tsuchida, R.: Bull. Chem. Soc. Japan *26,* 15 (1953)
81. Jφrgensen, C. K.: Adv. Chem. Phys. *5,* 33 (1963)
82. Fenske, R. F.: J. Am. Chem. Soc. *89,* 252 (1967)
83. Moeller, T., Ulrich, W. F.: J. Inorg. Nucl. Chem. *2,* 164 (1956)
84. Jφrgensen, C. K., Judd, B. R.: Mol. Phys. *8,* 281 (1964)
85. Reisfeld, R.: Structure and Bonding *13,* 53 (1973); *22,* 123 (1975)
86. Reisfeld, R., Velapoldi, R. A., Boehm, L.: J. Phys. Chem. *76,* 1293 (1972)
87. Sovers, O. J., Yoshioka, T.: J. Chem. Phys. *51,* 5330 (1969)
88. Judd, B. R.: J. Chem. Phys. *44,* 839 (1966)
89. Weinstock, B. L., Goodman, G. L.: Adv. Chem. Phys. *9,* 169 (1965)
90. Gruen, D. M., DeKock, C. W., McBeth, R. L.: Advances in Chemistry *71,* 102 (1967)
91. Gruen, D. M., DeKock, C. W.; J. Chem. Phys. *45,* 455 (1966)
92. Liu, C. S., Zollweg, R. J.: J. Chem. Phys. *60,* 2384 (1974)
93. Øye, H. A., Gruen, D. M.: J. Am. Chem. Soc. *91,* 2229 (1969)

94. Carnall, W. T., Fields, R., Wybourne, B. G.: J. Chem. Phys. *42,* 3797 (1965)
95. Choppin, G. R., Henrie, D. E., Buijs, K.: Inorg. Chem. *5,* 1743 (1966)
96. Karraker, D. G.: J. Inorg. Nucl. Chem. *31,* 2815 (1969) and *33,* 3713 (1971)
97. Bukietynska, K., Choppin, G. R.: J. Chem. Phys. *52,* 2875 (1970)
98. Choppin, G. R., Fellows, R. L.: J. Coord. Chem. *3,* 209 (1974)
99. Karraker, D. G.: Inorg. Chem. *6,* 1863 (1967)
100. Katzin, L. I., Barnett, M. L.: J. Phys. Chem. *68,* 3779 (1964)
101. Sinha, S. P., Mehta, P. C., Surana, S. S. L.: Mol. Phys. *23,* 807 (1972)
102. Coward, N. A., Kieser, R. W.: J. Phys. Chem. *70,* 213 (1966)
103. Birnbaum, E. R., Gomez, J. E., Darnall, D. W.: J. Am. Chem. Soc. *92,* 5287 (1970)
104. Nieboer, E.: Structure and Bonding *22,* 1 (1975)
105. Jørgensen, C. K.: Topics in Current Chemistry *56,* 1 (1975)
106. Pearson, R. G.: J. Am. Chem. Soc. *85,* 3533 (1963)
107. Pearson, R. G. (ed.): Hard and Soft Acids and Bases. Stroudsburg, Penn.: Dowden, Hutchinson and Ross 1973
108. Jørgensen, C. K.: Inorg. Chem. *3,* 1201 (1964)
109. Salzmann, J. J., Jørgensen, C. K.: Helv. Chim. Acta *51,* 1276 (1968)
110. Mason, S. F., Peacock, R. D., Stewart, B.: Chem. Phys. Letters *29,* 149 (1974)
111. Lagerwey, A. A. F., Blasse, G.: Chem. Phys. Letters *31,* 27 (1975)
112. Peacock, R. D.: Chem. Phys. Letters *35,* 420 (1975)
113. Gale, R., Godfrey, R. E., Mason, S. F., Peacock, R. D., Stewart, B.: Chem. Comm. (London) 329 (1975)
114. Romano, V., Bjerrum, J.: Acta Chem. Scand. 24 1551 (1970)
115. Jørgensen, C. K.: Halogen Chemistry (ed. V. Gutmann) *1,* 265. London: Academic Press 1967
116. Harrison, J. F.: Accounts Chem. Res. *7,* 378 (1974)
117. Jørgensen, C. K.: Acta Chem. Scand. *11,* 981 (1957)
118. Stepanov, B. I.: Zhur. Eksptl. Teoret. Fiz. *21,* 1158 (1951)
119. Orgel, L. E.: Discuss. Faraday Soc. *26,* 138 (1958)
120. Faucher, M., Caro, P.: J. Chem. Phys. *63,* 446 (1975)
121. Faucher, M., Caro, P.: J. Solid State Chem. *12,* 1 (1975)
122. Schneider, S. J., Roth, R. S.: J. Res. Nat. Bur. Stand. *64A,* 317 (1960)
123. Reisfeld, R., Mack, H., Eisenberg, A., Eckstein, Y.: J. Electrochem. Soc. *112,* 273 (1975)
124. Caro, P.: J. Solid State Chem. *6,* 396 (1973)
125. Caro, P., Schiffmacher, G., Boulesteix, C., Loier, C., Portier, R.: Defects and Transport in Oxides (ed. M. S. Seltzer and R. I. Jaffee) p. 519. New York: Plenum Press 1974
126. Caro, P.: Proceed. 5. Materials Research Symposium, p. 367. Nat. Bur. Stand. Special Publication no. 364. Washington D. C.: 1972
127. Felsche, J.: Structure and Bonding *13,* 99 (1973)
128. Caro, P., Derouet, J.: Bull. Soc. Chim. France 46 (1972)
129. Möbius, H. H.: Z. Chem. *2,* 100 (1962); *4,* 81 (1964)
130. Gutmann, V.: J. Chem. Educ. *47,* 209 (1970)
131. Jørgensen, C. K.: Chem. Phys. Letters *34,* 14 (1975)
132. Jørgensen, C. K.: Structure and Bonding *25,* 1 (1976)
133. Bunsen, R.: Liebig's Annalen *131,* 255 (1864)
134. Bahr, J. F.: Liebig's Annalen *135,* 376 (1865)
135. Mallory, W. S.: Phys. Rev. *14,* 54 (1919)
136. Ivey, H. F.: J. Luminescence *8,* 271 (1974)
137. Ives, H. E., Kingsbury, E. F., Karrer, E.: J. Franklin Inst. *186,* 401 and 585 (1918)

4. Energy Transfer

(References to this Chapter may be found p. 195, as Chapter 2, we consider here M^{3+} etc.)

As already suggested in Chapter 2, energy transfer from other absorbing species to rare earth ions may sensitize crystalline and glass lasers. In practice, the important energy donors have allowed transitions of high intensity. This is due to the fact that the narrow band absorption and emission in the lanthanides (which are essential for the laser activity) arise from parity-forbidden weak transitions. On the other hand, the parity-allowed transitions in Ce^{3+} and Tb^{3+} and in post-transition group ions such as Tl^{+}, Pb^{2+} and Bi^{3+} (isoelectronic with the neutral mercury atom) generally correspond to emission of broad bands, which generally are not suitable for laser action alone, but can be excellent for laser pumping by energy transfer processes. If weak transitions (such as the $3d^5$ transitions in Mn^{2+} or the exceptionally weak electron transfer band in UO_2^{2+}) have sufficiently long-lived excited states, they may work also. Energy transfer may increase the population of excited levels of lanthanides by one or two orders of magnitude compared with direct absorption of light into the rare earth ions. The interpretation of energy transfer in glasses may be complicated by inhomogeneous broadening in vitreous media with the result that the chances for a donor ion to be within a sufficiently short distance from an acceptor ion resulting in a rapid resonance transfer are rather low, and an additional factor responsible for the energy transfer must be considered.

Energy transfer between ions in a solid can be accomplished either radiatively or non-radiatively. Only the latter processes will be discussed in this Chapter. Radiative transfer (which is the trivial case of absorption by the acceptor of the light emitted by the donor) can be easily treated by measuring the absorption and emission characteristics of the ions involved, and by correcting for experimental geometry.

Non-radiative transfer from donors to acceptors (sometimes called *sensitizers* to *activators* in literature about luminescent solids) depletes the population of the excited state of the donor, decreases its life-time and makes its own luminescence weaken or disappear. The transfer is only significant and measurable if the rate is of the same order of magnitude (or higher) than the rate of the radiative transition in the donor ion [1]. It should be noted that the transfer probabilities obtained experimentally from the donor quantum efficiency in glass (as described in Ref. [2]) by Kraevskii *et al.* [3] and other authors often are several orders of magnitude higher than the transfer rates rates calculated by use of the classical resonance formula of Dexter [4] for the average transfer probability. Hence, additional mechanisms (to be discussed below) are needed to explain this discrepancy.

The microscopic behaviour of differing ions in dilute systems results from multipolar interactions. On the other hand, the multipole questions are absent from the rate equations used in evaluation of macroscopic data such as quantum efficiencies of fluorescence. Recently, the macroscopic treatment of energy transfer was performed independently by Fong and Diestler [5] and by Grant [6] who conclude that the concentration dependence of quantum efficiency reflects the number of interacting particles rather than the mechanism of interaction. While the microscopic insight into the mode of operation of energy transfer is mainly of academic interest, the measurements of quantum efficiency carried out by Reisfeld *et al.* [2], by Parke and Cole [7] and by Soules *et al.* [8] are of greater practical concern.

A. Transfer Probabilities

At first we shall review the existing theory of transfer probabilities and then extend this to phonon-assisted and diffusion-controlled energy transfer, discuss the influence of inhomogeneous broadening on the efficiency of transfer, and we shall also consider the possibilities of controlling the overlap between the donor and acceptor systems by electron transfer and Rydberg transitions, and wherever possible bring experimental examples, predict unexploited possibilities for energy transfer, and we shall finally discuss some infra-red to visible converters achieved by energy transfer processes.

a) Resonance Energy Transfer

Resonance transfer of energy between a donor (sensitizer) ion (S) and an acceptor (activator) ion (A) may occur when the energy differences between the ground and excited states of the S and the A systems are identical. If there exists a suitable interaction between the two electronic systems the energy absorbed by S may be transferred to A.

The coupling of adjacent ions in such a case can arise *via* exchange interaction if their wave-functions overlap directly, *via* super-exchange interactions involving intervening ions, or also *via* various electric or magnetic multipolar interactions.

Dexter [4] has derived the following expression for the probability of resonance transfer per unit time that energy will be transferred between the two ions:

$$P(R) = (2\pi/\hbar)|\langle \Psi_S^* \Psi_A | H_{SA} | \Psi_S \Psi_A^* \rangle|^2 \int g_S(E) g_A(E) dE \tag{4.1}$$

where Ψ_S^* and Ψ_S, Ψ_A^* and Ψ_A are the excited and ground state wave-functions, of the sensitizer S and the activator A ions respectively, and where H_{SA} is the interaction Hamiltonian. The integral is over the normalized emission band shape of the sensitizer and the absorption band of the activator in which the transitions $S \rightarrow S^*$ and $A \rightarrow A^*$ are represented by the line shape functions $g_S(E)$ and $g_A(E)$ each normalized in the sense $\int g(E)dE = 1$.

The ion-ion interactions differ in their dependence on donor-acceptor distance for various mechanisms. The radial dependence of the ion-pair transfer rate is derived from the square of the matrix element of H_{SA} in Eq. (4.1).

Recent reviews on energy transfer theory with special emphasis on rare earth ions are given by Reisfeld [2], Riseberg and Weber [9], Watts [10] and Auzel [11]. We shall now review in short the expressions for the matrix elements and transfer probabilities of various mechanisms and concentrate on the experimental findings which were mainly performed after the above-mentioned reviews appeared.

b) Exchange Interaction

Exchange interaction results from the overlap of the wave-functions of S and A and is consequently a very short-range interaction falling off exponentially with the distance between the ions. If the ions S and A have only a single active electron each, the initial and final states $|S^*A>$ and $|SA^*>$ must be antisymmetrized in the coordinates of the two electrons, and the exchange matrix elements will have the expression

$$< S^*(1)A(2) \left| -\left(\frac{e^2}{r_{12}}\right) P_{12} \right| S(1)A^*(2) > = < S^*(1)A(2) \left| -\frac{e^2}{r_{12}} \right| S(2)A^*(1) > \qquad (4.2)$$

P_{12} is the permutator which interchanges electrons 1 and 2.

For the general case where S, S*, A, A* are many-electron states the exchange matrix element is

$$< S^*A \left| -\sum_{p,t} (e^2/r_{pt}) P_{pt} \right| SA^* > \qquad (4.3)$$

p and t are indices referring to donor and acceptor electrons respectively. Dexter [4] has derived the following expression for the rate constant of energy transfer by the exchange mechanism,

$$p_{ex} = \frac{2\pi}{\hbar} Z^2 \int g_S(E) g_A(E) dE \qquad (4.4)$$

with $Z^2 = K^2 \exp(-2R/L)$, where K is a constant with the dimension of energy, and where L is the effective average Bohr radius. For exchange transfer, spin selection rules for conserving the total spin of the system, before and after the transfer must be satisfied ($S_{SA} = S_S + S_A$, $\Delta S_{SA} = 0$).

These spin selection rules were used by Antipenko and Ermolaev [12] in establishing the exchange transfer mechanism between Tb^{3+} and Eu^{3+} in solutions. However, it should be noted that because of the state admixing, exchange may be relatively free of selection rule restrictions.

Inokuti and Hirayama (I-H) [13] developed a quantitative theory of energy transfer by the exchange mechanism and predicted time dependence of fluorescence decay in such coupling. In the I-H approach the S ion is surrounded by a set of A ions at distances R_K. During the transfer process the environment of excited A ions changes with time, resulting in a nonexponential decay which for exchange coupling is of the form

$$\phi_{(t)} = \phi_0 \exp\left[-\frac{t}{\tau_A} - \gamma^{-3}\frac{C}{C_0}\, g\left(\frac{e^{\gamma}t}{\tau_A}\right)\right] \tag{4.5}$$

where C is the acceptor concentration, and τ_A the decay time of the pure donor,

$$C = 3N/(4\pi RV^3) \tag{4.6}$$

and C_0 is the critical transfer concentration, defined by $C_0 = 3/(4\pi R_0^3)$ where R_0 and γ are constants related to Dexter quantities by

$$\gamma = 2R_0/L \tag{4.7}$$

R_0 being the critical distance at which the probabilities for radiative and nonradiative transfers are equal.

In the I-H derivation it was assumed that the donor-donor interaction and the back transfer from the acceptors to the donor system is absent.

Soules *et al.* [8] attributed the energy transfer between antimony and manganese in the fluorophosphate phosphors to the exchange mechanism. These authors used order of magnitude arguments to show that the energy transfer rate by dipole-dipole and dipole-quadrupole interaction are too small to account for energy transfer between manganese and antimony. They fitted the decay curves of donor fluorescence to the theoretically calculated curves by the I-H theory and came to the conclusion that the exchange mechanism resembles the theoretical curve more than the other mechanisms.

Treadaway and Powell [14] have recently studied energy transfer between calcium tungstate and samarium and attributed the transfer to the exchange mechanism.

Blasse and Bril [15] came to the conclusion that the exchange interaction is active if the S emission band overlaps the 4f–4f absorption bands of A, and by electric multipole interaction if the S emission band overlaps allowed absorption bands of A. Their assumption was based mainly on the fact that the exchange interaction depends on the overlap integral only while the multipolar interaction depends on the absorption cross-section in addition to the overlap.

c) Magnetic Interaction

Another mechanism of paramagnetic ion coupling is the magnetic dipole-dipole interaction, the Hamiltonian of such interaction is of the form

$$H_{MDD} = \sum_{ij}\left[\frac{\vec{\mu}_i \cdot \vec{\mu}_j}{R^3} - \frac{3(\vec{\mu}_i R)(\vec{\mu}_j R)}{R^5}\right] \tag{4.8}$$

where $\vec{\mu}_i = \ell_i + 2\vec{s}_i$ and $\vec{\ell}_i + s_i$ are the orbital and spin operators for the i-th and j-th electrons of ions S and A respectively. The selection rules ΔS, ΔL, $\Delta J = 0 \pm 1$ for transition between $4f^N$ states are relaxed by *SLJ* state admixing. The MDD interaction has the same long range R^{-3} radial dependence as the electron dipole interaction which will be dealt with below.

Energy transfer from the point of view of ion-ion interaction in the ground state as measured by EPR has been discussed by Birgeneau *et al.* [16]. These types of interactions are not detected in optical measurements.

d) Electrostatic interaction

The electrostatic interaction is represented by

$$H_{es} = \sum_{i,j} \frac{e^2}{K|(\vec{r}_{S_i} - \vec{R} - \vec{r}_{A_j})|} \tag{4.9}$$

Here $\vec{r}_{S_i}$ and $\vec{r}_{A_j}$ are the coordinate vectors of electrons i and j belonging to ions S and A, respectively; $\vec{R}$ is the nuclear separation and K is the dielectric constant. The various multipolar terms appear from a power series expansion of the denominator. This expansion was expressed by Kushida [17] in terms of tensor operators. The leading terms are the electric dipole-dipole (EDD), dipole-quadrupole (EQD) and quadrupole-quadrupole (EQQ) interaction. These have radial a dependence of R^{-3}, R^{-4} and R^{-5} respectively.

In his calculation of the induced dipole-dipole and dipole-quadrupole processes of energy transfer Kushida [17] made use of the Judd-Ofelt expression for the forced electric dipole transition probability in the rare earths incorporated in solids.

The transfer rates for dipole-dipole, dipole-quadrupole and quadrupole-quadrupole processes in the rare earths as given by Kushida are of the form,

$$\bar{p}_{SA}^{(dd)} = \frac{1}{(2J_S+1)(2J_A+1)} \left(\frac{2}{3}\right) \left(\frac{2\pi}{\hbar}\right) \left(\frac{e^2}{R^3}\right)^2 [\Sigma_t \Omega_{tS} < J_S \| U^{(t)} \| J_S' >^2]$$

$$\times [\Sigma_t \Omega_{tA} < J_A \| U^{(t)} \| J_A' >^2] \bar{S} \tag{4.10}$$

$$\bar{p}_{SA}^{(dq)} = \frac{1}{(2J_S+1)(2J_A+1)} \left(\frac{2\pi}{\hbar}\right) \left(\frac{e^2}{R^4}\right)^2 [\Sigma_t \Omega_{tS} < J_S \| U^{(t)} \| J_S' >^2]$$

$$\times < 4f | r_A^2 | 4f >^2 < f \| C^{(2)} \| f >^2 < J_A \| U^{(2)} \| J_A' >^2 \bar{S} \tag{4.11}$$

$$\bar{p}_{SA}^{(qq)} = \frac{1}{(2J_S+1)(2J_A+1)} \left(\frac{14}{5}\right) \left(\frac{2\pi}{\hbar}\right) \left(\frac{e^2}{R^5}\right)^2 < 4f | r_S^2 | 4f >^2 < 4f | r_A^2 | 4f >^2$$

$$\times < f \| C^{(2)} \| f >^4 < J_S \| U^{(2)} \| J_S' >^2 < J_A \| U^{(2)} \| J_A' >^2 \bar{S} \tag{4.12}$$

Here Ω are the Judd-Ofelt intensity parameters which can be found in Section 2Hc; $<J||U^{(t)}||J'>$ is the matrix element of the transition between the ground and excited state of the sensitizer and activator respectively, (the calculation of these matrix elements in the intermediate-coupling scheme is now a well known procedure and may be found in Refs. [9] and [19]); $\bar{S}$ is the overlap integral; R is the interionic distance; $C^{(2)}$ is a numerical factor that depends on the orientation of the coordinate axis and $<4f|r_A^2|4f>^2$ is the expectation value of the radial integral of the 4f orbital.

Kushida presents an example for calculation of the transfer of energy between neighbouring ytterbium ions in crystals. The absorption band in ytterbium around 10^4 cm^{-1} arises from the $^2F_{7/2} \rightarrow {}^2F_{5/2}$ transition with oscillator strength of 3×10^{-6}.

The following values were inserted into Eq. (4.12). $R = 7a_0$ (a_0 is the hydrogen Bohr radius); for Yb $<4f|r^2|4f> = 0.691$ atomic units; the overlap integral $S = 10^{-3}$ and $<f^{13}\,{}^2F_{5/2}||U^{(2)}||f^{13}\,{}^2F_{7/2}>^2 = 6/49$. The transfer rates obtained in this way give $\bar{p}^{(qq)} = 1.4 \times 10^8$ sec^{-1} for the nearest-neighbour ions of ion separation $R = 7a_0$. As to the dipole-quadrupole (or quadrupole-dipole) and dipole-dipole transfer rates, inserting $\Sigma\Omega_t <f^{13}\,{}^2F_{5/2}||U^{(t)}||f^{13}\,{}^2F_{7/2}>^2 = 2 \times 10^{-20}$ cm^2 into Eqs. (4.10) and (4.11), Kushida obtained $\bar{p}^{(dq)} \sim 0.1\,\bar{p}^{(qq)}$ and $\bar{p}^{(dd)} \sim 0.2\,\bar{p}^{(dp)}$ for a pair of Yb ions of $R = 7a_0$. The dependence of the transfer rate on S-A separation may be displayed by writing:

$$P_{SA} = \alpha^{(6)}/R^6 + \alpha^{(8)}/R^8 + \alpha^{(10)}/R^{10} + \cdots \tag{4.13}$$

with $\alpha^{(6)}\alpha^{(8)}$ and $\alpha^{(10)}$ being constants for dipole-dipole, dipole-quadrupole and quadrupole-quadrupole transitions which may be evaluated from Eqs. (4.10), (4.11) and (4.12).

Using Eq. (4.13), Watts [10] plotted the energy transfer probabilities versus interionic separation for ytterbium-erbium transfer rate d-d [$\alpha(6)$], d-q [$\alpha(8)$] and q-q[$\alpha(10)$]. From his data it may be seen that the quadrupole-quadrupole mechanism is the highest and dipole-dipole the lowest at small interionic distances and reach the same order of magnitude at around 8 Å. At higher distances the situation is inverted, for example, at around 10 Å the dipole-dipole is highest and quadrupole-quadrupole lowest.

By taking different values of α (depending on the value of overlap integral which is inserted in the equation) a different distribution of the transfer rate may be obtained. However, it is clear that the behaviour of transfer rates with interionic separation may be difficult to use for estimation of the transfer mechanism.

Another estimate of transfer efficiency using spectral data may be obtained from Dexter's [4] formula for dipole-dipole resonance transfer,

$$P_{SA} = \frac{3\hbar^4 c^4}{64\pi^5 n^4} \quad \frac{Q_A}{\tau_S R^6} \cdot \int \frac{f_S(E) F_A(E) dE}{E^4} = \left(\frac{R_0'}{R}\right)^6 \cdot \frac{1}{\tau_S} \cdot \tag{4.14}$$

e) Statistical Aspects of Macroscopic Energy Transfer

In order to discuss energy transfer in macroscopic real systems a statistical analysis of the donor-acceptor distances is necessary. Only such an evaluation can be compared with experimentally measured quantities which are the time development of the sensitizer luminescence after pulse excitation and the relative yield of luminescence. The problem of time dependent luminescence was first treated by Forster [20] and by Inokuti and Hirayama [13]. Their development was based on the following consideration: When a random distribution of ions of type S is excited, each ion has the same probability of radiative decay resulting in a purely exponential fluorescence decay for the ensemble of ions. When the activator ions are also present and randomly distributed in the medium, some of the excited S ions will be sufficiently close to the A ion so that energy transfer will occur. Since the transfer probability depends on the interionic distance the decay rate of the individual S ions will vary as the distances to the A ion in a random way. Only ions with identical environments will have identical decay rates. Those donor ions which are near to the activator ions will decay rapidly, therefore the decay rates at short times after excitation will be large. After the ions near the quenching centres have decayed, the ions with no activator within a close distance will remain excited and their decay will approach the radiative rate, therefore energy transfer in absence of diffusion among the donor ions will be characterized by a non-exponential fluorescence decay with a rate approaching the purely radiative rate at long times.

Another approach to the macroscopic case of energy transfer is the use of rate equations which deal with the population of ions. The applicability of these equations has been discussed by Grant [6]. Based on the principle of quantum statistics Grant shows that the dependence of average transfer probability on concentration does not reflect the multipolar behaviour but rather the number of particles participating in the relaxation process. Recently, Fong and Diestler [5] on statistical consideration, also came to the conclusion that the transfer rate dependence on concentration reflects the many-body interactions. With the assumption of random mixing, the energy transfer rate at low concentration is linear with C as long as only pairwise interactions are considered. Quadratic or higher dependence on C can result from energy transfer between the sensitizer and 2, 3 or more activator ions. Fong *et al.* believe that the many-body transfer mechanisms are particularly important in the rare earth ions. Their results are similar to those of Grant.

The average yield of energy transfer in the case when transfer occurs between the nearest neighbours only,

$$\eta_{tr} = \int_0^\infty n(t)dt \tag{4.15}$$

where n(t) is the decaying population of the fluorescent state, η is related to the transfer probability in the following way,

$$\eta_{tr} = \int_{r_0}^\infty \frac{W_{tr}(r)}{W_D + W_{tr}(r)} \, e^{-4\pi r^3 n/3} 4\pi r^2 dr \tag{4.16}$$

where $W_{tr}(r)$ is the transfer probability and W_D the radiative transition probability. When the constraint caused by the transfer between the nearest neighbours is removed,

$$\eta_{tr} = \int_{r_0}^{\infty} \frac{W_{tr}(r)}{W_D + W_{tr}(r)} 4\pi r^2 dr \tag{4.17}$$

For concentration low enough and transfer probabilities small enough, the appropriate average for η due to Grant is

$$\eta_{tr} = 1 - \eta/\eta_0 = \langle W_{tr} \rangle / (W_D + \langle W_{tr} \rangle) \tag{4.18}$$

where $\langle W_{tr} \rangle$ is the transfer probability averaged over the system.

If it is assumed that the probability of the nearest activator being in a spherical shell $(v, v + dv)$ is $C(\exp(-cv))\,dv$, where $v = (4/3)(\pi R^3)$ is the volume, then the average transfer efficiency is given by,

$$\eta^{-1}(t) = \int_0^{\infty} \frac{ce^{-cv} P_{SA}}{P_{SA} + \tau_S^{-1}} dv = \int_0^{\infty} \frac{ye^{-yt}}{1 + t^2} dt \tag{4.19}$$

with $y = (4/3)(\pi c R_0^3)$. Thus, for any given y the efficiency may be calculated.

Experiments on energy transfer based on the resonance approach were performed by a large number of researchers. A method for calculating the average probability and efficiency of energy transfer between inorganic ions from the donor and acceptor luminescence and luminescence lifetime based on rate equations, was performed by Reisfeld *et al.* and is described in detail in Ref. [1].

When the absorption of the donor ion is much higher than the acceptor ion at the wavelength at which the system is excited (which is the practical case in optimum laser pumping), then the transfer efficiency η_t is given by a simple formula

$$\eta_t = 1 - \eta_d/\eta_d^0 \tag{4.20}$$

η_d being the fluorescence efficiency of the donor in the presence of an acceptor ion, and η_d^0 the fluorescence efficiency of the donor alone.

p_t, the average transfer probability, may be obtained from the formula,

$$p_t = \frac{1}{\tau_d}\left(\frac{\eta_d^0}{\eta_d} - 1\right) \tag{4.21}$$

τ_d being the measured lifetime of the donor.

B. Migration of Excitation

In many real systems the donor-donor transfer cannot be neglected. Because of the resonant condition, the S → S transfer may be even more rapid than the S → A transfer when the concentration of the two ions are comparable and especially in the rare earth ions where the Stokes shift is very small. Excitation energy may then be able to migrate among the sensitizer ions before passing to the activator thus decreasing the effective S–A distance. The migration of energy may be treated as a diffusion process or a hopping process. These are described in detail by Yokota and Tanimoto [22] and Watts [10]. Rapid energy diffusion can lead to a spatial equilibrium of excitation within the sensitizer systems. The rate will be independent of time since the distribution of excitation is always the same. The rate limiting step for the sensitizer relaxation will then be, either the energy transfer rate between the sensitizer and activator or the activator relaxation rate. In such a case a simple rate equation model for the donor system relaxation can be used which predicts a simple exponential decay. Mathematically the number of excited sensitizer ions as a function of time, is given by,

$$-\frac{dN_S}{dt} = \frac{N_S}{\tau} - N_S N_A P_{SA} \tag{4.22}$$

where $N_S(t) = \exp\{-\frac{1}{\tau} - N_A W_M\}t$, P_{SA} is the rate of transfer S → A and W_M is the energy transfer rate for those sensitizer ions which are closest to the activator ions.

When the diffusion is not fast enough to maintain the initial distribution of excitation the time dependence of population may be expressed as [22].

$$\frac{dN_S}{dt} = -\frac{N_S}{\tau} + D\Delta^2 N_S - \sum_i W[R_i(t)]N_S \tag{4.23}$$

In Eq. (4.23), D is the diffusion constant taken to be isotropic and $E_i(t)$ is the probability for energy transfer from the excited sensitizer to the n-th activator at position R_i.

Yokota and Tanimoto [22] have worked out the expected fluorescence decay when both quenching and diffusion are active and the diffusion is not fast enough to maintain the initial distribution of excitation. In such a case the decay function of the excited sensitizers is given by

$$N_S(t) = N_S(0)_I e^{-t/\tau} \left\langle \exp\text{-}t(-D\Delta + \sum_i \frac{\alpha}{r_i^6}) \right\rangle av \tag{4.24}$$

for a dipole-dipole interaction.

Under assumption of uniform distribution of sensitizers D is the diffusion constant and α a constant characterized by the dipole-dipole transfer interaction. The solution of this equation using the Pade approximation [22] is as follows:

$$N_S(t) = N_S(0)_e^{-t/\tau} \exp\left[-\frac{4}{3}\pi^{3/2}N_A(\alpha t)^{1/2}\left\{\frac{1+10.87x+15.50x^2}{1+8.743x}\right\}^{3/4}\right] \tag{4.25}$$

where $X = D\alpha^{-1/3}t^{2/3}$ and $W_{SA} = \alpha R^{-6}$.

In the hopping method W_{SA} is treated as a random variable. This situation is similar to the shift and broadening of the spectra of emitting atoms in a gas as a result of superposition effects between the perturbing ions as described by Anderson [23]. The latter mechanism was treated recently by Artamanova *et al.* [24] who performed a study of migration of electron excitation over Nd ions in glasses. These authors attributed the migration on Nd within the system to the stepwise mechanism analogous to the hopping mechanism. Both approaches of energy migration give similar results of behaviour of the excited population with time [10].

The decay function $\phi(t)$ is nonexponential at short times t where the migration is still unimportant. In this limit it approaches the I-H function for d-d interaction. For large t, $\phi(t)$ decays exponentially at a rate determined by the migration. As migration becomes more rapid, the boundary between these two regions shifts to shorter times until, for sufficiently fast migration, the decay appears to be a pure exponential. This long time behaviour is referred to as diffusion-limited relaxation. In the limit as t goes to infinity the fluorescence decay function [25] becomes,

$$\phi(t) = \exp\left\{-\frac{t}{\tau} - \frac{t}{\tau_D}\right\} \tag{4.26}$$

$$\frac{1}{\tau_D} = 0.51\,(4\pi N_A \alpha^{1/4} D^{3/4}) \tag{4.27}$$

The last equation characterizes the regime which is most easily investigated experimentally because the experimental conditions enable analyzing the behaviour of decay curves at longer times in a more tractable way.

Recently it was shown [26] that the diffusion limited decay depends on the concentration of sensitizer and activator ions by

$$\phi(t) = \exp\left\{-\frac{t}{\tau} - k_2 N_S N_A t\right\} \tag{4.28}$$

where k_2 is a constant proportional to α_S and α_A.

Experimental evidence for migration of energy has been given in many papers recently. Weber [25] investigated energy transfer between europium and chromium in phosphate glasses and analysed his data in view of the Yokota-Tanimoto theory. He verified that in the diffusion-limited relaxation the decay rate is proportional to $D^{3/4}$ as predicted by Eq. (4.27).

The effects of energy migration were studied by Krasutsky and Moos [27] in the system Pr^{3+} and Nd^{3+} in $LaCl_3$ at Pr^{3+} concentration 2 at % or more. A dipole-dipole interaction was assumed in the $Pr^{3+} - Pr^{3+}$ transfer and the exponential portion of the decay curves were used to determine the amount of diffusion. The largest diffusion constants obtained were for the $LaCl_3$: 20 – at % Pr (0.05 – and 0.1 – at % Nd) samples, with D ~ 5 x 10^{-9} cm^2sec^{-1}. This diffusion constant is much larger than that obtained by Weber of 6 x 10^{-10} cm^2sec^{-1} for 100% $Eu(PO_3)_3$ glass.

Migration effects of energy were found by Van der Ziel *et al.* [28] in terbium-doped aluminium garnets. The migration of energy is temperature sensitive and becomes dominant at 297 °K.

Watts and Richter [29] in their study of energy transfer between ytterbium and holmium in YF_3 have also found that diffusion in the ytterbium system may be important when the ytterbium concentration is increased. By variation of the ytterbium concentration they were able to vary the diffusion coefficient by three orders of magnitude and show consistence of their results with Eq. (4.25).

Bourcet and Fong [30] in their study of energy transfer between cerium and terbium in lanthanum phosphates performed an analysis of the dependence of the donor luminescence decay with temperature, and found that the diffusion in the cerium system plays an important role in the energy transfer process.

However, recent studies by Leah Boehm, Renata Reisfeld and Bernard Blanzat of energy transfer from cerium to terbium or thulium in phosphate glasses, show that the increased energy transfer in the former case as a function of increasing temperature is essentially due to the concomitant increase of the overlap between the absorption bands of Ce^{3+} and Tb^{3+}. On the other hand, energy transfer between a similar concentration of Ce^{3+} to Tm^{3+} is independent of temperature. It is well-known that the absorption spectrum of Tm^{3+} is much less affected by variation of temperature than Tb^{3+} with closely adjacent, excited levels belonging to $4f^75d$.

Madame F. Gaume organized in Villeurbanne (the campus of the University 1 of Lyon) an international colloquium "Spectroscopie des éléments de transition et des éléments lourds dans les solides" 28th June to 2nd July 1976, and the proceedings will be published by CNRS (Centre National de Recherches Scientifiques). Orbach presented there an important communication [188] about the "Spectral-Spatial Diffusion for Lorentzian Inhomogeneous Spectral Emission Profiles" where the time development of the optical emission profile is calculated numerically for phonon-assisted energy transfer processes which fall off as the inverse square of the energy mismatch. The emission after a sharp line excitation consists of a sharp line which broadens in time, (in contrast to the case of Gaussian equilibrium profile) while the remainder of the equilibrium emission (assumed to be Lorentzian) increases at the expense of the sharp line. These calculations were compared by Orbach with experiments on the ruby, where there is a critical concentration of active sites below which excitation localization occurs effectively, whereas energy migration is perceived above this threshold concentration. Chang Hsu and Powell have observed analogous effects with Sm^{3+} in $CaWO_4$ crystals.

Events beyond our control prevented one of us from presenting her communication [189], entitled "Radiative and Non-radiative Processes and Energy Transfer in Vitreous States," at this Colloquium.

Studies of energy transfer between manganese and erbium in MnF_2 performed by Flaherty and DiBartolo [31] revealed energy migration within the manganese system via excitation type process. Hopping migration of excitons in calcium tungstate crystals to the samarium centers was also observed by Treadaway and Powell [14].

Soules *et al.* [8] have found that the energy transfer between antimony and manganese is a few orders of magnitude greater than that predicted for d-d coupling. Their results can be interpreted by migration of energy within the manganese system.

Kraevskii *et al.* [3] found that energy transfer between Yb-Er, Nd-Yb, Eu-Dy and Sm–Eu in phosphate glasses cannot be explained by resonance transfer and made an attempt to correlate their results with the diffusion of energy in the sensitizer system. However, they found that this additional mechanism is inadequate in explaining their results and that it is necessary to consider contributions from other mechanisms, *e.g.* superexchange.

Energy migration in glass is mentioned by Pant *et al.* [32] in connection with energy transfer in sodium borate glasses. A study of the former relation existing among the recent theories of excitation transfer in a two-molecule system is given by Kenkre [33] and by Golubov and Konobeev [34]. Both these theories have so far not been applied experimentally to the inorganic ions.

C. Inhomogeneous Broadening

The emission lines of rare earth ions due to f–f transitions in glassy hosts are usually rather broad in comparison with crystal hosts as a result of the considerably large inhomogeneous broadening originating from the site to site variation of the crystal field acting on the rare-earth ion in the glassy state [1]. The inhomogeneous broadening from rare earth ions may amount to 100 cm^{-1} (*e.g.*, $^5D_0 - {}^7F_0$ line of europium where the crystal field splitting is absent). The effect of inhomogeneous broadening is to destroy the resonance, that is, to decrease the value of the overlap integral.

As will be shown below, using narrow bandwidth laser excitation it is possible to excite only those ions in an inhomogeneous distribution which are resonant with the laser line. In the absence of diffusion those sites which are not in resonance with the laser are not excited. The emission from the selectively excited ion is narrow, considering the homogeneous line width of the ions.

Recently, Kushida and Takushi [35] determined homogeneous spectral widths from the inhomogeneously broadened transition $^7F_0 - {}^5D_0$ of Eu^{3+} ion in $Ca(PO_3)_2$ glass using monochromatic dye laser excitation. In their experiments the homogeneous width was determined from the width of the resonance fluorescence line under monochromatic light excitation. Care was taken to eliminate the diffusion by use of small concentrations of europium. Under the dye laser excitation at 578 nm the resonance fluorescence line was narrowed to 6 ± 0.6 $cm^{-1}\,\delta\nu$. This line shape could be described by a Lorentzian and was assigned to the homogeneous width. When the concentration of rare earth ions is increased, after a short laser pump, diffusion of the excitation from the selectively excited ions to all the other ions in the inhomogeneous distribution will cause the initial narrow fluorescence to broaden gradually until the full inhomogeneous line is observed. The characteristic time of the spectral diffusion is the same as the time for spatial diffusion [36].

Spectral migration by this very elegant method has been observed for trivalent europium by Motegi and Shionoya [37] and by Yen *et al.* [38]. Spectral diffusion of ytterbium in silicate and phosphate glasses was demonstrated by Weber *et al.* [39] from the gradual broadening of the time-resolved 978.5 nm fluorescence with time. Hence, the initial narrowed line-emission decreases and a broad emission of inhomogeneously broadened bands grows up.

Selzer *et al.* [40] and Flach *et al.* [41] have directly observed the migration of energy with the excited state of $^3P_0$ in $Pr_xLa_{1-x}F_3$ by time-resolved fluorescence line-narrowing.

Before we move from the description of energy migrations having (at least roughly) probabilities proportional to R^{-k}, where k is a number higher than 3 and R the internuclear distance, we would like to mention the (mathematically rather trivial) fact that a statistical distribution with equal probability of a lanthanide nucleus to occur in each volume element produces a *divergent* infinite value of the probability of energy transfer because R^{-k+2} (after multiplication by $4\pi R^2 dR$) cannot be integrated to a lower limit zero. Hence, it is imperatively needed to introduce a cut-off minimum distance R_0 in such descriptions, and this has no obvious value in a glass, and may be perceptibly influenced by thermal vibrations and by fluctuations of the concentration of the lanthanide.

D. Phonon-assisted Energy Transfer

As shown in the preceding section, the probability that the resonance energy transfer will occur is very small because the inhomogeneous broadening decreases the possibility of the sensitizer and activator ions which are in resonance will be at short distances. However, a non-resonant transfer can easily occur if the energy difference between the interacting ions is given up or taken from the lattice vibrations. Phonon-assisted energy transfer was first described by Orbach [42] and is discussed in detail in Ref. [43].

In the case of phonon-assisted energy transfer the basic Eq. (4.14) of the resonance transfer of Dexter applies, however, there is a need of modification. The interaction Hamiltonian must contain an electron-phonon part. The initial and final states must include the initial and final phonon states which will differ by a number of phonons whose total energy is ΔE. The line-shape factors must include the phonon side-bands. If one phonon of energy $\hbar\omega = \Delta E$ is created in the process of energy transfer the transfer rate is

$$W_{SA} = (4\pi^2/h)|< S^*A|H_{SA}|SA^*>|^2 c[n(\omega) + 1] \int g_S(E) g_A(E - \hbar\omega) dE \qquad (4.29)$$

C is the function of electron-phonon coupling parameters and n (ω) is the number of phonons present of energy $\hbar\omega$. For the absorption of phonons the square bracket is replaced by $n(\omega)$ and the sign changed before $\hbar\omega$.

When the energy of one phonon is sufficient for energy conservation the probability of a transfer assisted by a single acoustic phonon between the i–th and j–th ions in the case of q–q interaction is given by Orbach [42]. His formula includes $\hbar\omega_q$, the energy of the acoustic phonon involved, H_1, the ion-ion interaction Hamiltonian, V_1, the ion-vibration interaction, ρ, the density of the host material and v the velocity of sound.

In the Orbach process the transfer rate is composed of a temperature-dependent part associated with the phonon-occupation number n_q and a temperature independent part corresponding to energy transfer accompanied by spontaneous phonon emission.

Experimental studies of energy transfer between ytterbium ions in zinc borate glasses based on the Orbach model were performed recently by Speed *et al.* [44]. These authors also performed an averaging over all possible separation distances of the two ions appearing in Orbach's equation. Such treatment is essential for correlation of the measured emission efficiency and decay times which arise from many different sites. Speed *et al.* [44] have measured the energy migration between Yb^{3+} ions in zinc borate glasses by studying the dependence of emission efficiency and the lifetime of Yb^{3+} on temperature and concentration, they found a good agreement of the results with Orbach model in which the acoustic phonons assist in the energy transfer. Many real systems which exhibit ion-pair interaction involve considerable energy mismatches requiring the participation of several phonons for energy conservation.

When the energy transfer occurs between the levels of a donor and an acceptor in which the mismatch of energy of several thousand cm^{-1} multiphonon phenomena must be considered. This was done by Miyakawa and Dexter [45]. In their theoretical analysis of multiphonon processes, Miyakawa and Dexter derived a comparative relaxation analogue of the multiphonon gap dependence. According to their theory, the probability of phonon-assisted energy transfer is expressed by:

$$W_{PAT}(\Delta E) = W_{PAT}(0)e^{-\beta\Delta E} \tag{4.30}$$

where ΔE is the energy gap between the electronic levels of donor and acceptor ions and β is a parameter determined by the strength of electron-lattice coupling, as well as by the nature of the phonon involved. The above equation has the same form as that for the energy gap dependence of the multiphonon relaxation rate, which is also given by the Miyakawa-Dexter theory as:

$$W_{MPR}(\Delta E) = W_{MPR}(0)e^{-\alpha\Delta E} \tag{4.31}$$

It is further indicated that the parameter α is given by:

$$\alpha = \frac{1}{\hbar\omega} [\ln\{N/g(n+1)\} - 1] \tag{4.32}$$

and α and β are connected with each other as

$$\beta = \alpha - \gamma$$

and

$$\gamma = \frac{1}{\hbar\omega} \ln(1 + g_S/g_A) \tag{4.33}$$

Here g is the electron-lattice coupling constant, suffixes S and A are sensitizer and activator ions respectively, n is the number of phonons excited at the temperature of the system, $\hbar\omega$ is the phonon energy which contributes dominantly to these multi-phonon processes and N is the number of phonons emitted in the processes, namely,

$$N = \Delta E/\hbar\omega \tag{4.34}$$

Non-resonant phonon-assisted energy transfer between various trivalent rare-earth ions in yttrium oxide crystals were thoroughly studied by Yamada *et al.* [46]. In their experiments the energy gap between the sensitizer and activator system varied in a wide range of energies up to 4,000 cm^{-1}. The probability of phonon-assisted transfer was observed in order to obtain the exponential dependence on energy gap predicted by the Miyakawa-Dexter theory. It was revealed that the phonons of about 400 cm^{-1} which produce the highest intensity in the vibronic side bands of yttrium oxide contribute dominantly to the phonon-assisted process. Table 25 presents the energy transfer rate between various pairs of sensitizers and activators [46].

Table 25. Calculated rates [46] of phonon-assisted energy transfer for various combinations of donor (sensitizer, S) and aceeptor (activator, A) trivalent lanthanide ions in Y_2O_3 at 77 °K

		Transition of energy transfer		Energy gap	Transfer rate
S	A	S	A	cm^{-1}	sec^{-1}
Sm	Eu	$^4G_{5/2} \rightarrow {}^6H_{5/2}$	$^7F_0 \rightarrow {}^5D_0$	600	1,700
Eu	Yb	$^5D_0 \rightarrow {}^7F_6$	$^2F_{7/2} \rightarrow {}^2F_{5/2}$	1,670	17
Eu	Yb	$^5D_1 \rightarrow {}^5D_0$	$^2F_{7/2} \rightarrow {}^2F_{5/2}$	1,225	40
Eu	Yb	$^5D_2 \rightarrow {}^5D_1$	$^2F_{7/2} \rightarrow {}^2F_{5/2}$	1,935	1,100
Tb	Yb	$^5D_4 \rightarrow {}^7F_0$	$^2F_{7/2} \rightarrow {}^2F_{5/2}$	4,200	2.3
Ho	Sm	$^5S_2 \rightarrow {}^5I_4$	$^6H_{5/2} \rightarrow {}^6H_{13/2}$	190	59,000
Ho	Tm	$^5S_2 \rightarrow {}^5I_7$	$^3H_6 \rightarrow {}^3H_4$	480	18,000
Ho	Yb	$^5I_4 \rightarrow {}^5I_8$	$^2F_{7/2} \rightarrow {}^2F_{5/2}$	2,610	460
Er	Yb	$^4S_{3/2} \rightarrow {}^4I_{13/2}$	$^2F_{7/2} \rightarrow {}^2F_{5/2}$	1,070	8,400
Tm	Yb	$^1G_4 \rightarrow {}^3H_5$	$^2F_{7/2} \rightarrow {}^2F_{5/2}$	1,840	530

Recently, Auzel [47] obtained results similar to Yamada, Shionoya and Kushida [46] with multi-phonon excited Stokes and anti-Stokes fluorescence. Various rare-earth ions were excited by a c.w. dye laser at energies just below and just above the energy of their electronic and one phonon vibronic absorption range. The energy gap ΔE involved between the excitation and the electronic level was larger than $\hbar\omega$, the highest phonon energy of the host. From this fact and the exponential dependence of the probability for anti-Stokes and Stokes excitation with the energy gap Auzel concluded that the process was a multiphonon one. In his work Auzel connects the parameter α (in his notation α_s) for Stokes excitation with α_{as} anti-Stokes by the relation $\alpha_{as} = \alpha_s + 1/KT$. The experiments described by Auzel lead to a new method for studying non-radiative and energy transfer parameters because such methods enable monitoring of the energy gaps at desired intervals.

Multi-phonon energy transfer in glasses was first proposed, to our knowledge, in Ref. [48]. The probability of phonon-assisted energy transfer between Eu and Yb, and between Tb and Yb in borosilicate, phosphate and germanate glasses was measured recently by Komiyama [49]. It was between 75 and 650 °K and correlated with the composition of the host glass. That the phonon-assisted energy transfer process occurs between Pr ions is concluded in Ref. [40] from the dependence of transfer probability on temperature.

E. Selected Examples and Suggestions

It is obvious that we cannot treat here all the conceivable cases of a donor (sensitizer) system S transferring energy to an acceptor (activator) A in a way possibly permitting laser action. We restrict ourselves to reported cases of lanthanide acceptors, to a relatively rare category of energy transfer from a lanthanide donor to species containing a partly filled 3d shell, and to a few characteristic cases which have not yet been scientifically exploited. Since the theoretical and general discussion above is centered around energy transfer *between* trivalent lanthanides, in Section 4Ef we shall take up this question again. It may once more be remembered that energy transfer well below the melting point of vitreous and crystalline solids is not accompanied by diffusion of material particles (distinctly not nuclei, and electrons only in very special cases of luminescent semi-conductors) whereas, it is not so easy to make a sharp-cut distinction between energy transfer (as discussed here) and optical consequences of the formation of collision complexes in liquids of low viscosity and in gaseous mixtures.

a) Donors with Spin-forbidden 3d Transitions

Certain excited states of iron group compounds having the total spin quantum number S, one unit lower than the groundstate, are known to transfer energy to lower-lying J-levels of trivalent lanthanides. At first, it might seem unlikely that the very weak spin-forbidden transitions in the partly filled 3d shell would be suitable for pumping of a laser, but they may be the terminal step of feeding higher states (including spin-allowed 3d transitions and electron transfer bands) rapidly showing radiationless decay. Whereas, the relatively long life-time of the lowest spin-forbidden transition by itself enhances the possibility of energy transfer, the general situation that Russell-Saunders coupling is nearly always a good approximation in the sense that S is well-defined [50] may increase the probability of energy transfer in the combined system (having a considerable internuclear distance R) with $S = S_1 \oplus S_2$ obtained by vector coupling according to Eq. (5.5) being the same for the excited donor (with a given S_2) and acceptor groundstate, as it is for the groundstate of the donor ($S_2 + 1$) and the excited state of the acceptor, to which energy transfer takes place. This argument connects our subject with antiferromagnetic coupling discussed in Section 5F.

The fluorescence of the ruby $Cr_xAl_{2-x}O_3$ was already mentioned in the "Tales of 1,001 Nights" and it is not surprising that attempts have been made to transfer energy from the lowest doublet ($S = 1/2$) level of the octahedral chromophore $Cr(III)O_6$ possessing quartet ($S = 3/2$) groundstate. Weber [51] demonstrated that energy transfer occurs between chromium on the octahedral sites of the perovskite $YAlO_3$ to trivalent neodymium, holmium, erbium, thulium or ytterbium incorporated on the yttrium sites. The energy is carried to one of the two spin-allowed transitions at 555 or 410 nm, and narrow emission lines, (accompanied by vibronic structure) due to the spin-forbidden transition can be observed close to 725 nm. At low Cr concentration, the lifetime is 54 msec at 77 °K and 34 msec at 295 °K. At higher Cr concentration, pair interactions result in a modified emission spectrum and a faster, non-exponential decay. In $Y_{1-x}Nd_xAl_{1-y}Cr_yO_3$ at 300 °K, the energy transfer to neodymium $^4F_{3/2}$ was found to be more efficient than in a comparably doped garnet $Y_3Al_5O_{12}$. Holmium in $YAlO_3$ has $^5I_4$ only 260 cm^{-1} below the $^2E(O_h)$ level of Cr(III), and it probably decays in a radiationless way to $^5I_6$ from which the fluorescent transition to the $^5I_8$ groundstate was observed. By the same token, $^4I_{9/2}$ of erbium is 1,060 cm^{-1} below ^2E, but the fluorescence observed is $^4I_{11/2} \rightarrow {}^4I_{15/2}$. In the case of thulium $^3F_4$ (which we prefer to call $^3H_4$) is 860 cm^{-1} below ^2E, and its fluorescence to the $^3H_6$ groundstate was observed [51]. Actually, Cr^{3+} is known to sensitize Tm^{3+} laser action in $YAlO_3$. Also energy transfer to ytterbium (producing the $^2F_{5/2} \rightarrow {}^2F_{7/2}$ emission in the near infra-red) was observed. Sharp, Miller and Weber [52] detected effective energy transfer from chromium to ytterbium in a silicate glass, where the life-times are 2 msec (at 295 °K) for $^2F_{5/2}$ of Yb^{3+} taken alone, and as short as 0.02 msec for Cr^{3+} alone.

Whereas, the lowest spin-forbidden transition in $Cr(III)O_6$ essentially remains within the same sub-shell (MO) configuration as the groundstate (with concomitant narrow bands) the lowest spin-forbidden transition from the quartet $^4T_1$ to sextet $^6A_1$ groundstate of both octahedral $Mn(II)X_6$ and tetrahedral $Mn(II)X_4$ contain one anti-bonding electron less in the upper sub-shell (but a much higher interelectronic repulsion [53]) than in the groundstate. Hence, the vibronic broadening according to the Franck-Condon principle corresponds to a *decreased* equilibrium internuclear distance R_e in the excited quartet compared with R_0 in the sextet groundstate. Orgel [54] pointed out that the Stokes shift should be larger and the emission bands much broader, of octahedral $3d^5$ compared with tetrahedral systems. Empirically [114], mixed oxides containing $Mn(II)O_6$ are known sometimes to emit orange fluorescence, and tetrahedral sites (especially in sulphides) green light. Solutions or solids containing water or other hydrogen-containing molecules generally do not fluoresce, much in analogy to lanthanides, and seem to relax by multi-phonon processes to the groundstate. The first monomeric, tetrahedral complex shown [55] to be strongly fluorescent is $MnBr_4^{2-}$ and later, this was also found [56] for tetrahedral $MnCl_4^{2-}$ and MnI_4^{2-}. Recently, we measured the luminescence [57] of solid $[N(C_2H_5)_4]_2MnBr_4$ having a quantum yield close to 1, an exponential decay at 80 and at 290 °K with $\tau = 0.8$ msec, an excitation spectrum with much detail, corresponding to the narrow $3d^5$ excited levels, and an emission band at 19,500 cm^{-1} with a shoulder at 18,700 cm^{-1}. We shall continue with work on possible energy transfer from this anion.

MnF_2 is a rutile containing octahedral chromophores. Hegarty and Imbusch [58] pointed out that the mobility of the excitation is so pronounced that it is very difficult to avoid the trapping of energy at crystalline defects (*e.g.* due to the presence of

10^{-6} magnesium) and emitted at slightly decreased wave-numbers. Hence, it is not surprising that efficient energy transfer to 10^{-3} europium is observed. The splitting in several fine lines shows that the Eu^{3+} is at a site with low symmetry (not octahedral) and the two strongest line groups are the magnetic dipole transition $^5D_0 \rightarrow {}^7F_1$ and $^5D_0 \rightarrow {}^7F_4$. The transfer rate is about $5 \cdot 10^{10}$ sec^{-1}. Energy transfer to Er^{3+} was also observed [58] whereas, Yb^{3+} quenched, but without its own luminescence. Flaherty and DiBartolo [31, 59] also reported many details about the behaviour of Er^{3+} in MnF_2. The levels $^4F_{9/2}$, $^4I_{11/2}$ and $^4I_{13/2}$ are all three able to fluoresce, with $\tau = 0.25$ msec, 9.4 msec and 17 msec at 77 °K, but τ does not decrease dramatically with increasing temperature. Parke and Cole [7] studied energy transfer from Mn^{2+} in phosphate glasses to Nd^{3+}, Ho^{3+} and Er^{3+}, and Kumar [147] to Nd^{3+} in barium borate glass.

It is well-known that less than 10^{-6} iron can quench the luminescence of many solids. It seems plausible that this quenching really is energy transfer to $^4T_1$ of Fe^{3+} followed by fluorescence in the near infra-red. It might be worthwhile to look for energy transfer from iron(III) to *J*-levels of lanthanides situated below 8,000 cm^{-1}. Though electron transfer bands with high oscillator strength might pump the lowest quartet, they may also suffer from the difficulty pointed out by Blasse [60] that the potential surfaces expressing the energy of a given electronic state as a function of the (3N–6) variables, indicating the positions of N nuclei may cross in a highly unexpected fashion far away from the regions compatible with the Franck-Condon principle for absorption bands. Octahedral chromophores such as $Co(II)F_6$ and $Ni(II)O_6$ to be studied [53] in glasses, or in crystals such as $KCoF_3$, $NiTiO_3$ or $Ni_xMg_{1-x}O$ might also produce energy transfer below 7,000 cm^{-1} but the excited $^4T_2$ or $^3T_2$ are short-lived and have expanded R_e. It is not absolutely true that a necessary (though not sufficient) condition for energy transfer is that the lowest excited state of the donor shows luminescence, but it is very rare that a competiting energy transfer would be even faster.

The $4d^6$ rhodium(III) is known [61] to fluoresce with τ close to 1 msec from $^3T_1$ (to the groundstate $^1A_1$) around 14,700 cm^{-1}, whereas the spin-allowed transitions occur in absorption to $^1T_1$ at 25,400 cm^{-1} and to $^1T_2$ at 31,200 cm^{-1} (to be compared with 25,500 and 32,800 cm^{-1} in the hexaqua ion [62]) in $Rh_xAl_{2-x}O_3$. The former, spin-forbidden transition may conceivably transfer energy in mixed oxides.

b) Post-transitional (Mercury-like) Donors with Laporte-allowed Transitions

The groundstate $^1S_0$ of the mercury atom constitutes the closed-shell configuration $[Xe]4f^{14}5d^{10}6s^2$ (in the following abbreviated to $6s^2$). The first excited state $^3P_0$ at 37,645 cm^{-1} belongs to the configuration 6s6p. In spite of this state having odd parity, the transition from $J = 0$ to the groundstate with even parity and $J = 0$ is forbidden, not only as electric dipolar transitions but also as electric or magnetic multipolar transitions of any order, as long the mercury atom conserves its spherical symmetry. The configuration 6s6p comprises the three other levels of odd parity $^3P_1$

at 39,412 cm^{-1}, $^3P_2$ at 44,043 cm^{-1} and $^1P_1$ at 54,069 cm^{-1}. Since the triplet and singlet characteristics are rather heavily mixed by relativistic effects for the two levels with $J = 1$, they both represent a high oscillator strength for transitions to the ground-state and correspond to the two "resonance lines" in the ultra-violet.

It has been attracting attention for a long time [63, 64] that complexes in solution of the isoelectronic series thallium(I), lead(II) and bismuth(III) show a strong and a very strong absorption band at positions dependent on the neighbour atoms (the ligands) but roughly comparable to $^3P_1$ and $^1P_1$ in the mercury atom. This identification was first made by Seitz [65] for octahedrally coordinated thallium(I) incorporated in crystalline alkaline metal halides. Though it is beyond doubt that here are two strongly allowed transitions, it is not perfectly certain [50] that the description of the excited configuration as 6s6p remains a good approximation. The crystallographic time-average picture of the stereochemistry of such species is not only frequently highly distorted (as if they contained a lone-pair located at one side, representing a mixture of 6s and 6p character) but the much shorter time-scale inherent in the optical spectra in the visible and in the ultra-violet produce an "instantaneous picture" of lower symmetry.

It has become customary to describe the decrease of inter-shell transition wave-numbers, such as 6s → 6p or 4f → 5d, as a function of decreasing ligand electronegativity, by a *nephelauxetic ratio* β times the corresponding energy difference in the gaseous ion (such as Pb^{2+}, Bi^{3+}, Ce^{3+} . . .) though this parameter β was originally introduced [53] for describing the decrease of the phenomenological parameters of interelectronic repulsion for transitions within the *same* electron configuration containing only *one* partly filled shell. In particular, Duffy and Ingram [67, 68] have been quite active in extending the concept originating [53, 63] in the β of d^q systems that, it is possible to write as a good approximation

$$1 - \beta = \mathsf{hk} \tag{4.35}$$

where h is a property of the ligands (generally chosen 1.00 for the aqua ion) and k a (rather small) decimal fraction characterizing the central atom (and generally increasing strongly as a function of increasing oxidation state z in a given isoelectronic series). Reisfeld and Boehm [68] determined β for Sn^{2+}, Sb^{3+}, Tl^{+}, Pb^{2+} and Bi^{3+} in various glasses and pointed out that Eq. (4.35) shows a certain interpendence of h and k (as is definitely the case for hexacyano complexes [53] of the 3d group). Following studies of post-transition group ions in molten sulphates and halides [69], Duffy and Ingram [70, 71] introduced the concept of *optical basicity*, connected not only with the variation of h, but also the molar refractivities and the vibrational spectra and rationalized the influence of the kind and amount of differing net-work modifying cations in glasses. It should noted that the transitions in Sn^{2+} and Sb^{3+} described in this approximation as $5s^2 \rightarrow 5s5p$ produce a strong absorption band at the highest wave-number to $^1P_1$ but that the spin-forbidden character of the transition to $^3P_1$ shows up as much weaker bands [63, 64]. Actually, several weak bands are frequently observed, and they may correspond to $^3P_2$ components or to a separation (at least on an instantaneous picture [50]) of the three sub-levels formed from $^3P_1$ in low symmetry.

Corresponding to the $^3P_1$ emission, fluorescence spectra of Sn^{2+} and Sb^{3+} in borate, phosphate and germanate glasses [72] show a non-exponential decay at 293 °K which can be resolved in two τ close to $8 \cdot 10^{-8}$ sec and $2 \cdot 10^{-6}$ sec, whereas the decay at 87 °K corresponds to a single exponential in borate glass, $\tau = 11.3 \cdot 10^{-6}$ sec for Sn^{2+} and $6.5 \cdot 10^{-6}$ sec for Sb^{3+}. This behaviour can be explained by a much slower decay of $^3P_0$ which is situated (presumably) some 1,000 to 2,000 cm^{-1} below $^3P_1$ (it is not very clear in what region of the (3N – 6) dimensional variation of the internuclear distances involving (N – 1) nearest neighbour atoms the two potential surfaces are at the shortest distance) and from which the thermal energy repopulate the rapidly decaying $^3P_1$. Similar behaviour [73] was previously found in Tl^+, Pb^{2+} and Bi^{3+}. At 87 °K, τ of Bi^{3+} in borate glass is 0.019 msec and in germanate glass 0.017 msec, whereas both these two cases and Pb^{2+} in germanate glass show a decay curve at 293 °K which can be resolved in two exponentials with τ close to $3 \cdot 10^{-8}$ sec and $3 \cdot 10^{-7}$ sec. As seen below in Eq. (4.36), the quantum yields are quite moderate. Pedrini [74] discussed the observable effects of the thermal repopulation of the emitting $^3P_1$ from the long-lived $^3P_0$ of Bi^{3+} incorporated in metaantimonates (such as $CaSb_2O_6$ and $SrSb_2O_6$) and succeeded in cooling below a rather sharp transition

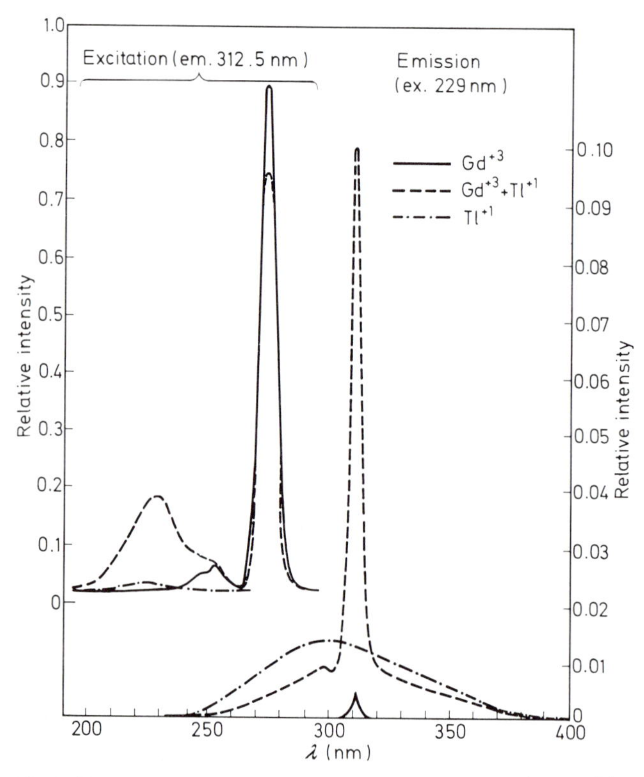

Fig. 7. Excitation spectra (monitored at 312.5 nm) of Tl^+ or Gd^{3+} alone in a borate glass, and of simultaneously present thallium and gadolinium, as well as their emission spectra excited by irradiation at 229 nm
[Repetition of Figure 3 of R. Reisfeld: Structure and Bonding *30* (1976) 65 on page 76]

temperature ~ 60 °K where the emission band shifts some 400 cm^{-1} toward lower energy, corresponding to $^3P_0$ emission. Jacquier [75] gave an extensive MO treatment of BiOCl and Bi^{3+} incorporated in Y_2O_3, LaOCl and YOCl showing a rather complicated luminescent behaviour. It is quite clear that the 12 mutually orthogonal states constituting $^3P_0$, $^3P_1$, $^3P_2$ and $^1P_1$ of the configuration 6s6p in spherical symmetry are mixed to a considerable extent in the compounds, and furthermore, that in sites lacking a centre of inversion, the orbitals show mixed 6s and 6p character. The variation of the energy differences as a function of the coordinate representing the Bi–O distance in the pure breathing mode of the chromophore was investigated [75]. In view of the very complicated stereochemistry of such systems this may be an over-simplification.

Characteristic data for the lead [76] and bismuth [73] in three glasses may be compared with $^3P_1$ 64,391 cm^{-1} above the groundstate of gaseous Pb^{2+} and 75,926 cm^{-1} in Bi^{3+}. It may be noted that $^3P_0$ is situated 4,963 cm^{-1} below the latter level, whereas the only directly determined vibronic origin of Bi^{3+} in a compound is 26,051 cm^{-1} (only 1,188 cm^{-1} below the origin of $^3P_1$) when incorporated [77] in CaO. By the way, this material is known to show a striking chemiluminescence producing violet light when heated with a few parts per million of bismuth in a hydrogen flame. δ is the one-sided half-width in cm^{-1} and ϕ the quantum yield of the fluorescence, and P the oscillator strength of the absorption (with maximum at σ to be compared with σ_{em} of the emission, both in cm^{-1}):

	borate	phosphate	germanate	
Pb^{2+}, σ	42,700	45,600	~37,000	
δ	2,700	2,400	–	
P	0.09	0.10	–	
σ_{em}	25,000	32,400	22,000	
δ	4,200	3,600	3,350	
ϕ	0.01	–	0.044	(4.36)
Bi^{3+}, σ	41,300	43,000	~37,000	
δ	2,600	3,100	–	
P	0.06	0.11	–	
σ_{em}	24,000	27,400	22,700	
δ	2,500	2,100	2,650	
ϕ	0.0014	$3 \cdot 10^{-5}$	0.019	

The most striking aspect of these result is perhaps the unusually large Stokes shift between 13,000 and 17,000 cm^{-1}. This fact combined with the rather large δ may suggest motion along other normal modes of vibration than the totally symmetric breathing mode.

Table 26 summarizes the recent studies of energy transfer from ions with strong inter-shell transitions to excited levels of $4f^q$ in trivalent lanthanides. Both the efficiency η and the rate of probability of the energy transfer are given. The quite striking case [1, 78] of Tl^+ to Gd^{3+} is also illustrated on Figure 7. The energy transfer from lead to Eu^{3+} was studied by Reisfeld and Lieblich-Sofer [79] in germanate glass, where the presence of 0 to 7 weight percent Eu_2O_3 produces a regularly decreasing

Table 26. Efficiency η and probability p (in the unit 10^6 sec^{-1}) of energy transfer between donors (S) having intense inter-shell transitions, to trivalent lanthanides (A). Concentrations are given in weight percent oxides

S	A		p	η
Tl^+	Gd^{3+}			
0.01	1.0	borate	28.3	0.85
0.01	3.0	borate	38.3	0.89
0.01	5.0	borate	47.6	0.91
0.01	7.0	borate	61.7	0.93
Pb^{2+}	Eu^{3+}			
1.0	1.0	germanate	0.33	0.09
1.0	2.0	germanate	1.14	0.25
1.0	3.0	germanate	2.57	0.41
1.0	5.0	germanate	5.00	0.59
1.0	7.0	germanate	7.32	0.68
Bi^{3+}	Sm^{3+}			
1.0	1.0	borate	0.69	0.19
1.0	0.5	germanate	0.63	0.28
1.0	1.0	germanate	2.13	0.43
Bi^{3+}	Eu^{3+}			
1.0	0.5	borate	3.45	0.50
1.0	1.0	borate	0.64	0.18
1.0	0.5	germanate	1.14	0.29
1.0	1.0	germanate	0.72	0.20
Ce^{3+}	Tb^{3+}			
0.025	1.0	borate	2.0	0.07
0.025	2.0	borate	6.7	0.19
0.025	2.5	borate	8.7	0.23
0.025	3.0	borate	13.7	0.32
Ce^{3+}	Tm^{3+}			
0.13	0.05	borate	2.4	0.09
0.13	0.15	borate	4.8	0.17
0.13	0.25	borate	7.0	0.25
0.13	0.50	borate	11.6	0.41
0.13	0.75	borate	14.0	0.49
0.13	1.00	borate	14.9	0.52

intensity of the broad lead emission at 445 nm to about a-quarter of the original value. Correspondingly, the longer lasting component of the non-exponential decay (at room temperature) shows τ decreasing from $2 \cdot 6 \cdot 10^{-7}$ sec to 10^{-7} sec. As seen from Table 26, the europium emission increases regularly in these samples. Comparable results [80] were obtained for energy transfer from trivalent bismuth to Sm^{3+} and to Eu^{3+} in borate and germanate glasses. Recent studies [81] of temperature dependence of energy transfer between Bi^{3+} and Eu^{3+} in glasses show that the probability of energy transfer depends on the quantum efficiency of the sensitizer ion, and in addition indicate that the non-radiative relaxation within the sensitizer system is much faster than the rate-determining step of energy transfer to the lanthanide acceptor. We feel that this is true in general. The systems studied sofar perform non-radiative relaxation in the donor (most frequently phonon-assisted) orders of magnitude more rapidly than the transfer.

In the mercury atom, the excited state $^3P_0$ cannot be expected to have an infinitely long life-time, even in a very dilute gas. When one of the nuclei (such as mercury 201 with $I = 3/2$) have positive I, the admixture of different J-levels presenting the same $F = I \oplus J$ produce life-times of the order of some minutes, and if the breakdown of parity in our Universe produces oscillator strengths P close to 10^{-12}, τ is expected to have values of about a few hours. However, in laboratory experiments, a much more obvious reason for desactivation is the collision with wall surfaces or with gaseous molecules. Mercury atoms adsorbed in zeolites were found by Prener, Hanson and Williams [82] still to show the resonance line at 39,200 cm^{-1}, but broadened to a certain extent, showing broad emission around 27,000 and 23,000 cm^{-1}. A quite interesting possibility of energy transfer is that one can make zeolites (cage-structure "molecular sieves" consisting usually of sodium aluminium silicates) containing large amounts of lanthanides (which are used as highly effective catalysts for methyl migration in the "cracking" of hydrocarbons) and ultra-violet radiation may allow a build-up of a stationary concentration of $^3P_0$ conceivably able to transfer energy to rare earths.

Like the first excited (electron transfer) state of the uranyl ion discussed in Section 1 D is sufficiently long-lived to have characteristic chemical properties (a strong oxidant), the $^3P_0$ state of the isolated mercury atom reacts with many gaseous molecules. Thus, the complex $HgNH_3$ (having a radiative $\tau = 1.8 \cdot 10^{-6}$ sec) was shown by Callear and Freeman [83] to give a moderately broad emission band in the near ultra-violet at 345 nm (29,000 cm^{-1}), whereas sufficiently low ammonia pressures (below 0.1 torr) allows the detection of a comparable band at 300 nm due to the complex $HgNH_3$ in the $^3P_1$ state.

A related problem is *excimers,* diatomic molecules formed from excited states, where for instance [84], the Hg_2 excimer produces two broad emission bands with maxima close to 440 and 600 nm due to transitions from the *same* potential curve formed from an odd level with $\Omega = 0$ (originating in one $6s^2 : {}^1So$ and one $6s6p : {}^3P_0$ atom) down to a repulsive even groundstate, also with $\Omega = 0$. If the same transition takes place from the minimum of the excited curve, it occurs at 510 nm.

A related excimer is TlHg discussed by Drummond and Schlie [85] with a repulsive ground potential curve derived from the lowest level $^2P_{1/2}$ (belonging to the configuration $6s^2 6p$) of the thallium atom, whereas $^2P_{3/2}$ has a minimum of energy, and an excited state can emit a fairly narrow band at 656 nm by a transition to the latter state, producing effectively a four-level laser system. Actually, such excimers as TlHg and the analogous TlXe seem to be very good candidates for laser activity [85].

c) Electron Transfer Bands in Early Transition Elements as Donors

Whereas inter-shell transitions such as $6s \rightarrow 6p$ in the mercury isoelectronic series, or $4f \rightarrow 5d$ in the lanthanides, are centered around a single atom (and hence quite acceptable to solid-state physicists), the idea of *electron transfer spectra* [64, 86] (where an electron is transferred from a molecular orbital situated on one strongly reducing ligand, or delocalized on several adjacent atoms, to an empty or partly filled shell of the central atom) has been ignored frequently, as being too distant from the concepts of atomic spectroscopy. The parallel concept in energy-band theory is a transition from the valence band to the conduction bands which are empty in stoichiometric closed-shell compounds such as Ta_2O_5 and WO_3 and contain electrons in metallic ReO_3 and the isotypic tungsten bronzes Na_xWO_3 or La_xWO_3. These two descriptions are *not equivalent* because the d or f shells characterizing the transition group compounds have much lower electron affinities than ionization energies. There is no doubt that the electron transfer bands in the visible and in the near ultra-violet correspond to transitions limited to the penta-atomic anion in MnO_4^- and CrO_4^{2-} and it has also been established [87] that the loosest bound MO have the symmetry t_1 (in the point-group T_d) in the LCAO approximation consisting exclusively of oxygen 2p, whereas the lowest empty MO are 3d-like and have the symmetry type e. It is striking [64] that the oscillator strength P of this transition increases from 0.04 in MnO_4^- and 0.09 in CrO_4^{2-} to 0.15 in VO_4^{3-} (existing in strongly alkaline solutions with pH above 13). Whereas, the regularities in electron transfer spectra (discussed in Section 1 D) described by optical electronegativities in Eq. (1.44) work well in octahedral MX_6^{+z-6} and tetrahedral and quadratic MX_4^{+z-4} complexes of the halides X^- but not as well (as far goes band positions) for the corresponding MO_4^{+z-8}. This difficulty is connected with strong π-anti-bonding and a specific kind of inductive effects [87] (making the central atom look far less oxidizing than in the corresponding fluoride) and it also shows analogies with the rather unexpected behaviour of oxide glasses studied by Duffy [67] and by us [68]. The early transition-group elements with a low-lying empty d shell most frequently show the coordination numbers $N = 4$ and 6. Under equal circumstances, the monomeric tetrahedral complexes in *higher* oxidation states z are closer to the general treatment of electron transfer bands, whereas lower z, $N = 6$ and in particular bridging of each oxide by two (as in the perovskite $SrTiO_3$), three (as both in rutile and in anatase TiO_2) or by four (as in C-type Sc_2O_3) transition-group atoms, seem to be closer to an energy band description.

There is another interesting problem for monomeric tetroxo complexes. The MO configuration $(t_1)^5(e)^1$ described above contains two partly filled shells producing 20 states which combine to the four terms $^3T_1$, $^3T_2$, $^1T_1$ and $^1T_2$ (normally, such term distances are quite negligible and do not produce observable energy separations in electron transfer spectra [86]). There is no doubt that $^1T_1$ occurs in MnO_4^- some 3,000 cm^{-1} below the 20 times more intense (symmetry-allowed) transition to $^1T_2$ (and in our opinion, at least one of two triplets is about 1,000 cm^{-1} below $^1T_1$ but does not fluoresce). In chromate and orthovanadate, the weak transitions (before the first $^1T_2$) are reduced to uncharacteristic shoulders. The situation is more extreme in monomeric MoO_4^{2-} and WO_4^{2-} in (weakly alkaline) solution where a plot of log ϵ as a function of the wave-number produces a parabola [corresponding to the Gaussian error curve in Eq. (1.37)] even down to very low ϵ. Hence, the weak transitions are too close to $^1T_2$ to be detected, much in the same way as $^3P_0$ is too close to $^3P_1$ in the mercury-like atoms.

Yttrium orthovanadate crystallizes in the zircon ($ZrSiO_4$) type with $N = 8$ for

Y^{3+}. This crystal was of great importance for developing the Eu^{3+} cathodoluminescence used in red colour television, and fluorescence in near-ultra-violet excitation occurs of incorporated Sm^{3+}, Eu^{3+}, Dy^{3+}, (weakly Er^{3+}) and Tm^{3+} on the yttrium sites. It was rather enigmatic that Tb^{3+} does not fluoresce but Draai and Blasse [88] resolved this problem by studying a large number of samples of $CaSO_4$ where 10^{-4} to 10^{-2} of the calcium is replaced by trivalent lanthanides and of the sulphate by orthovanadate anions. The energy transfer to M = Pr, Sm, Eu, Tb and Dy is quite extended, with emission lines producing a strong orange-red, orange, red, green and yellow luminescence, respectively. It seems that the quenching of luminescence in $Y_{1-x}Tb_xVO_4$ is due to a competiting electron transfer process going to $4f^7$ Tb^{4+} and $3d^1V^{4+}$, expected [60] to have highly variable potential surfaces. In the weakly doped $Ca_{1-x}M_xS_{1-x}V_xO_4$ one can detect the vanadate emission [88] below 200 °K as a broad band (with maximum σ_{em} in cm^{-1}) and originating in an excitation spectrum (with maximum σ_{exc}) as a function of M^{3+}:

M =	La	Gd	Y	Lu	Sc	
σ_{exc}	40,300	40,300	39,200	38,200	38,900	(4.37)
σ_{em}	20,600	21,400	21,500	21,200	19,400	

showing a distinct interaction between the (colourless) M^{3+} and VO_4^{3-}, possibly due to a non-statistical distribution in the calcium sulphate. If the vanadate concentration is increased above 0.1 percent a new band in the excitation spectrum grows up at 32,800 cm^{-1} probably indicating the formation of clusters with many Y^{3+} and VO_4^{3-}. The solubility of YVO_4 in $CaSO_4$ is limited but many-phase behaviour was not observed before 10 percent. It is noted that the Stokes shift is large, between 17,000 and 19,500 cm^{-1}. At 77 °K the luminescence of the sample containing 0.1 percent YVO_4 does not correspond to an exponential decay but can be described reasonably well with two components having τ_1 = 1.18 and τ_2 = 0.21 microsecond. The numerical value of τ_1 and the strong influence on undiluted yttrium vanadate $Y_{1-x}M_xVO_4$ of temperature (where cooling quenches the M^{3+} emission and promotes the broad vanadate emission) suggests low-lying, long-lived (say $^3T_1$ or $^1T_1$) levels which can thermally repopulate higher, short-lived levels in analogy to $^3P_0$ and $^3P_1$ of 6s6p. The quenching is not due [88] to a temperature variation of the probability of energy transfer from vanadate to the lanthanide, but of the energy transfer between adjacent vanadate anions. Another interesting kind of system is $Y_{1-x}M_xP_{1-y}V_yO_4$ (pure YPO_4 is the same crystal type) where the number of V-P and V-V closest distances is statistically determined, decreasing the rate of energy transfer from one vanadate group to another when y decreases. We may finally mention the surprising observation [88] that $CaSO_4$ containing a little Tm^{3+} and VO_4^{3-} emits not only the $^1G_4 \rightarrow {}^3H_6$ lines at 20,900 cm^{-1} but also $^1D_2 \rightarrow {}^3H_6$ as a weaker line at 27,400 cm^{-1}, well above the emission band of vanadate in Eq. (4.37). A (perhaps somewhat naive) explanation would be that energy transfer takes place from the vanadate to the thulium well before most of the Stokes shift has had the time to take place, again suggesting a plurality of adjacent vanadate states.

We have not heard about any energy transfer from chromates or permanganates to lanthanides, and this may be partly due to the fact that near infra-red lines have not been looked for, but also because the migration of energy between anions may be far less pronounced than in solid vanadates.

Whereas niobate is not readily obtained in aqueous solution as monomeric NbO_4^{3-} but rather as isopolyniobates such as $Nb_6O_{19}^{8-}$ (somewhat comparable to paramolybdate $Mo_7O_{24}^{6-}$ obtained by a weak decrease of pH of MoO_4^{2-}). Minerals such as fergusonite are isotypic with the phosphate monazite, being also used as a source of thorium (with the composition $M_{1-x}Th_xP_{1-x}Si_xO_4$) and synthetic orthoniobates, such as $YNbO_4$ and corresponding tantalates, such as $LaTaO_4$. Since niobium(V) is less oxidizing than vanadium(V), and tantalum(V) even less, both excitation and emission maxima occur at higher wave-numbers than in Eq. (4.37). $N = 6$ is more frequent in solid compounds of niobium and tantalum, not only in the stoichiometric perovskites such as $KNbO_3$ and $RbTaO_3$ but also in many mixed oxides studied by Blasse [89]. However, both these cases and molybdates ($N = 4$) have not been exploited to a great extent for energy transfer. Some materials, such as $PbMoO_4$ or the pyrochlore $Pb_2Nb_2O_7$ might combine interesting aspects of 6s → 6p inter-shell transitions and electron transfer bands of the d^0 systems.

Tungstates are the classical examples [90] showing energy transfer to trivalent lanthanides, where $M'_{0.98}Na_{0.01}M_{0.01}WO_4$ (M' = Ca, Sr, Ba, Cd and Pb), $Na_{0.5}M_{0.01}Y_{0.49}(WO_4)_2$ and many other scheelites prepared in the laboratory, transfer energy quite efficiently to Eu^{3+} and to Tb^{3+}. It was also suggested as a technique of analytical detection [91] that a solution of 0.6 M Na_2WO_4 (containing 0.04 M borax $Na_2B_4O_7$) at pH = 6 can keep trivalent lanthanides in solution, and 10^{-7} to 10^{-5} molar Sm^{3+}, Eu^{3+}, Tb^{3+} and Dy^{3+} produce perceptible line emission with 265 nm excitation. The constitution of such solutions is not known, but may be related to the salts [92] of $MW_{10}O_{35}^{7-}$ which served as an example in Table 24 in Section 3 B. Van Uitert [93] has continued the theoretical treatment of solid tungstates.

Blasse and Bril [94] pointed out that mixed oxides containing tungsten(VI) show highly different luminescent behaviour according to whether $N = 4$ or $N = 6$ (oxygen bridges) occur. Further theoretical and experimental studies were performed by Van Oosterhout [95] and it seems that these chromophores are very promising for sensitizing lanthanide luminescence. As an example of how ionic radii can be much more important than the oxidation state for determining solid-state chemistry one can cite the lead-containing [96] perovskite $Sr_{1-x}Pb_xLaLiWO_6$, where lanthanides may occupy the sites with $N = 12$, whereas the small lithium(I) and tungsten(VI) use the octahedral sites. A comparable [96] perovskite is $Ba_{2-x}Pb_xMgWO_6$. Alberda and Blasse [97] reported energy transfer from tungstate and uranate groups substituting the tellurate in the garnet $Y_3Li_3Te_2O_{12}$ to lanthanides (such as Sm^{3+}, Eu^{3+} and Dy^{3+}) replacing the yttrium. As in YVO_4, Tb^{3+} does not fluoresce. The tungstate produces a blue-green luminescence with a broad band around 20,000 cm^{-1}.

It is well-known that both metalloids (such as boron, silicon, phosphorus and arsenic) and trivalent metals (such as chromium, iron, cobalt and lanthanides [92]) form heteropolytungstates, such as $SiW_{12}O_{40}^{4-}$ and $PW_{12}O_{40}^{3-}$. In the latter case, it is possible to obtain definite species [98, 99] by adding 1, 2, 3, 4 . . . electrons having strong dark blue colours due to high absorption bands in the near infra-red. The reduced species can at most transfer energy to lanthanide J-levels in the region below 8,000

cm^{-1}. The non-reduced species have very broad absorption bands extended over nearly the whole ultra-violet.

Reisfeld *et al.* [100] studied the phosphotungstate glasses and their energy transfer to Eu^{3+}, both in cases where the samples are heterogeneous with microcrystallites suspended in a vitreous phase, and in clear-cut homogeneous cases.

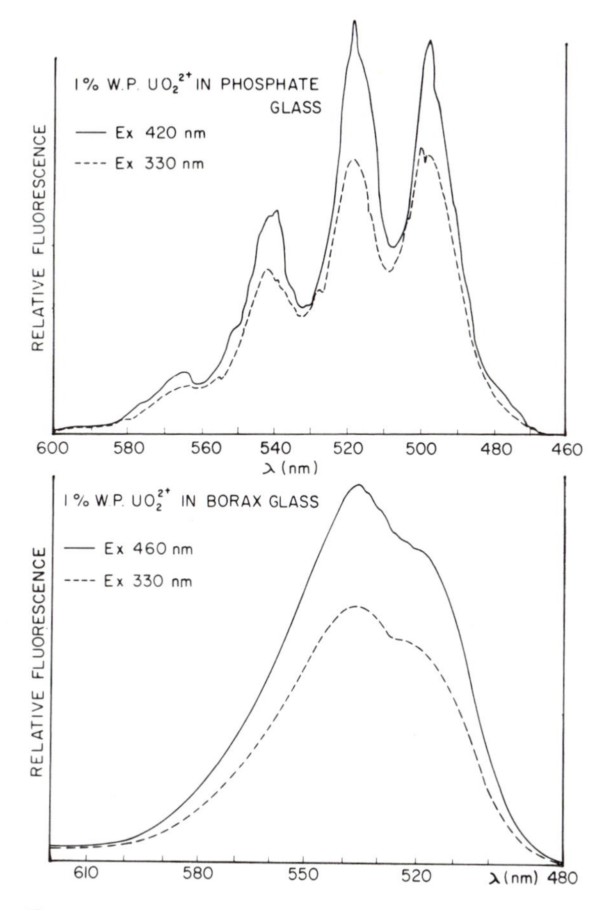

Fig. 8. Luminescence spectra (excited at 460, 420 or 330 nm) of uranyl ions in phosphate and borate glasses

d) Energy Transfer from the Uranyl Ion to Lanthanides

The first excited state of the uranyl ion UO_2^{2+} is electron transfer like the first state of tetrahedral vanadate or tungstate. However, a major difference is that even in aqueous solution, and in glasses, the uranyl ion is known since 1833 to fluoresce. In certain solids, such as $RbUO_2(O_2NO)_3$ or even $UO_2(NO_3)_2{,}6H_2O$ the quantum yield is quite high, and frequently above 0.7. As already mentioned in Section 1 D, we maintain [101] that this state (frequently having a life-time τ of about 10^{-4} sec) in contrast to most other chemical systems does not present a well-defined S, but even parity (like the groundstate) and the quantum number Ω (replacing J in linear symmetry [50]) equal to 4 or 1. This state is chemically very oxidizing [102] and is extraordinarily effective for abstracting hydrogen atoms from organic compounds, which may have important consequences for the photochemistry exploiting solar energy and the significant concentrations of uranyl carbonate complexes present in the ocean and in river and lake water [103, 104].

Phosphate glass [60, 105] containing (say, 1 weight percent) uranyl ions produce an absorption spectrum above 20,500 cm^{-1} and an emission spectrum below this wave-number, with several equidistant vibrational components, very much like the uranyl aqua ion. Such a glass shows a non-exponential decay (two representative τ values at 300 °K are 0.13 and 0.50 msec) and it is not certain whether this is due to highly non-equivalent uranyl sites in the glass or rather to the closely adjacent excited levels, which must be formed [101] by some of the 56 states belonging to the configurations consisting of one of the four electrons in the π_u orbitals (consisting almost exclusively of oxygen 2p in the LCAO approximation) being transferred to the uranium 5f shell. Professor Minas Marcantonatos, University of Geneva, kindly informed us that he has evidence for non-exponential decay of uranyl fluorescence in aqueous solution which might be connected with the formation of a quenching, dimeric excimer.

Borate and germanate glass [105] gives a broad, irregular structure of the emission band with a maximum at 18,500 and a shoulder at 19,200 cm^{-1} similar to that found by Blasse *et al.* [106, 107], in perovskites, such as $Ba_2MgW_{0.997}U_{0.003}O_6$, rather than a vibrational structure like the phosphate glass. Then, at least the crystallographic time-average picture, shows regular octahedra with $N = 6$. The excited state of these octahedral chromophores (at a wave-number slightly below the uranyl ion, hardly dependent on the equatorial ligands at long distances) corresponds to electron transfer from an orbital with odd parity, most probably t_{1u} (and not t_{1g} as in 4d and 5d group hexahalides [86]). This question [103] is intimately connected with the unexpected low wave-number 24,600 cm^{-1} of a very weak band [108, 109] of gaseous and solid UF_6, whereas we believe that the shoulder at 38,000 cm^{-1} is due to the Laporte-forbidden electron transfer band (from t_{1u}) whereas the strong maximum at 46,800 cm^{-1} most probably is from t_{1g} in good agreement with an extrapolation from the corresponding strong band of gaseous NpF_6 at 38,800 and of PuF_6 at 31,700 cm^{-1}. A straightforward extrapolation from UCl_6 apparently having its first band in the red, and Eq. (1.44), suggests the first such band of UF_6 close to 42,000 cm^{-1}. It may be noted that introduction of one oxo ligand [110] in red $UOCl_5^{2-}$ suffices to shift the first electron transfer band to 24,000 cm^{-1}, almost at the same position as UCl_6^-. According to Blasse, τ of the electron transfer band of the perovskite $Ba_2CdW_{0.997}U_{0.003}O_3$ is 0.35 msec independent of the temperature, when below 100 °K. This shows a quite low upper limit to P from Einstein's formula from 1917, Eq. (2.30).

The uranyl phosphate glass [80, 105] shows efficient energy transfer to the $^5D_0$ state of Eu^{3+} when present in concentrations around a few percent. It must be considered as a quite normal energy transfer from an excited state (at 3,000 cm^{-1} higher energy) and does not seem to have much to do with the strongly oxidizing character (high electron affinity) of the excited uranyl ion. This situation might be different in conducting inorganic glasses where the energy transfer might be accompanied by electron transfer, such as tin(IV) oxide doped with antimony(V) or glasses simultaneously containing Ce^{3+} and Eu^{3+} of which Ce^{4+} and Eu^{2+} is an optically accessible electron transfer state. Organic vitreous polymers might also be combined with characteristics allowing electron transfer, as the presence of substituted *p*-phenylenediamine $R_2NC_6H_4NR_2$, known [53] for readily loosing one electron and forming strongly coloured intermediates.

Flash photolysis experiments [102] can transform a large part of a given concentration of the uranyl ion to its relatively long-lived, first excited state. This state forms a blue solution with a strong absorption band at 17,000 cm^{-1} presumably due to an electron jumping from π_g to the vacant position in π_u.

This type of experimentation reminds one of the pulse radiolysis of aqueous solutions. Faraggi and Feder [111] used the linear accelerator at the Hebrew University to bombard aqueous lanthanide sulphate solutions with 5 MeV electrons arriving in microsecond pulses. In the case of praseodymium(III), the OH radicals liberated provided an oxidation to an aqua ion of Pr(IV) not previously known, having a broad electron transfer band at 20,000 cm^{-1} and a much stronger band at 30,000 cm^{-1} with $\epsilon \sim 1{,}000$ and $P = 0.03$. These positions are not very different from Pr(IV) in mixed oxides reviewed by Blasse [60]. It was argued [111] that Pr^{4+} aqua ions (much like Th^{4+}) are formed in acidic solution, and deprotonate with pK = 3.2 to monomeric $PrOH^{3+}$. This behaviour is in striking contrast to cerium(IV) always present (at least for a large part) as polymeric hydroxo complexes (even at pH = 0) in absence of strongly complexing anions. However, the differences may be due to the very short duration and low concentration of the Pr(IV) formed. By the way, the electron transfer spectrum of this species (lasting for about a millisecond) might suggest a sulphate complex with x_{opt} close to 3.2.

Another effect of pulse radiolysis is the liberation of hydrated electrons (existing for a much shorter time than the solvated electrons in the blue solutions of reactive metals in liquid ammonia) and Faraggi and Tendler [112] demonstrated that samarium, europium and ytterbium sulphate was reduced to divalent aqua ions. Their 4f→5d inter-shell transitions were discussed in Chapter 1 C. Sm(II) has the weaker band at 17,700 cm^{-1} followed by a complicated structure with maximum at 32,200 cm^{-1}; Eu(II) two bands at 31,300 and 38,100 cm^{-1} and Yb(II) two at 28,500 and 38,500 cm^{-1}, which were compared [112] with these ions in CaF_2, SrF_2 and BaF_2 crystals.

Since we discuss energy transfer without diffusion of matter, we do not generally mention these aspects of photochemistry and radiochemistry in this book. However, they may become significant by building up quasi-stationary concentrations of strongly coloured intermediates in lasers, and the rapid reactions may intervene in streaming solutions in liquid lasers.

e) Energy Transfer Between Lanthanides

By far the most important energy transfer between lanthanides comes from the 4f→5d excitation of cerium(III) discussed in Section 1C. Weber [113] studied $Ce_xY_{1-x}AlO_3$ showing four or five absorption bands starting at 33,000 cm^{-1}. The distribution of these transitions shows that (at least) the instantaneous picture is far removed from the cubic point-group O_h of an idealized perovskite. As in fluorite crystals the 5d one-electron energies tend to separate for unknown reasons. At 300 °K, the fluorescence of this crystal ($\tau = 1.6 \cdot 10^{-8}$ sec) corresponds to a rather broad band peaking at 28,500 cm^{-1} which was not resolved in the two components (to $^2F_{7/2}$ and to the groundstate $^2F_{5/2}$) first reported for Ce^{3+} by Kröger [114]. This may be explained by the large separation of sub-levels of $^2F_{7/2}$ found [113] between 2,085 and 3,250 cm^{-1} (gaseous Ce^{3+} 2,253 cm^{-1}). Weber found four lines in the region between 10,500 and 11,800 cm^{-1} in the crystal $Y_{0.85}Ce_{0.15}Al_{0.99}Cr_{0.01}O_3$ corresponding to emission of Ce^{3+} and Cr^{3+} pairs (^2E) terminating at $^2F_{7/2}$ sub-levels. Since the Ce^{3+} emission in $YAlO_3$ overlaps absorption bands of both Nd^{3+} and Cr^{3+}, Weber [113] studied $Y_{0.987}Ce_{0.0005}Nd_{0.0125}Al_{0.9985}Cr_{0.0015}O_3$ and the transfer from the cerium 5d to chromium ^2E is less than a-tenth as intense as from $^4T_1$ of Cr^{3+}, and the transfer from Ce5d to Nd^{3+} is even less effective. Previously, Blasse and Bril [115] observed energy transfer from Ce^{3+} to Cr^{3+}, Sm^{3+}, Eu^{3+}, Tb^{3+} and Dy^{3+} in $Y_3Al_5O_{12}$ having the first 4d → 5d transition already at 22,000 cm^{-1} and Weber [116] discussed the influence of the temperature on this state and the 4f5d states (starting at 35,000 cm^{-1}) of Pr^{3+} showing radiationless decay to the $4f^2$ state $^3P_0$.

The energy transfer from Ce^{3+} to Nd^{3+} in a lithium calcium silicate glass was studied by Jacobs *et al.* [117]. The 4f → 5d absorption starts at 29,000 cm^{-1} and the emission peaks at 28,000 cm^{-1} corresponding to a Stokes shift about 3,700 cm^{-1}. If hundred times more Nd^{3+} is present than Ce^{3+} the ^4D absorption bands around 28,000 cm^{-1} are stronger than the Ce^{3+} band at higher energy. Some of the energy transfer is radiative but a part was shown to be non-radiative.

As summarized in Table 26, Reisfeld and Hormodaly [118] studied energy transfer from Ce^{3+} to Tb^{3+} in borate glass. The emission spectrum of 0.025 weight percent Ce^{3+} alone is asymmetric and can be resolved in two Gaussian components with maxima at 25,600 and 28,100 cm^{-1}, corresponding to final the states $^2F_{7/2}$ and $^2F_{5/2}$, respectively. The natural (radiative) life-time was calculated to be $7.1 \cdot 10^{-8}$ sec, and the quantum yield close to 0.46. Whereas the luminescence of Tb^{3+} alone shows an excitation spectrum (monitored at 540 nm $^5D_4 \rightarrow {}^7F_5$) with a strong peak at 230 nm and weak peaks at 370 and 350 nm. In the presence of Ce^{3+}, an additional broad, intense peak appears in the excitation spectrum at 315 nm, corresponding to energy transfer, most probably of the dipole-dipole type.

Energy transfer from cerium to thulium was studied by Reisfeld and Eckstein [119] in borate and phosphate [123] glasses. Figure 9 gives the excitation spectra monitored at 455 nm ($^1G_4 \rightarrow {}^3H_6$) with and without cerium present. In Table 26 it can be seen that the 4f → 5d band is indeed quite effective for the energy transfer.

Though the feeding of the fluorescent level $^5D_4$ of Tb^{3+} by its own $4f^8 \rightarrow 4f^75d$ transitions in the ultra-violet are not, strictly speaking, energy transfer, it is quite interesting for our purposes that Hoshina [120] studied the detailed structure of this excitation band of Tb^{3+} incorporated in MBO_3 (M = Sc, Y, Gd, In), $Y_3Ga_5O_{12}$, $ScPO_4$, YPO_4, Sc_2O_3 and Y_2O_3. Loriers and Heindl [121] studied energy transfer from Ce^{3+} to Tb^{3+} incorporated in $Y_{1-x}Ce_xPO_4$ and $Ce_2Si_2O_7$ obtaining in both cases a very effective cathodoluminescence. Blanzat, Denis and Reisfeld [122]

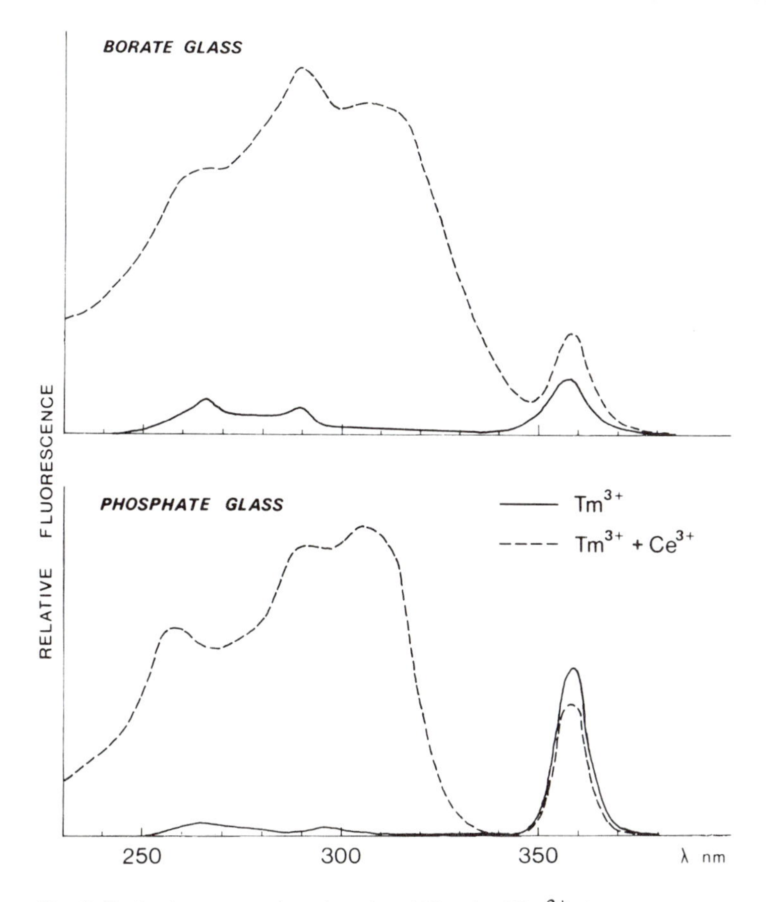

Fig. 9. Excitation spectra (monitored at 455 nm) of Tm^{3+} alone, or in simultaneous presence of Ce^{3+}, in borate and phosphate glasses
[Repetition of Figure 2 of R. Reisfeld: Structure and Bonding *30* (1976) 65 on page 75]

recently studied energy transfer from Ce^{3+} to Tb^{3+} in ultraphosphates, such as $Ce_xTb_{1-x}P_5O_{14}$ monitored at 538 nm and excited at the cerium 4f → 5d band at 307 nm, obtaining $\tau = 3.7$ msec independent of x in the whole interval from 0.1 to 0.9. The decay of pure CeP_5O_{14} cannot be described by one exponential, but rather $\tau_1 = 1.4 \cdot 10^{-8}$ sec and $4 \cdot 10^{-8}$ sec. These values decrease smoothly to $8 \cdot 10^{-9}$ sec and $3 \cdot 10^{-8}$ sec as a function of increasing terbium concentration (x down to 0.1). It may finally be mentioned that Boehm, Reisfeld and Blanzat [123] have studied energy transfer from Ce^{3+} (excited at 305 nm) to Tb^{3+} and Tm^{3+} in phosphate glass. The comparison between the results for ultraphosphate crystals and phosphate glasses shows that the energy transfer is far more rapid in glasses which may be due to a combination of the higher disorder (removing any symmetry restrictions) and the possibility of some (perhaps a minor part) of the inter-lanthanide distances being shorter in the glass (not restricted by a repeating unit cell) increasing the average value of R^{-k} for high k values.

Goldschmidt, Stein and Würzberg [124] used 25 nanosecond pulses of a frequency-quadrupled Nd laser at 265 nm on terbium dissolved as perchlorate in D_2O or in borate glass to show fluorescence from the higher $^5D_3$ (to $^7F_5$ at 418 nm) with $\tau \sim 10^{-5}$ sec and the corresponding grow-up and subsequent decay of $^5D_4$ in the glass. The effect of phonon frequences and energy gaps were thoroughly discussed.

As another type of Ce^{3+} luminescence may be mentioned, $La_{1-x}Ce_xMgAl_{11}O_{19}$ and $CeMgAl_{11}O_{19}$, studied by Sommerdijk *et al.* [125] in all cases showing an exponential decay (over at least two decades) with $\tau = 2 \cdot 10^{-8}$ sec. These crystals can be activated with Tb^{3+} or Eu^{3+}.

Energy transfer *between trivalent* lanthanides was discovered in scheelites [126] already in 1962. These crystals have highly mobile excitations migrating between the tungstate anions, like the vanadates discussed above. Energy transfer from the $^5D_4$ level of terbium to the $^5D_0$ level of europium [126] could be detected in $M_{0.5}Tb_{0.5-x}Eu_xWO_4$ (where M = Li, Na, K, Rb, Cs) in samples with x varying from 10^{-4} to 0.5. Later [127] the energy transfer from Er^{3+} to Tm^{3+} and to Ho^{3+} was found in crystals of $CaMoO_4$. A quite striking phenomenon in these scheelite-type crystals is the *concentration quenching.* There exists [128] an optimum of total yield of fluorescence somewhere between 0.01 and 0.1 europium/78 Å^3 (the unit cell volume). The energy transfer between two different trivalent lanthanides may also show the aspect of quenching, if the acceptor decays without radiation (or by infrared luminescence going undetected). Axe and Weller [129] investigated the influence of all the other lanthanides M^{3+} on the luminescence of $Y_{1.96}Eu_{0.02}M_{0.02}O_3$. The M fall in three groups: La, Sm, Gd and Yb (having no resonant levels) with a yield between 0.9 and 1 relative unit, Dy, Ho, Er and Tm (with two or three resonant levels) with yields between 0.4 and 0.7, and Nd with the very small yield 0.05 of Eu^{3+} luminescence.

Chrysochoos [130] ascribed the concentration quenching of Nd^{3+} luminescence in silicate glasses to the transformation (by energy transfer) of one excited $^4F_{3/2}$ and one groundstate $^4I_{9/2}$ to two Nd^{3+}, both in the state $^4I_{15/2}$. Chrysochoos compared his experimental results with the Inokuti-Hirayama [13] equations and concluded in a predominance of dipole-quadrupole and quadrupole-quadrupole interactions. It is worthwhile to note that the non-exponential behaviour of the decay curves (ascribed to multipolar coupling in the I–H theory) is observed only in cases where the energy is transferred from a sensitizer ion in which the absorption and emission are due to transitions within $4f^q$. When the sensitizer has allowed transitions, such as the transfer from Ce^{3+} to Nd^{3+} the decay of the sensitizer remains purely exponential, even at the beginning. This exponential behaviour in Ce^{3+} cannot be explained by the diffusion of energy in the donor system, because the relaxation in the 5d states is much faster than the transfer rate $\sim 10^7$ sec^{-1}, followed by a large Stokes shift preventing resonance between neighbour cerium ions with concomitant diffusion of energy.

Energy transfer between Gd^{3+} (the ^6I levels close to 36,000 cm^{-1}) and Sm^{3+} in borate glasses was demonstrated by Reisfeld, Greenberg and Biron [131] by the fact that excitation at 273 nm increased the intensity of the samarium fluorescence (originating from $^4G_{5/2}$ at 18,000 cm^{-1}) by an order of magnitude. Quantitative treatment indicates that the transfer actually takes place between $^6P_{7/2}$ of Gd^{3+} at 31,950 cm^{-1} to ^6P levels of Sm^{3+} around 25,000 cm^{-1} (*cf.* Table 23) and that the energy gap is bridged by phonons in the glass, and it seems that the transfer has dipole-dipole character.

In glasses similar energy transfer has been noted [132] between Sm^{3+} and Eu^{3+} and [133] between Tm^{3+} and Er^{3+}. Energy transfer between Tb^{3+} and Sm^{3+} in phosphate crystals was studied by Hirano and Shionoya [134] and in other crystals by Van Uitert, Dearborn and Rubin [135] and Blasse and Bril [136].

A fascinating example of energy transfer was described by Chun Ka Luk and Richardson [137] as complexes of (optically active) multidentate carboxylates such as malate, aspartate, glutamate, . . . containing at least two lanthanide ions, showing very effective energy transfer from $^5D_4$ of terbium to $^5D_0$ of europium. By far the strongest emission line of the latter ion is the hypersensitive (Ω_2 dependent) pseudo-quadrupolar transition $^5D_0 \rightarrow {}^7F_2$. The circular dichroism was also studied [137].

f) Processes of Infra-red to Visible Up-conversion

We already discussed in Section 2Hh that certain vitroceramic materials [138] (having heterogeneous compositions on a very small scale) are able to perform rather unexpected optical processes corresponding to a special kind of energy transfer. The donor is normally the $^2F_{5/2}$ level of Yb^{3+} having rather unique properties. Both terbium, holmium, erbium and thulium can be excited from levels below 12,000 cm^{-1} to produce visible fluorescence after being provided with 10,000 additional cm^{-1}. Auzel [139] has given an excellent review of such infra-red detectors. The processes involved can be 1) stepwise energy transfer, 2) cooperative sensitization of luminescence, and 3) cooperative luminescence.

As an example of stepwise energy transfer we may consider a system containing Er^{3+} and Yb^{3+}. The latter ion absorbs 970 nm (10,300 cm^{-1}) photons, and in phonon-assisted energy transfer an erbium ion is excited to its $^4I_{11/2}$ state. Then a second photon is absorbed by another ytterbium ion and produces $^4I_{11/2} \rightarrow {}^4F_{7/2}$ in the excited erbium by resonance energy transfer. This successive double energy transfer accounts for the fact that the excitation spectrum of such a system agrees with the reflection spectrum of Yb^{3+} and that the intensity of the resulting green emission $^4S_{3/2} \rightarrow {}^4I_{15/2}$ of erbium is proportional to the square of the flux of infra-red photons.

Two excited ($^2F_{5/2}$) ytterbium ions can cooperatively excite a third ion of another lanthanide, *e.g.* terbium, if the sum of the energies of the donor system corresponds to the excited level of the acceptor system, as has been experimentally verified by Livanova *et al.* [140] and by Ostermayer and Van Uitert [141]. It shall be further discussed in Section 5F that Yb_2O_3 and YbOF have recently been shown [142] to have weak absorption bands starting 490 nm due to simultaneous excitation of two (presumably oxygen-bridged) Yb^{3+}.

Cooperative luminescence is the opposite process of cooperative absorption and can be described as two excited ions that simultaneously make transitions downwards, emitting one photon having the sum of the two energy differences. Nakazawa and Shionoya [143] observed this process by obtaining a green emission from $YbPO_4$ exposed to a high intensity of near infra-red radiation. A similar case where an ion emits one photon and simultaneously exciting another ion was found by Feofilov and Trofimov [144] in $Yb_{2-x}Gd_xO_3$. We have already discussed that such a system might conceivably constitute a 4-level laser, whereas Gd^{3+} alone at best may work as a 3-level laser. This phenomenon occurs [145] also in $Yb_{1-x}Gd_xPO_4$. An inverse process was observed [146] in a silicate glass containing both Yb^{3+} and Tb^{3+}. An extensive theory of the processes discussed above, with experimental comparisons, is given by Kushida [17].

A related phenomenon is the conversion of a single ultra-violet photon in two visible photons [148, 149] with the result that the quantum efficiency can be higher than 1. If 0.1 percent Pr^{3+} is incorporated in YF_3, LaF_3, $NaLaF_4$ or $NaYF_4$ and excited with the mercury spectral line at 185 nm and the 4f5d states formed decay non-radiatively to the $^1S_0$ ($4f^2$) state. This system is able to generate two visible photons (the measured quantum yield in $Y_{0.99}Pr_{0.01}F_3$ is 1.4 ± 0.15) by $^1S_0 \rightarrow {}^1I_6$ followed by non-radiative decay to the closely adjacent $^3P_0$ and then another photon is emitted by transitions to one of the six J-levels of ^3H or ^3F. A condition for this cascade process is that the nephelauxetic effect (for inter-shell transitions) is sufficiently weakly pronounced for the lowest 4f5d state to be above $^1S_0$. Hence, oxidic lattices probably cannot provide this situation, in contrast to fluorides.

An excellent review on upconverting phosphors published before 1974 can be found in the Chapter [187] by Thomson. Most commercial samples consist of $La_{1-x-y}Er_xYb_yF_3$, but Auzel's vitroceramics [138] have twice as high a conversion efficiency.

g) Energy Transfer from Organic Ligands to Lanthanides

It was already noted by Weissman [150] (at a time when the J-values of the term ^5D had not been satisfactorily established) that organic compounds of Eu^{3+} enhance the emission of spectral lines in the orange, and in several cases, the quantum yield approaches 1 at liquid air temperature. This effect attracted general attention when Schimitschek and Schwarz [151] in 1962 reported liquid lasers containing such compounds. The first two years of parallel effort in a large numbers of laboratories concentrated on the neutral complexes of three β-diketonate anions $Eu(RCOCHCOR)_3$ soluble in many organic solvents. In 1961, Crosby, Whan and Alire [152] described the mechanism by which near ultra-violet radiation excites the strong singlet→singlet absorption bands of the ligand. In the case of M = La, Gd and Lu, a broad emission is observed, which may be due to fluorescence of the lowest excited singlet state, or in other cases, the longer-lived "phosphorescence" where the conjugated organic system has arrived by radiationless processes at the lowest triplet (producing spin-forbidden emission). All the cases of M = Eu and most of Tb gives line emissions due to $^5D_0 \rightarrow {}^7F_J$ levels and $^5D_4 \rightarrow {}^7F_J$, respectively. Some of the compounds of M = Sm, Dy, Tm and Yb (in the near infra-red) give lines, frequently accompanied by the broad band emission. These results were explained [152] as energy transfer from the lowest triplet state of the ligand to the partly filled 4f shell. These observations were confirmed [153, 154] and a quenching influence of air oxygen (with triplet groundstate) noted [154]. The life-times τ were determined [155–158] usually close to 0.03 msec for $Sm(^4G_{5/2})$, 0.5 msec for $Eu(^5D_0)$ but only $3 \cdot 10^{-6}$ sec for $Eu(^5D_1)$,

0.6 msec for $Tb(^5D_4)$ and 0.015 msec for $Dy(^4F_{9/2})$. For comparison it may be noted that the acetylacetonates Rh aca_3 and In aca_3 show "phosphorescence" with $\tau = 0.31$ and 0.07 sec, and these values are undoubtedly shortened by relativistic effects (spin-orbit coupling) in rhodium and indium from a value of several seconds obtaining without this "heavy-atom perturbation". The stereochemistry of the rare earth β-diketonates is far more complicated than the classical octahedral tris-complexes, and in particuliar, water molecules and organic solvents are retained tenaciously [158] and many salts have $N = 8$ in tetrakis-anions $Eu(RCOCHCOR)_4^-$, actually favouring laser performance [160]. This behaviour was utilized by Halverson, Brinen and Leto [161] to enhance luminescence by adding bulky organic substituents R and by forming synergic complexes with ligands containing extended chromophores. Actually, this approach was later used for the development of proton magnetic resonance "shift-reagents" and for biological studies [162].

Many conjugated ligands can be used for energy transfer to lanthanides. Thus, the chelates of *o*-hydroxyacetophenone [150] and of *o*-hydroxybenzophenone [163] are excellent. Carboxylates of aromatic ring systems, such as phthalates and naphthalates, were studied by Sinha, Jørgensen and Pappalardo [164] and Sm, Eu, Tb and Dy shown to produce line emission in some cases. At this time, test-tube experiments showed that methanol (freezing sharply at −100 °C) and ethanol (becoming highly viscous at −190 °C) allow energy transfer from rather unexpected species (such as fluoresceinate, umbelliferone, pyrene, and even the quinine protonated cation) to europium. The complex chemistry and spectroscopic properties of many lanthanides are discussed in the two books [165, 166] by Sinha. An interesting case of two-step energy transfer in solution was reported by Gallagher, Heller and Wasserman [167] where 4,4′-dimethoxybenzophenone does not produce line emission of europium alone, but in presence of terbium, both Tb (546 nm) and Eu (613 nm) lines are seen, demonstrating energy transfer from the Tb adduct to Eu. The 546 nm emission is quenched by the presence of Pr, Nd, Sm, Ho, Er or Tm, but not by Dy.

Perhaps one of the most spectacular cases of large organic molecules bound with an oxygen atom to the lanthanide is the hexakis (antipyrene) complex, where *antipyrene* is the trivial name for 1-phenyl-2,3-dimethyl-5-pyrazolone $C_{11}H_{12}ON_2$. In the iodide of this trivalent cation, the M–M closest approach is 13.8 Å with the result that such crystals are used for studies needing extensive magnetic dilution. Correspondingly, the τ are very large [168] and above 1 msec (up to 5 msec) for all the lanthanides studied. Mixed crystals $Tb_xM_{1-x}(antipyrene)_6I_3$ with x = 0.1, 0.25, 0.5 and 1 were studied [169] with 253.7 nm excitation, and the intensity of the terbium 548 nm emission measured as function of M, as in the case [129] of $Y_{2-x}Eu_xO_3$ the M fall in three distinct groups with respect to this quenching. If the standard intensity 1 obtains for M = Gd, Y and La, the series Sm, Pr, Yb, Ce, Dy, Eu and Tm decreases from 0.2 to 0.05, and Ho, Er and Nd show very low values [169] between 0.02 and 0.01.

Some years ago it was believed that rare earths do not normally bind nitrogen-containing ligands. This was disproved by Kononenko and Poluektov [170] extracting the neutral complexes M $phen_2sal_3$ (where phen is the bidentate heterocyclic ligand 1,10-phenanthroline and sal^- the salicylate anion) in benzene, showing strong line emission of Eu and Tb (which can be used for analytical detection) and the absorption spectra of the extracted species shows strong (about 5 times stronger than the aqua ions) hypersensitive pseudoquadrupolar transitions, for M = Nd at 583 nm, Ho at 451 and Er at 521 nm. That these complexes are not loose adducts has been abundantly shown by later crystallographic studies [171] to contain two nitrogen atoms from each 1,10-phenanthroline or 2,2'-dipyridyl among, the typically, $N = 10$ or 9 coordinated atoms. Sinha [172] also prepared complexes of the terdentate terpyridyl. In the meantime, Moeller [173] has prepared complexes of aliphatic diamines such as $M\,en_4^{3+}$ (of ethylenediamine) stable in perfectly anhydrous acetonitrile, showing that the difficulty in preparing nitrogen-containing complexes of rare earths is essentially a problem of competition of water precipitating hydroxides and transferring protons to the nitrogen-containing ligand acting as a Brønsted base (much like solutions of trivalent iron). The lanthanide complexes of nitrogen-containing heterocyclic molecules are quite effective [164, 166] for line emission of Sm, Eu, Tb and Dy. The lowest triplet of phenanthroline [174] is situated at 22,213 cm^{-1} and shows $\tau = 1.6$ sec in solution.

Much equipment [175] has been constructed for studying spectra in the visible and the ultra-violet at extremely short time scales and Wild [176, 177] discussed how such measurements can help with the understanding of triplet states in organic molecules. Though oxygen shows slighthly larger spin-orbit coupling than carbon, the life-time of traces of fluorescein dispersed in molten boric acid is about 20 seconds at room temperature, producing a beautiful green phosphorescence.

The classical cases of *inverted electron transfer* from a partly (or completely) filled d shell to low-lying empty, delocalized MO in conjugated ligands, as is known from iron(II) and copper(I) complexes of phenanthroline and dipyridyl. The first cases to be extensively classified [63] were iridium(III) complexes, having one to four pyridine molecules among the six unidentate ligands. Recently the interest in the utilization of solar energy has intensified studies [178] of the photophysics of ruthenium(II) complexes of substituted dipyridyl derivatives and their photochemical reactions with other species in solution. There are not any cases known of inverted electron transfer spectra of trivalent lanthanides, and the colourless cerium(III) and terbium(III) complexes of phenanthroline and similar ligands show that the 4f→5d transitions are not intermixed either to any significant extent. One would expect divalent lanthanides to be far more effective in this respect but the complexes formed seem to be extremely weak and lose the competition with solvent molecules and anions, besides the chemical fact that the stronger bonding in the trivalent case makes the standard oxidation potential E^0 more negative.

Many other systems than those discussed here are able to transfer energy to lanthanides. Semi-conductors with most or all of the visible region below the "absorption edge" are known, such as NaCl-type SrS containing [179] each of the lanthanides. Though the familiar line emissions can be recognized, it is surprising that even Lu^{3+} provide four emission bands, and these results illustrate the great experimental difficulties [114] in the work with these "Bologna phosphors". Actually, the (more or less) exponential decay with τ of several hours is related to the defective trapping of energy which is very slowly released, and forms an intermediate category from "thermoluminescence" where energy from cosmic rays or natural radioactivity is stored in crystals (such as fluorites) for centuries, and released by strong heating, but generally below the perceptibly red emission of standard opaque bodies above 500 °C. The classical cathodoluminescent material ZnS also produces lanthanide line fluorescence, such [180] as Nd, Tb, Dy and Tm. There is strong evidence for pairing of adjacent lanthanide ions in such materials, which are quite fascinating, and the base for the recent technology [181] of *electroluminescence.*

Energy transfer from mixed oxides of titanium(IV) might very well have been treated in 4Ec. because the strong absorption in the near ultra-violet is intimately connected with the empty 3d shell. However, it also shows aspects comparable to CaS and ZnS. An interesting case is Sm^{3+} in the perovskite $BaTiO_3$ studied by Makishima *et al.* [182] which can occupy both the barium ($N = 12$) and titanium ($N = 6$) sites, and emit from the $^4G_{5/2}$ to ^6H levels. When passing through the ferroelectric phase transition, or going toward rhombohedral and orthorhombic lattices by cooling, these lines split to a small extent. We might also have treated the work [183] on sensitized luminescence of Tb^{3+} in $Sr_{2.5}Mg_{0.3}(PO_4)_2$ in 4Eb., but the fact that Cu^+ is also a sensitizer, producing at the same time a broad band, is reminescent of typical semi-conductors [114]. Tl^+ alone produces broad emission bands at 340 and 390 nm, but if both Tb^{3+} and Tl^+ are present in this double phosphate, narrow lines originating from $^5D_3$ are visible, like in the high-energy transfer by cathodoluminescence or by short light pulses [124].

h) Energy Transfer from Lanthanides to Other Species

The main reason why this process is not so frequently observed, is that most other species have higher molar extinction ϵ at higher wave-numbers than their first fluorescent level compared with the $4f^q$ transitions. This argument would not necessarily be directed against $4f \rightarrow 5d$ transitions in Ce^{3+} nor in Sm^{2+}, Eu^{2+} and Yb^{2+}.

Actually, Blasse and Bril [115] observed efficient energy transfer from Ce^{3+} to Cr^{3+} in $Y_3Al_5O_{12}$ and in $YAl_3B_4O_{12}$. There is some evidence [51] of energy transfer from $Ho^{3+}(^5S_2)$ to Cr^{3+} in $YAlO_3$. Van der Ziel and Van Uitert [184, 185] studied $EuAl_{1-x}Cr_xO_3$ and found evidence for exchange coupling between ^2E and the ^7F *J*-levels, and for cooperative absorption [185] to the simultaneously excited state $^2E + {^7F_1}$. One would regard energy transfer to Mn^{2+} to be quite feasible, but no work seems to have been reported.

The long-lived (τ frequently 2 msec) lowest excited level $^6P_{7/2}$ of Gd^{3+} should offer excellent possibilities for energy transfer to d group ions. We have already mentioned the radiative transition [144, 145] in $Yb_{2-x}Gd_xO_3$ and in $Yb_{1-x}Gd_xPO_4$ leaving Yb^{3+} in its excited state $^2F_{5/2}$. It would be interesting for the operation of a four-level laser if one would find a d group ion replacing the Yb^{3+}. Since the pale-yellow $Cr(CN)_6^{-3}$ has ^2E at about 13,000 cm^{-1} but the first quartet $^4T_2$ at 26,600 cm^{-1} (and $^4T_1$ at 32,200 cm^{-1}), it is conceivable that $GdCr(CN)_6$, rather than just transfer energy might emit photons around 18,000 cm^{-1}. The very low intrinsic probability of such transfer must be compensated by the coupling of the sextet state of Gd^{3+} with the quartet state of Cr^{3+} to give states having less spin-forbidden transitions. The $3d^6$ system Fe^{2+} (with quintet groundstate) might offer other possibilities, either by the broad $^5T_2 \rightarrow {}^5E$ transition around 10,000 cm^{-1} (the only one having ϵ above 1) replacing the Yb^{3+} in the example above (of course, this does not give a sharp line, but it might still work like an adjustable dye laser) or one might hope that the sharp transitions [186] to triplet levels (situated in $Fe(H_2O)_6^{2+}$ at 19,800, 21,100 and 22,000 cm^{-1}) might produce perceptible radiative transitions from Gd^{3+}. If exciting sources above 40,000 cm^{-1} are available, all of these arguments apply also to energy transfer from Tb^{3+}.

If the first level ^7S of the excited configuration $3d^5 4s$ producing photo-conductivity of solid iron(II) salts [21] is not thermally excited to ^5S, it might also transfer energy to other species.

References

1. Reisfeld, R.: Structure and Bonding *22,* 123 (1975) and *30,* 65 (1976)
2. Reisfeld, R.: Structure and Bonding *13,* 53 (1973)
3. Kraevskii, S. L., Rudnitskii, Yu. P., Sverchkov, E. I.: Opt. Spectrosc. *36,* 662 (1974)
4. Dexter, D. L.: J. Chem. Phys. *21,* 836 (1953)
5. Fong, F. K., Diestler, D. J.: J. Chem. Phys. *56,* 2875 (1972)
6. Grant, W. J. C.: Phys. Rev. *B4,* 648 (1971)
7. Parke, S., Cole, E.: Phys. Chem. Glasses *12,* 125 (1971)
8. Soules, T. F., Bateman, R. L., Hewes, R. A., Kreidler, E. R.: Phys. Rev. *B7,* 1657 (1973)
9. Riseberg, L. A., Weber, M. J.: Relaxation Phenomena in Rare Earth Luminescence. In: Progress in Optics (ed. E. Wolf) Vol. 14, in press
10. Watts, R. K.: Optical Properties of Solids (ed. R. DiBartolo) New York: Plenum Press 1975
11. Auzel, F. E.: Proceedings of the IEEE *61,* 758 (1973)
12. Antipenko, B. M., Ermolaev, V. L.: UDC 535, 372 2, 758 (1968) (p. 415 in translation)
13. Inokuti, M., Hirayama, F.: J. Chem. Phys. *43,* 1978 (1965)
14. Treadaway, M. J., Powell, R. C.: Phys. Rev. *B11,* 862 (1975)
15. Blasse, G., Bril, A.: J. Chem. Phys. *47,* 1920 (1967)
16. Birgeneau, R. J., Hutchings, M. T., Baker, J. M., Riley, J. D.: J. Appl. Phys. *40,* 1070 (1969)
17. Kushida, T.: J. Phys. Soc. Japan *34,* 1318, 1327 and 1334 (1973)
18. Peacock, R. D.: Structure and Bonding *22,* 83 (1975)
19. Wybourne, B. G.: Spectroscopic Properties of Rare Earths. New York: Interscience, John Wiley 1965
20. Forster, T.: Z. Naturforsch. *4a,* 321 (1949)
21. Sulzer, J., Waidelich, W.: Physica Status Solidii *3,* 209 (1963)
22. Yokota, M., Tanimoto, O.: J. Phys. Soc. Japan *22,* 779 (1967)
23. Anderson, P. W.: Phys. Rev. *86,* 809 (1952)
24. Artamanova, M. V., Briskina, C. M., Burshtein, A. I., Zusman, L. D., Skleznev, A. G.: Sov. Phys. JEPT *35,* 457 (1972)
25. Weber, M. J.: Phys. Rev. *B4,* 2932 (1971)
26. Layne, C. B.: Ph. D. Thesis, University of California, 1975
27. Krasutsky, N., Moos, H. W.: Phys. Rev. *B8,* 1010 (1973)
28. Van der Ziel, J. P., Kopf, L., Van Uitert, L. G.: Phys. Rev. *B6,* 615 (1972)
29. Watts, R. K., Richter, H. J.: Phys. Rev. *B6,* 1584 (1972)
30. Bourcet, J. C., Fong, F. K.: J. Chem. Phys. *60,* 34 (1974)
31. Flaherty, J. M., DiBartolo, B.: Phys. Rev. *B8,* 5232 (1973)
32. Pant, T. C., Bhatt, B. C., Pant, D. D.: J. Luminescence *10,* 323 (1975)
33. Kenkre, V. M.: Phys. Rev. *B11,* 1741 (1975)
34. Golubov, S. I., Konobeev, Y. V.: Physica Status Solidii *56b,* 69 (1973)
35. Kushida, T., Takushi, E.: Phys. Rev. *12,* 824 (1975)
36. Riseberg, L. A.: Phys. Rev. *A7,* 671 (1973)
37. Motegi, N., Shionoya, S.: J. Luminescence *8,* 1 (1973)
38. Yen, W. M., Sussman, S. S., Paisner, J. A., Weber, M. J.: UCRL Report 76481, Lawrence Livermore Laboratories
39. Weber, M. J., Paisner, J. A., Sussman, S. S., Yen, W. M., Riseberg, L. A., Brecher, C.: J. Luminescence *12,* 729 (1976)
40. Selzer, P. M., Hamilton, D. S., Flach, R., Yen, W. M.: J. Luminescence *12,* 737 (1976)
41. Flach, R., Hamilton, D. S., Selzer, P. M., Yen, W. M.: Phys. Rev. Letters *35,* 1034 (1975)
42. Orbach, R.: Optical Properties of Ions in Crystals (eds. H. M. Crosswhite and H. W. Moos), p. 445. New York: Interscience 1967
43. Orbach, R.: "Relaxation and Energy Transfer" in Optical Properties of Solids (ed. R. DiBartolo). New York: Plenum Press 1975
44. Speed, A. R., Garlick, G. F. J., Hagston, W. E.: Physica Status Solidii *27a,* 477 (1975)
45. Miyakawa, T., Dexter, D. L.: Phys. Rev. *B1,* 2961 (1970)
46. Yamada, N. S., Shionoya, S., Kushida, T.: J. Phys. Soc. Japan *32,* 1577 (1972)

47. Auzel, F.: J. Luminescence *12,* 715 (1976)
48. Reisfeld, R., Eckstein, Y.: J. Non-crystalline Solids *11,* 261 (1973)
49. Komiyama, T.: J. Non-crystalline Solids *18,* 107 (1975)
50. Jǿrgensen, C. K.: Modern Aspects of Ligand Field Theory. Amsterdam: North-Holland Publishing Co. 1971
51. Weber, M. J.: J. Appl. Phys. *44,* 4058 (1973)
52. Sharp, E. J., Miller, J. E., Weber, M. J.: J. Appl. Phys. *44,* 4098 (1973)
53. Jǿrgensen, C. K.: Oxidation Numbers and Oxidation States. Berlin–Heidelberg–New York: Springer 1969
54. Orgel, L. E.: J. Chem. Phys. *23,* 1958 (1955)
55. Jǿrgensen, C. K.: Acta Chem. Scand. *11,* 53 (1957)
56. Cotton, F. A., Goodgame, D. M. L., Goodgame, M.: J. Am. Chem. Soc. *84,* 167 (1962)
57. Reisfeld, R., Greenberg, E., Jǿrgensen, C. K.: in preparation
58. Hegarty, J., Imbusch, G. F.: Colloque CNRS no. 255, Lyon 1976, (ed. F. Gaume), p. 199, Paris 1977
59. Flaherty, J. M., DiBartolo, B.: ibid., p. 191
60. Blasse, G.: Structure and Bonding *26,* 43 (1976)
61. Blasse, G., Bril, A.: J. Electrochem. Soc. *114,* 1306 (1967)
62. Jǿrgensen, C. K.: Acta Chem. Scand. *10,* 500 (1956)
63. Jǿrgensen, C. K.: Absorption Spectra and Chemical Bonding in Complexes. Oxford: Pergamon Press 1962
64. Jǿrgensen, C. K.: Adv. Chem. Phys. *5,* 33 (1963)
65. Seitz, F.: J. Chem. Phys. *6,* 150 (1938)
66. Duffy, J. A., Ingram, M. D.: Inorg. Chim. Acta *7,* 594 (1973)
67. Duffy, J. A.: Structure and Bonding *32,* 147 (1977)
68. Reisfeld, R., Boehm, L.: J. Non-crystalline Solids *17,* 209 (1975)
69. Duffy, J. A., Ingram, M. D.: J. Inorg. Nucl. Chem. *36,* 43 (1974)
70. Duffy, J. A., Ingram, M. D.: J. Non-crystalline Solids *21,* 373 (1976)
71. Duffy, J. A., Ingram, M. D.: J. Inorg. Nucl. Chem. *38,* 1831 (1976)
72. Reisfeld, R., Boehm, L., Barnett, B.: J. Solid State Chem. *15,* 140 (1975)
73. Reisfeld, R., Boehm, L.: J. Non-crystalline Solids *16,* 83 (1974)
74. Pedrini, C.: J. Chem. Phys. *63,* 3085 (1975)
75. Jacquier, B.: J. Chem. Phys. *64,* 4939 (1976)
76. Reisfeld, R., Lieblich, N.: J. Non-crystalline Solids *12,* 207 (1973)
77. Runciman, W. A., Manson, N. B., Marshall, M.: J. Luminescence *12,* 413 (1976)
78. Reisfeld, R., Morag, S.: Appl. Phys. Letters *21,* 57 (1972)
79. Reisfeld, R., Lieblich-Sofer, N.: J. Electrochem. Soc. *121,* 1338 (1974)
80. Reisfeld, R., Lieblich-Sofer, N., Boehm, L., Barnett, B.: J. Luminescence *12,* 749 (1976)
81. Reisfeld, R., Kalisky, J., Boehm, L., Blanzat, B.: Proceed. 12. Rare Earth Conference (Denver, Colorado) p. 378
82. Prener, J. S., Hanson, R. E., Williams, F. E.: J. Chem. Phys. *21,* 759 (1953)
83. Callear, A. B., Freeman, C. G.: Chem. Phys. Letters *45,* 204 (1977)
84. Schlie, L. A., Guenther, B. D., Drummond, D. L.: Chem. Phys. Letters *34,* 258 (1975)
85. Drummond, D., Schlie, L. A.: J. Chem. Phys. *65,* 3454 (1976)
86. Jǿrgensen, C. K.: Progress Inorg. Chem. *12,* 101 (1971)
87. Müller, A., Diemann, E., Jǿrgensen, C. K.: Structure and Bonding *14,* 23 (1973)
88. Draai, W. T., Blasse, G.: Physica Status Solidii *21a,* 569 (1974)
89. Blasse, G.: J. Inorg. Nucl. Chem. *33,* 4356 (1971)
90. Van Uitert, L. G.: J. Chem. Phys. *37,* 981 (1962)
91. Alberti, G., Massucci, M. A.: Analyt. Chem. *38,* 214 (1966)
92. Stillman, M. J., Thomson, A. J.: J. Chem. Soc. Dalton 1138 (1976)
93. Van Uitert, L. G.: J. Luminescence *4,* 1 (1971)
94. Blasse, G., Bril, A.: Z. physik. Chem. *57,* 187 (1968)
95. Van Oosterhout, A. B.: Luminescence of Tungstates. Thesis, University of Utrecht 1976
96. Bleijenberg, K. C., Blasse, G.: J. Luminescence *11,* 279 (1975)

97. Alberda, R. H., Blasse, G.: J. Luminescence *12,* 687 (1976)
98. Prados, R. A., Meiklejohn, P. T., Pope, M. T.: J. Am. Chem. Soc. *96,* 1261 (1974)
99. Altenau, J. J., Pope, M. T., Prados, R. A., Hyunsoo So: Inorg. Chem. *14,* 417 (1975)
100. Reisfeld, R., Mack, H., Eisenberg, A., Eckstein, Y.: J. Electrochem. *122,* 273 (1975)
101. Jørgensen, C. K., Reisfeld, R.: Chem. Phys. Letters *35,* 441 (1975)
102. Burrows, H. D., Kemp, T. J.: Chem. Soc. Rev. (London) *3,* 139 (1974)
103. Jørgensen, C. K.: Revue de chimie minérale (Paris) *14,* 127 (1977)
104. Jørgensen, C. K.: Colloquium "Procédés photographiques non-argentiques" September 1976. Association pour l'étude et la caractérisation des émulsions photo sensibles, Paris
105. Lieblich-Sofer, N., Reisfeld, R., Jørgensen, C. K.: Inorg. Chim. Acta, submitted
106. Van der Steen, A. C., De Hair, J. T. W., Blasse, G.: J. Luminescence *11,* 265 (1975)
107. DeHair, J. T. W., Blasse, G.: Chem. Phys. Letters *36,* 111 (1975)
108. McDiarmid, R.: J. Chem. Phys. *65,* 168 (1976)
109. Lewis, W. B., Asprey, L. B., Jones, L. H., McDowell, R. S., Rabideau, S. W., Zeltmann, A. H., Paine, R. T.: J. Chem. Phys. *65,* 2707 (1976)
110. Bagnall, K. W., DuPreez, J. G. H., Gellatly, B. J., Holloway, J. H.: J. Chem. Soc. Dalton 1963 (1975)
111. Faraggi, M., Feder, A.: J. Chem. Phys. *56,* 3294 (1972)
112. Faraggi, M., Tendler, Y.: J. Chem. Phys. *56,* 3287 (1972)
113. Weber, M. J.: J. Appl. Phys. *44,* 3205 (1973)
114. Kröger, F. A.: Some Aspects of the Luminescence of Solids. Amsterdam: Elsevier 1948
115. Blasse, G., Bril, A.: J. Chem. Phys. *47,* 5139 (1967)
116. Weber, M. J.: Solid State Comm. *12,* 741 (1973)
117. Jacobs, R. R., Layne, C. B., Weber, M. J., Rapp, C. F.: J. Appl. Phys. *47,* 2020 (1976)
118. Reisfeld, R., Hormodaly, J.: J. Solid State Chem. *13,* 283 (1975)
119. Reisfeld, R., Eckstein, Y.: Appl. Phys. Letters *26,* 253 (1975)
120. Hoshina, T.: J. Chem. Phys. *50,* 5158 (1969)
121. Loriers, J., Heindl, R.: Colloque CNRS No. 255, Lyon 1976 (ed. F. Gaume), p. 265. Paris 1977
122. Blanzat, B., Denis, J. P., Reisfeld, R., in preparation
123. Boehm, L., Reisfeld, R., Blanzat, B.: Chem. Phys. Letters *45,* 441 (1977)
124. Goldschmidt, C. R., Stein, G., Würzberg, E.: Chem. Phys. Letters *34,* 408 (1975)
125. Sommerdijk, J. L., Van der Does de Bye, J. A. W., Verberne, P. H. J. M.: J. Luminescence *14,* 91 (1976)
126. Van Uitert, L. G., Soden, R. R.: J. Chem. Phys. *36,* 1289 (1962)
127. Johnson, L. F., Van Uitert, L. G., Rubin, J. J., Thomas, R. A.: Phys. Rev. *133A,* 494 (1964)
128. Van Uitert, L. G., Jida, S.: J. Chem. Phys. *37,* 986 (1962)
129. Axe, J. D., Weller, P. F.: J. Chem. Phys. *40,* 3066 (1964)
130. Chrysochoos, J.: J. Chem. Phys. *61,* 4596 (1974)
131. Reisfeld, R., Greenberg, E., Biron, E.: J. Solid State Chem. *9,* 224 (1974)
132. Reisfeld, R., Boehm, L.: J. Solid State Chem. *4,* 417 (1972)
133. Reisfeld, R., Eckstein, Y.: J. Non-crystalline Solids *9,* 152 (1972)
134. Hirano, H., Shionoya, S.: J. Phys. Soc. Japan *30,* 1343 (1971)
135. Van Uitert, L. G., Dearborn, E. F., Rubin, J. J.: J. Chem. Phys. *45,* 1578 (1966)
136. Blasse, G., Bril, A.: J. Chem. Phys. *47,* 1920 (1967)
137. Luk, C. A., Richardson, F. S.: J. Am. Chem. Soc. *97,* 6666 (1975)
138. Auzel, F., Pecile, D., Morin, D.: J. Electrochem. Soc. *122,* 101 (1975)
139. Auzel, F. E.: Proceed. IEEE *61,* 758 (1973)
140. Livanova, L. D., Satokulov, I. G., Stolov, A. L.: Fiz. Tverdgo Tela *11,* 918 (1969); english translation: Soviet Physics, Solid State *11,* 750 (1969)
141. Ostermayer, F. W., Van Uitert, L. G.: Phys. Rev. *B1,* 4208 (1970)
142. Schugar, H. J., Solomon, E. I., Cleveland, W. L., Goodman, L.: J. Am. Chem. Soc. *97,* 6442 (1975)
143. Nakazawa, E., Shionoya, S.: Phys. Rev. Letters *25,* 1710 (1970)
144. Feofilov, P. P., Trofimov, A. K.: Optika i Spektroskopia *27,* 538 (1969)

145. Nakazawa, E.: J. Luminescence *12,* 675 (1976)
146. Bilak, V. I., Zverev, G. M., Karapetyan, G. O., Onischenko, A. M.: Zh. Exper. Theor. Fiz. *14,* 301 (1970); english translation: Soviet Physics JETP Letters *14,* 199 (1971)
147. Kumar, R.: Chem. Phys. Letters *45,* 121 (1977)
148. Piper, N. W., DeLuca, J. A., Ham, F. S.: J. Luminescence *8,* 344 (1974)
149. Sommerdijk, J. L., Bril, A., Jager, A. W.: J. Luminescence *8,* 341 (1974)
150. Weissman, S. I.: J. Chem. Phys. *10,* 214 (1942)
151. Schimitschek, E. J., Schwarz, E. G. K.: Nature *196,* 832 (1962)
152. Crosby, G. A,, Whan, R. E., Alire, R. M.: J. Chem. Phys. *34,* 743 (1961)
153. Whan, R. E., Crosby, G. A.: J. Mol. Spectr. *8,* 315 (1962)
154. El-Sayed, M. A., Bhaumik, M. L.: J. Chem. Phys. *39,* 2391 (1963)
155. Nardi, E., Yatsiv, S.: J. Chem. Phys. *37,* 2333 (1962)
156. Freeman, J. J., Crosby, G. A.: J. Phys. Chem. *67,* 2717 (1963)
157. Bhaumik, M. L., Lyons, H., Fletcher, P. C.: J. Chem. Phys. *38,* 568 (1963)
158. Samelson, H., Lempicki, A.: J. Chem. Phys. *39,* 110 (1963)
159. De-Armond, M. K., Hillis, J. E.: J. Chem. Phys. *49,* 466 (1968)
160. Brecher, C., Lempicki, A., Samelson, H.: J. Chem. Phys. *41,* 279 (1964)
161. Halverson, F., Brinen, J. S., Leto, J. R.: J. Chem. Phys. *41,* 157 and 2752 (1964)
162. Nieboer, E.: Structure and Bonding *22,* 1 (1975)
163. Crosby, G. A., Whan, R. E., Freeman, J. J.: J. Phys. Chem. *66,* 2493 (1962)
164. Sinha, S. P., Jørgensen, C. K., Pappalardo, R.: Z. Naturforsch. *19a,* 434 (1964)
165. Sinha, S. P.: Complexes of the Rare Earths. Oxford: Pergamon Press 1966
166. Sinha, S. P.: Europium (in English). Berlin–Heidelberg–New York: Springer 1967
167. Gallagher, P. K., Heller, A., Wasserman, E.: J. Chem. Phys. *41,* 3921 (1964)
168. Peterson, G. E., Bridenbaugh, P. M.: J. Opt. Soc. Amer. *53,* 494 (1963)
169. Van Uitert, L. G., Soden, R. R.: J. Chem. Phys. *36,* 1797 (1962)
170. Kononenko, L. I., Poluektov, N. S.: Russ. J. Inorg. Chem. *7,* 965 (1962)
171. Sinha, S. P.: Structure and Bonding *25,* 67 (1976)
172. Sinha, S. P.: Z. Naturforsch. *20a,* 164 (1965)
173. Forsberg, J. H., Moeller, T.: Inorg. Chem. *8,* 883 (1969)
174. Brinen, J. S., Rosebrook, D. D., Hirt, R. C.: J. Phys. Chem. *67,* 2651 (1963)
175. Tuan, Vo Dinh; Wild, P.: Appl. Optics *12,* 1286 (1973) and *13,* 2899 (1974)
176. Wild, U. P.: Chimia (Aarau) *27,* 421 (1973) and *30,* 382 (1976)
177. Wild, U. P.: Topics in Current Chemistry *55,* 1 (1975)
178. Lin, C. T., Böttcher, W., Chou, M., Creutz, C., Sutin, N.: J. Am. Chem. Soc. *98,* 6536 (1976)
179. Keller, S. P.: J. Chem. Phys. *29,* 180 (1958)
180. Anderson, W. W., Razi, S., Walsh, D. J.: J. Chem. Phys. *43,* 1153 (1965)
181. Degenhardt, H.: Naturwiss. *63,* 544 (1976)
182. Makishima, S., Yamamoto, H., Tomotsu, T., Shionoya, S.: J. Phys. Soc. Japan *20,* 2147 (1965)
183. Mizuno, H., Masuda, M.: Bull. Chem. Soc. Japan *37,* 1239 (1964)
184. Van der Ziel, J. P., Van Uitert, L. G.: Phys. Rev. *180,* 343 (1969)
185. Van der Ziel, J. P., Van Uitert, L. G.: Phys. Rev. *B8,* 1889 (1973)
186. Jørgensen, C. K.: Acta Chem. Scand. *8,* 1502 (1954)
187. Thomson, A. J.: Electronic Structure and Magnetism of Inorganic Compounds *4,* p. 149. London: Chemical Society Specialist Periodical Reports 1976
188. Holstein, T., Lyo, S. K., Orbach, R.: Colloque CNRS No. 255, Lyon 1976, (ed. F. Gaume) p. 185, Paris 1977
189. Reisfeld, R.: *ibid.,* p. 149

5. Applications and Suggestions

(References to this chapter are found p. 218)

The two subjects of this book are on the one hand the (trivalent) lanthanides containing a partly filled 4f shell and emitting narrow spectral lines (to a certain extent imitating the behaviour of monatomic entities in a dilute gas) and on the other hand the stimulated emission of coherent laser light. Obviously, we can refer only very superficially to laser technology (giving a few references), but we want to devote a few pages to the discussion of future possibilities, especially those related to very high energy densities and to communications at very long distances. Finally, we shall make a few remarks about the distinct characteristics of spectroscopy of condensed matter (as compared with isolated atoms).

A. The Standard Emission Spectrum of Opaque Objects (in Atomic Units)

We mentioned at the end of Chapter 3 that a standard continuous spectrum is emitted by opaque objects (traditionally called "black bodies"). It is worthwhile repeating that opaqueness is not a distinct property of a given compound (as for instance the molar extinction coefficient ϵ) but rather dependent on the thickness of the sample and by geometrical conditions (such as the laboratory device consisting of a hole in a uniformly heated crucible). Empty space in equilibrium with objects heated to the absolute temperature T °kelvin shows an energy density (in erg/cm^3 or joule/m^3) of electromagnetic radiation with the frequency (in hertz = sec^{-1}) between ν and $(\nu + d\nu)$

$$\rho(\nu) = 8\pi h\nu^3 d\nu / c^3 \{\exp(h\nu/kT) - 1\} \qquad (5.1)$$

which is (4/c) times larger than the total emissive effect (power) from the surface of the opaque object (in units such as erg/cm^2 sec) first suggested by Planck. The latter expression is equivalent to Eq. (3.13) because $\nu = c/\lambda$ with the consequence that $d\nu = -c d\lambda/\lambda^2$. The photons do not have half-numbered spin (unlike electrons, neutrinos, protons, neutrons, . . .) and hence do not follow Fermi-Dirac statistics with the exclusion principle, but rather Bose-Einstein statistics where each state can be occupied by many particles. Verifiable consequences are found in the behaviour of liquid helium 3 and helium 4 which follow the two kinds of statistics. Actually, the bracket in the denominator of Eq. (5.1) is the reciprocal value of the population number of the Bose-Einstein states, of which the number (as function of the frequency) is proportional to $\nu^2 d\nu$.

The integral of Eq. (5.1) over all ν values gives the total energy density of electromagnetic radiation

$$\rho_{tot} = \left(\frac{4}{c}\right) \sigma_{S.B.} \, T^4 \qquad (5.2)$$

where the Stefan-Boltzmann constant $\sigma_{S.B.}$ given by

$$\sigma_{S.B.} = 2\,\pi^5 k^4/15\,c^2 h^3 \qquad (5.3)$$

is $5.6696 \cdot 10^{-5}$ erg cm^{-2} sec^{-1} $(°K)^{-4}$ and the Boltzmann constant $k = 1.3806 \cdot 10^{-16}$ erg/(°K). Since the gas constant R_0 is the product of Avogadro's number $N_0 = 6.022 \cdot 10^{23}$ and k, the heat content (at constant volume) of normal matter is a small multiple (3/2 for monatomic gases, 5/2 for diatomic gases, to which a T-dependent vibrational contribution is added, and 3 for solids following the law of Dulong and Petit) of nR_0T, where n is the number of moles. The heat capacity of most samples does not vary greatly with T. This is not at all true for the electromagnetic field present also in empty space, since the differential quotient of Eq. (5.2) with respect to T is a constant ($16\,\sigma_{S.B.}/c$) times T^3. At temperatures such as T = 1,000 °K, a gas does not need to be diluted below the feasible limits of vacuum technology to contain less kinetic energy of the gas atoms than stocked in the "black-body" radiation $7.56 \cdot 10^{-3}$ erg/cm^3. Indeed, the two energy densities are identical for $1.6 \cdot 10^{-10}$ mole/cm^3 of a monatomic gas, corresponding to a pressure of 0.01 torr (a torr is very nearly 1 mm Hg). There are important astrophysical and cosmological consequences in the fact that a dilute gas is carrier of two energy densities, one of which is proportional to T^4 and independent of the density of matter. This was particularly true for the first hours after the singularity of extremely high density, from which the Universe developed some 10^{10} years ago. It may be noted that certain crystals cooled to such a T that kT is well below nearly all the phonon frequences also may show heat capacities rather varying with the second or third power of T.

McWeeny [1] recently discussed the atomic units of many physical quantities, obtained by using 1 bohr as unit of length, 1 hartree as unit of energy, the electric charge + e of a proton, and $2.4189 \cdot 10^{-17}$ sec as unit of time. Some of the atomic units derived are quite comparable with conventional Laboratory conditions, such as $6.62 \cdot 10^{-3}$ ampère for electric current, while other atomic units are quite formidable, such as $5.142 \cdot 10^9$ volt/cm for electric field strength. Previously, there has been no attempt to introduce atomic units of molarity nor of several other quantities closer related to classical chemistry than to chemical physics.

Since the atomic unit of volume is a cubic bohr, the mole N_0 per 1,000 cm^{-3} (the definition of the present-day litre) corresponds to $8.924 \cdot 10^{-5}$ particles per cubic bohr, and the *atomic unity of molarity* is 11,206 M. The general tendency for internuclear distances to be some 4 to 5 bohr restrict molarities at normal pressure from exceeding one or two percent of an atomic unit. Thus, metallic aluminium, silver, gold and the compounds LiF and NiO are all between 98 and 100 M, metallic sodium 42 M, NaCl 37 M and ThO_2 38 M as compared with liquid water 55 M and methanol 25 M. The highest molarities known do not correspond to particularly high densities, but to low atomic weights. Thus, metallic beryllium is 205 M and diamond 293 M, and the highest molarity of a compound may be 120 M of BeO.

The only reasonable definition of the *atomic unit of temperature* T_a is, that kT_a is one hartree, and hence $T_a = 315{,}700$ °K. It is worth noting that room-temperature is a tiny bit less than one thousandth atomics unit, and a tungsten lamp with a filament heated to 3,157 °K one percent. On the otherhand, the interior of the Sun (where the thermonuclear transmutation of hydrogen to helium takes place) has about 50 units of atomic temperature. Seen from this point of view. Eq. (5.2) can be re-written (1 hartree/bohr^3 = $2.94 \cdot 10^{14}$ erg/cm^3):

$$\rho_{tot} = 2.57 \cdot 10^{-9} (T/T_a)^4 \text{ hartree/bohr}^3 \qquad (5.4)$$

where the dimension-less constant $2.57 \cdot 10^{-9}$ is only $2.88 \cdot 10^{-5}$ times the reciprocal value of the atomic unity of molarity. It is not to be expected that an easily identified constant would be obtained in Eq. (5.4) because such a relation would imply connections with the unit of atomic weight, one-twelfth of carbon 12. Whereas, the energy m_0c^2 corresponding to the rest-mass m_0 of an electron is exactly 137.036^2 hartree ($c = 137.036 \pm 0.001$ atomic units of speed), then one would rather expect $m_0 = 5.486 \cdot 10^{-4}$ unit of atomic weight to enter Eq. (5.4).

B. Induced Thermonuclear Reactions

Though considerable development took place in the study and utilization of fission of uranium and (synthetic) trans-uranium elements between 1939 and 1945, there is no doubt that the most frequent nuclear reaction in the present Universe is the fusion of hydrogen to helium 4, and in spite of a certain heat evolution in the Earth's crust due to the spontaneous radioactivity of potassium 40, thorium 232 and uranium 238 and their descendents, all fossil fuel and all meteorological phemenomena (excluding tidal waves) derive their energy from thermonuclear reactions in the Sun. Though certain prodigalous stars transmute 10^{-8} of their hydrogen atoms per year, the energy production $4 \cdot 10^{26}$ watt (joule/sec = 10^7 erg/sec) of the Sun corresponds to a transmutation rate of $9 \cdot 10^{-12}$ per year, since 1 g of hydrogen produces $7 \cdot 10^{11}$ joule by the formation of helium, and the solar mass is $2 \cdot 10^{33}$ g. This fortunate aspect of a reaction slowed down (by what would be an enormous activation energy in Arrhenius' equation) also suggests that it is rather difficult to maintain a quasi-stationary, non-explosive thermonuclear reaction in small volumes. The typical stars have masses of 10^5 to 10^6 times of the Earth, and most physicists and engineers have argued that the most plausible technique of confinement are specific magnetic fields. There is general agreement that isotopes other than ordinary hydrogen 1 are more suitable though the most profitable end product of fusion is helium 4 as in the stars. One of the reasons is the advantage of forming two nuclei (rather than a simple absorption) as in chemical reactions in gases. Indeed, the presence of approximately 10^{-4} deuterium in the ocean and lake water and lithium 6 and lithium 7 in the Earth's crust is surprising enough because their concentration is far lower in stellar atmospheres. There are several reasons for believing that they are primordial in the sense that they never have been incorporated in a star where they would have been transmuted rapidly. Fortunately for the inorganic chemist, the elements from $Z = 30$ to 92 have roughly constant abundances (10^{-4} to 10^{-7}) in the Earth's crust (suggesting an origin of at least 0.2 percent as residues of supernova explosions) whereas, the logarithm of the abundance of elements in normal stars decreases roughly linearly with Z from carbon and oxygen on [2–4]. Another isotope of great interest for relatively easy thermonuclear reactions is tritium (hydrogen 3) emitting an electron and a neutrino with a half-life of 12 years, which can be made in large quantities by irradiating lithium 6 in a nuclear reactor with a high flux of neutrons. Besides the attempts at confined plasma conservation, there has been recently some experimentation with injection of energy more closely related to beams impinging on a target, as in Van der Graaf linear accellerators and cyclotrons.

The Livermore laboratories [5, 6] in California have developed neodymium glass lasers providing about $2 \cdot 10^{12}$ watt. However, this cannot be achieved in a continuous operation but in pulses lasting about 10^{-9} sec (a nanosecond) every minute or so, and each containing an energy of 2,000 joule. One of the technological problems with such short lengths of time is that light only propagates 30 cm in air or empty space during a nanosecond, and $(30/n)$ cm in a medium with the refractive index n. Hence, designs of beam-splitting and coincidence of two beams have to be worked out carefully. The energy content in such a pulse corresponds to about 10^4 joule/cm^3 if the cross-section is 1 mm^2, and hundred times more (corresponding to 0.03 atomic unit or $T = 45\ T_a$ in Eq. (5.4)) if the cross-section of the beam is 100 times smaller. The corresponding flux in 1 mm^2 is $2 \cdot 10^{14}$ watt/cm^2. The original experiments were conducted with strongly cooled pellets of the solid consisting of the molecule DT, and where the intended nuclear reaction between deuterium and tritium forms a neutron and a helium 4 nucleus. The best technique for a high thermonuclear yield is to concentrate simultaneously, say ten, laser beams from all directions on the pellet, producing an implosion shock wave from the surface inwards. Though most of the pellet evaporates too rapidly, perhaps one percent is trapped by the implosion shock wave and allows the emission of, say 10^9 neutrons from 10^{-14} g reacted material, it is not possible to ascribe a definite temperature to the hot centre. A more advantageous procedure is to fill microspheres (glass or plastic bubbles) with gaseous DT. In this case the thickness of the surface of the microspheres is a very critical parameter, too thin bubbles rapidly loosing their gas content and too thick layers absorbing too much of the energy. It is conceivable that a minor but significant amelioration of this technique would be to fill deuterated paraffin (polyethylene) spheres with tritium T_2 gas, or the other way round, microspheres of tritiated polymers filled with D_2 thus diminuishing the danger of escape of the biologically rather objectionable tritium. Nuclear explosions initiated by fission bombs can occur in crystalline (102 M) lithium deuteride or tritide, and it is conceivable that such solid particles might replace the microspheres.

Obviously, the final goal of the Livermore experiments is to produce more energy from the thermonuclear fusion than is spent on the laser excitation. This goal still seems very far off, and one of the aspects which one may hope to improve would be the use of the fusion energy in the plasma to obtain directly the light needed to pump the laser rather than to adapt the ackward method of first making electric current for this purpose. We shall discuss in the next section a few observations of laser operation (with very low yield) from rapid fission products. At the moment one of the major practical difficulties of the Livermore project is an unexpected variation (10^{-6} to 10^{-5} at 10^{10} W/cm^2) of the refractive index n of the neodymium glass [7] in the presence of very high light intensities. The consequence is a tendency of self-focusing of the beam during the operation, distorting and melting the glass rods in an unpredictable way. An empirical search is at present being conducted for glasses having the smallest possible coefficients of this undesirable second-order effect. A system SHIVA is now designed [8] for operation at the end of 1977, it contains 20 laser beams delivering once every second more than 10^4 joule in less than 10^{-9} sec, corresponding to a focusable power $2 \cdot 10^{13}$ W (20 terawatt).

It is not perfectly clear whether the heating of the pellets or microspheres containing deuterium and tritium is a genuine effect of non-linear optics. The choice of near infra-red radiation from the neodymium glass is mainly determined by the optimal characteristics of this four-level laser, and not by any property of the colourless target. At this point, it is worth speculating about several extreme properties of the electromagnetic field. For instance, photon energies above 2 (137.036^2) hartree are capable of producing an electron and a positron. This exchange between two electrically charged particles, each possessing the same rest-mass m_0, and a photon (moving with the speed c relative to any observer, disregarding effects of strong fields and the refractive index n of condensed matter) is reversible, since the electron and the positron can also "annihilate" with formation of one (or more) photons. The temperature where λ_{max} of Eq. (3.14) would be comparable to this photon energy is about 7,500 T_a or $2.36 \cdot 10^9$ °K. Under such circumstances the thermonuclear reactions would rapidly establish thermodynamic equilibrium for all elements (not only hydrogen and helium) with high proportions of iron, cobalt and nickel being formed. However, an energy density of the "black-body" radiation Eq. (5.4) of 137^2 hartree/bohr^3 would already occur for $T \sim 200\ T_a$ or about $6 \cdot 10^7$ °K which is within the astrophysical range of stellar interiors. The density of photons between ν and $(\nu + d\nu)$ can be obtained by dividing Eq. (5.1) by the photon energy $h\nu$. If we are going to discuss a "dimerization" of two identical photons it is necessary to take into account the width acceptable as a physical replacement of $d\nu$. The situation at $T = 200\ T_a$, where the average photon energy is some 500 hartree, corresponds to a density of about 40 photons (of whatever energy) per cubic bohr[1]. Though we are singularly lacking in

---

[1] As discussed in Section 5 E, the spherical average of stellar light in the part of the Galaxy where the Sun is situated, has the order of magnitude 10^{10} photons/cm^2 sec with the result that the photon density is roughly one per cm^3. The corresponding equivalent mass according to Einstein is 10^{-33} g/cm^3. For practical purposes space is an irreversible sink for light; very little gets re-absorbed under existing conditions. Hence, the paradox first pointed out by Olbers that the night sky has a luminosity 10^{13} times weaker than the average stellar surface can be explained away [33] as the stars have been shining 10^{10} years only and not the 10^{23} years needed to establish equilibrium (which could never happen, because even a complete annihilation of matter, disregarding the baryon conservation rule, would only supply light for 10^{13} years at the present rate).

experimental information about such conditions, it is qualitatively obvious that the multi-photon processes are highly enhanced. Whereas, the second-order effects studied in non-linear optics have exponentially decreasing probabilities of involving 3, 4, . . . rather than 2 photons, it is evident that there exists a limit of high photon densities where the probability of multi-photon processes is divergent in the sense that it is not less likely that 100 photons "collide" rather than 10. The corresponding up-grading of infra-red or visible radiation might have unexpected effects, even in the regime of thermonuclear reactions. A density of several photons within a sphere having the corresponding wave-length λ as diameter (λ is (2π) 137 bohr or 456 Å, divided by the photon energy in hartree) is quite outside the domain of conventional linear optics, and it is quite significant that the lowest wave-length of a stimulated emission known at present [9] is the $\lambda = 182$ Å (corresponding to 5 rydberg) $3d \rightarrow 2p$ transition (for this one-electron system, the energy does not depend on l, but only on the principal quantum number) in C^{+5}. The population inversion started 10^{-9} sec after a 0.5 joule pulse (lasting $1.4 \cdot 10^{-10}$ sec) from a neodymium(III) glass laser. It is conceivable that other four-level lasers may become technologically prominent in time, and we discussed in Chapter 4 why, for instance energy transfer from the lowest excited level $^6P_{7/2}$ of gadolinium(III) at 32,100 cm^{-1} above the groundstate to $^2F_{5/2}$ of ytterbium(III) at 10,100 cm^{-1} above its groundstate may yield 22,000 cm^{-1} laser radiation.

C. Cathodoluminescence and Energy Dissipation from Rapid Elementary Particles and Fission Products

In the opinion of Lavoisier the *elements* cannot be transmuted and have a definite atomic weight (which might very well be an infinite decimal fraction like π) whereas, the *principles* do not possess mass and are essentially present everywhere. Only a few principles were recognized, such as light and heat, and the question whether there are two electric principles or only an excess or deficit of one electricity was still actively debated in the time of Benjamin Franklin. Actually, most electric currents are due to the motion of electrons, but it is known from electrolysis of salt solutions and molten salts, and from the conductivity of oxide ions in Nernst's lamp that both positively and negatively charged ions may contribute. The correct identification by Lavoisier of metallic elements and their oxides (and the prediction of magnesia, lime, strontia and baryta as oxides of metals not yet isolated) is generally taken as the break-down of the phlogiston theory. However, this break-down was by far not as complete as supposed by most text-books, and in particular, the idea that redox (reduction-oxidation) reactions [10] are the transfer of a principle from the reducing to the oxidizing agent has been fully verified. Though there had been some suspicion that phlogiston might have negative mass, and though the Swedish chemist Berzelius had suggested that all chemical bonding is connected with the attraction between positive and negative charges it came as a general surprise that the principle of redox reactions at the same time is the negatively charged elementary particle (the lightest and the only non-radioactive and stable in the presence of positive nuclei), the electron.

Though the electron certainly is a principle in the sense of being present in all material samples (and this characteristic has been accentuated in quantum mechanics, where the wave-functions are anti-symmetric in such a way as to make all electrons indiscernible) it has a definite rest-mass m_0 though it is 1836 times smaller than a hydrogen atom. The first studies of the electron in empty space were performed by Crookes, showing that potential differences above 10,000 V producing sparks at atmospheric pressure (as well as the diffuse discharges characterizing neon lamps and other Geissler tubes at 100 times lower pressures) are able to emit free electrons from the negatively charged electrode below 10^{-6} atm. The speed of these electrons is considerable, the non-relativistic expression (well below c) is the square-root of the kinetic energy in eV, multiplied by 593.2 km/sec. When they hit the walls of the Crookes tube they emit X-rays, as found later by Röntgen, both a continuous spectrum of "Bremsstrahlung" showing a maximum between 0.3 and 0.6 times the kinetic energy of the electrons, and sharp emission lines characterizing the elements in the wall or "anti-cathode". Electromagnetic radiation is also emitted of much lower frequences and in particuliar, visible luminescence. Crookes demonstrated that colourless materials (such as CaO and $CaSO_4$) emit broad very weak bands when bombarded with rapid electrons. When small traces of trivalent lanthanides are present they emit narrow, intense bands corresponding to transitions between different J-levels of $4f^q$. Possibly, because of energy transfer processes, it is not possible to draw valid conclusions about relative concentrations since very small amounts may dominate the emission spectrum of the cathodoluminescence. In fact, Urbain [11] demonstrated in 1909 that two of the elements (victorium and incognitum) postulated by Crookes to be obtained by fractional crystallization of rare earths represent two modes of emission of trivalent europium discovered in 1901 by Demarcay. Though the red lines of such cathodoluminescence are the usual transitions from $^5D_0$ to $^7F_1$ and $^7F_2$, known from conventional luminescence, many of the green and blue lines which occur in cathodoluminescence are only due to transitions from several higher J-levels of the terms ^5D, ^5L, ^5G, . . . Fifty years later the red cathodoluminescence of Eu(III) proved to be of great technological interest for colour television, first in the orthovanadate [12] $Y_{1-x}Eu_xVO_4$ and later [13] in the oxysulphide $Y_{2-x}Eu_xO_2S$ producing a deeper saturated red light because of two combined effects: the Judd-Ofelt parameter Ω_2 is larger, corresponding to a more intense pseudoquadrupolar emission $^5D_0 - {}^7F_2$ in the more polarizable oxysulphide (isomorphous with A-La_2O_3) where there is a (weak) nephelauxetic shift relative to the vanadate. The pale blue colour of so-called black-and-white television is due to the cathodoluminescence of ZnS, and it is possible to obtain other colours corresponding to relatively broad emission bands, such as green (when incorporating traces of copper and chloride in zinc sulphide) or yellow or dull orange from other semi-conducting crystals, such as $ZnS_{1-x}Se_x$ or $CdS_{1-x}Se_x$. Many materials emit in the near infra-red.

The physical mechanism behind the cathodoluminescence is not at all well understood. It is beyond doubt that the rapid electron produces a large number of secondary electrons during the passage in the solid. This situation influenced the early studies of radioactivity where it was established that each α-particle is a helium 4 nucleus carrying two positive charges, but that the number of negative charges transported by each β-particle was by one to two orders of magnitude too high. The additional electrons were for some time called δ-particles. High-energy photons (X-rays) and rapid electrons ionize inner shells (with high ionization energy I) with relatively higher probability [14] than loosely bound electrons. The great difference between nuclei and electronic systems is that a given nucleus has an affinity with an additional neutron close to 8 MeV (except deuterium and beryllium 9 which can photo-dissociate with photons having a quarter of this energy) whereas, I of electrons vary by a factor above 10^4, and I of each of the two strongest bound 1 s electrons in a typical atom is close to Z^2 rydberg to be compared with the energy of formation $-Z^{2.4}$ rydberg of the atom from the nucleus and Z electrons at mutually large distances. The main origin of the secondary electrons is the *Auger process* where an atom, initially ionized in an inner shell, changes to another state where an electron is emitted with a definite kinetic energy which is determined by the energy difference between the final and the initial state and where two electrons are lacking in the same or two different shells (both having I much lower than I of the shell initially lacking an electron). Hence, it is possible to study the terms distributed roughly in Russell-Saunders coupling of the partly filled shell $3d^8$ not only in the krypton atom [15] but also in metallic copper and zinc, and in copper(I) compounds [16]. Whereas, Auger spectrometry is traditionally performed on gaseous or metallic samples, a photo-electron spectrometer using 1,253.6 or 1,486.6 eV photons can be used [17, 18] for the detection of Auger signals of isolating compounds (especially of elements having I of a given inner shell slightly below the photon energy) and interesting information [16, 19, 20] can be obtained about the chemical bonding. Coming back to cathodoluminescence, it seems that the major part of the energy transfer to luminophores is done by electrons with kinetic energies in moderate excess of the lowest I value, say between 100 and 20 eV. However, it is quite conceivable that even slower electrons during the process of capture in the solid excite states, in analogy to the optically accessible electron transfer states, discussed by Blasse [21], are intermediates in the luminescence produced by ultraviolet radiation. It would be interesting to investigate whether oxidizing species with high electron affinity of the partly filled shell show a high propensity for cathodoluminescence. It may also be mentioned that the luminescence induced by strong radioactivity was first noted by Maria Sklodowska and Pierre Curie in radium salts (presumably due to the anions) and later recognized [22] in curium(III) compounds as the transition at 16,900 cm^{-1} between the two lowest levels (both having $J = 7/2$) of $5f^7$. The α-particles lose most of their energy by multiple ionization processes at the final part of their trajectory, and again, their energy transfer seems to be connected with secondary Auger electrons. The glimpse of green light produced by an individual α-particle in zinc sulphide or in willemite Zn_2SiO_4 activated by traces of manganese(II) had a great impact on early nuclear physics. For completeness we may also mention the blue light (with a continuous spectrum rising in the ultra-violet) emitted by charged particles moving with a velocity higher than that of light in a medium with the refractive index n (and hence in the interval between (c/n) and c) discovered by Cerenkov. Though the yield of these "optical shock-waves" is low, it is conceivable that it may contribute to luminescence by energy transfer.

Several people have elaborated on the idea that high levels of radioactivity or a quasi-stationary nuclear reaction may sustain laser action. One such case would be the luminescence of the lowest (electron transfer) state of the uranyl ion in a homogeneous reactor [22] of uranyl sulphate (enriched in uranium 235) in deuterium oxide. However, the first experimental confirmation of a laser pumped with the kinetic energy of fission fragments involved [23] the far infra-red vibrational lines at about 1,900 cm^{-1} in gaseous CO. Later the pumping of a helium-xenon laser emitting at 3,508 nm was also reported [24]. As discussed above, the use of lasers in a fusion reactor would be helped considerably if a feed-back of energy to the laser were feasible. This is one reason why the coupling between high-intensity electromagnetic fields and plasma containing rapidly moving electrons and positive ions is not exclusively of academic interest but may enhance the available energy supply.

D. Communications and Holography

The use of mirrors to deflect the parallel light beams from the Sun as a transmitted signal (optical telegraphy) has been known for millenaries, one of its inherent technical difficulties is the curvature of the Earth. In cloudy weather or at night, the use of lamps as light sources for this purpose meets the major difficulty that the light intensity (neglecting additional effects of absorption and scattering) distributes equally on a sphere with surface $4\pi R^2$ at a distance R from the light source at the centre, with the result that "passive" (non-directive) reflection (like solar light reflected from the planets at larger distances than Mars) produce light intensities proportional to R^{-4}. This difficulty is circumvented to a certain extent by using parabolic mirrors with the light source at the focus, and in principle, "active" reflection from planar mirrors of such collimated beams (or reflection from curved mirrors of other types of beams) do not involve a necessarily weakened intensity for large R. Comparable results, although losing a large part of the mis-directed intensity can be obtained with lenses. However, when the laser became available, it turned out to be a far more stylish way of obtaining a non-diverging light beam. The angular extension is frequently given in millirad ($360°/2{,}000\ \pi$) corresponding to 3.44 minutes (or 206 seconds) of arc (about a-ninth of the apparent diameter of the Sun or the Moon). At a distance R, the light-beam with diameter 1 millirad produces a spot with the diameter 10^{-3} R (hence 380 km on the Moon). The surface $4\pi \cdot 10^6$ millirad^2 of uniformly irradiated light show that such a beam is 16 million times more concentrated on the spot (with area $\pi/4$ squared millirad). It may be noted that technical descriptions of lasers generally use half of the angle (the radius of the light spot) with a certain "convention about fixing the edge of the spot, and call this half-angle the divergence".

The laser signals are not only sent through the atmosphere (with unavoidable small fluctuations of the refractive index due to variations of temperature and pressure) and the interplanetary space (where charged particles in the solar wind might produce weak diffraction) but also through vitreous fibers (for instance in underground cables). This type of "light pipes" has been much elaborated in recent years, one of the main goals being to obtain the lowest possible energy loss/km. It is not easy to fabricate glasses or organic polymers, which are really colourless and free from light-scattering defects such as small bubbles, dust, and heterogeneous compositions. We refer to two recent books [25, 26] on the technological applications of lasers, and shall only make a few more detailed remarks about the possibilities of communications and related subjects.

The coherent laser beam has a variety of free parameters not available to a light-house lamp. The time dependence of a sequence of light and dark periods (perhaps of two different durations) such as morse signals can transform information very rapidly, especially in the case of pulse-operated lasers, but it is also possible to superpose oscillations corresponding to trigonometric functions with differing periods. Hypothetically, the density of information in such a modulated laser beam is sufficient to allow the simultaneous transmission of millions of telephone calls but it has proved very difficult to exploit these potentialities in practice. Using a laser with an organic dye-stuff, it is also possible to vary the photon energy hundreds of times the spectral line-width, which may serve purposes discussed in the next section. Quite generally, it may be interesting to study the pattern designed in a Minkowski four-dimensional space by a laser with a rapid variation with time.

This brings us to a problem whether the unconventional pattern can be stored in stationary objects, and brought to realization by a coherent light beam. The most spectacular instance of such a device is the *hologram* invented by the Hungarian physicist Gabor. Perhaps the most surprising aspect of this fixation of a three-dimensional image in a transparent matrix is that each small fragment is carrier of the same (though less sharp) image, reminding one of some of the less understood facts about storage of information in the brain. It is possible to have pocket-size holograms containing many volumes of typical encyclopedic texts, and it may very well replace traditional libraries and the storage of information in units each allowing two alternatives, such as tape containing crystallites of ferromagnetic CrO_2. The miniaturiazation and diversification of memory devices is one of the major expressions of progress in contemporary Society, and may well be a second important application of lasers complementary to the exploitation of nuclear fusion as an energy source.

E. Geodesy and Trigonometry in the Solar System

Since the Apollo expeditions the general public is accustomed to the time delay of 2.5 seconds inherent in telephone communications with people standing on the Moon. The measurement of distances between fixed points or to moving objects has been refined considerably with recent electronic equipment. We are not going to discuss airplanes [25, 26] and satellites but rather the interesting example [27] of determining the orbit of the Moon with a precision of 42 cm (1 part in 10^9) using the reflectors for laser beams installed on the Moon. The purpose of this series of measurements was to see whether the general theory of relativity is valid, including the contribution to the mass of the Moon due to gravitational self-energy (and it seems indeed to be so). Previous experiments [28] of the Eötvös type have abundantly shown that inertial and gravitational masses of objects in the laboratory are equivalent within 1 part in 10^{13}.

It should be mentioned that Radar echo determinations of distances in the 10^8 km range, using electromagnetic wave-lengths of some cm, have already previously been carried to Venus and Mars as passive reflectors but obviously with a lower precision. It will be of great interest to deploy laser ranging equipment in the future exploration of the Solar system, in particular for the study of the finer details of the general theory of relativity and the minor deviations from Euclidean geometry due to the presence of the massive bodies. Actually, the Keppler ellipses representing the planetary orbits with the centre of gravity of the Solar system (close to the Solar surface) at one focus, are geodesic curves of free fall in the coordinate systems chosen in the general theory of relativity.

The Sun is not only the dominating mass in this system but creates obvious problems for equipment using weak intensities of rather special light. The radius R_0 of the Sun being 695,950 km, the surface area is $6.14 \cdot 10^{18}\,m^2$ and the energy production $3.86 \cdot 10^{26}$ W hence is $6.29 \cdot 10^7\,W/m^2$ corresponding to Eq. (5.2) for $\sigma_{S.B.}T^4$ with T = 5,770 °K (or 5,500 °C). This radiation density is only attenuated by the division of $(R/R_0)^2$ at the distance $R = 1.496 \cdot 10^8$ km of the Earth, being $1{,}360\,W/m^2$ (corresponding to the "solar constant" = 1.94 $gcal/cm^2 min$), just outside our atmosphere. The observed attenuation close to sea level has been discussed by Moesta [29]. As known from our almost black sky at night, there is fortunately very little scattering of light from suspended dust and comets (which would constitute the equivalent of fog in atmospheric observations) and there is no reason to doubt the R^{-2} dependence goes out to many light-years away. We have already discussed [30] the fact that the number of green, blue and violet photons (able to excite the uranyl ion) in Solar light here is about $8 \cdot 10^{16}$ $photons/cm^2 sec$ perpendicular on the light beam. Consequently, about 30 Einstein (moles of photons) arrive per m^2 sea surface per day. The number of photons at night is far more difficult to evaluate. The full Moon has a luminosity $4 \cdot 10^5$ times smaller than the Sun, and hence, about 10^{12} visible photons arrive per $cm^2 sec$. For practical purposes, the luminosity of the greater planets can be added to the stars. The scale of magnitude of the stars is made in such a way that the decadic logarithm of the visible (or the photographic) luminosity falls 0.4 unit per unit of magnitude. Hence, a star of the magnitude 6.0 (*i.e.* the limit of visibility to the naked eye) is 100 times less luminous than of magnitude 1.0, and a star (for large telescopes) of the magnitude 16 is 10,000 times less luminous than of 6. The number of stars of the lower magnitudes (*i.e.* higher magnitude classifiers) increases rapidly, and there is no obvious way of integrating the total luminosity. Furthermore, our galaxy produces a diffuse belt of light (it is known that the light from the numerous stars closer to the centre of our galaxy is heavily attenuated by clouds of dust). The order of magnitude of $photons/cm^2 sec$ is $8 \cdot 10^6$ for magnitude 0, 800 for magnitude 10, and a rough integration shows that each magnitude class from 1 to 8 has a total contribution increasing from $3 \cdot 10^7$ to $1.3 \cdot 10^8$ $photons/cm^2 sec$ suggesting the order of magnitude of a few times 10^9 $photons/cm^2 sec$ of all stars. The total luminosity [31] of 1 square degree is about $5 \cdot 10^5$ $photons/cm^2 sec$ (like a star of magnitude 3) but it seems that only about a-fifth is due to non-scattered stellar light. Then the stars of the whole sky provide $4 \cdot 10^{10}$ $photons/cm^2 sec$. Of course, most observers are cut off from half of this luminosity by a horizon. These values pertain to non-directional detection. If the photon detector selects a narrow angular region, the statistics of varying light intensities becomes erratic, because the region selected may contain a bright star. In practice the geometrical orientation of satellites or interplanetary probes is frequently performed by the detection of the brightest stars, such as Sirius or Canopus.

Among the obvious parameters affecting the strategy of detecting (necessarily weak) laser signals is the narrowness of the laser line. If it is 1 cm^{-1}, it represents about 1 in 10^4 of the prevailing photon wave-numbers. However, it is not easy to have detectors so monochromatically selective. Interference filters can be readily deviced for selecting some 100 cm^{-1} representing 1 percent of the pertinent spectrum. The detector would be "clever" by comparing photon counts in the laser line with counts in closely adjacent wave-numbers, thus helping to avoid coincidence with Fraunhofer absorption (or the rare emission) lines. It is very difficult to have confidence in photon counts below 10 (though a stable latent image is obtained in silver bromide crystallites with four photons, for specific reasons [32]) and we compare this limit with the original output of the laser, where 1 W green (20,000 cm^{-1}) light corresponds to $2.5 \cdot 10^{18}$ photons/sec. Extremely sensitive low-level photon counters may have to be cooled to well below 100 °K, for instance by the surrounding space, in order not to show thermal noise (*e.g.* when they are electron photo-multiplier devices).

Though a large part of modern astronomy is based on radio observations around some cm wave-length, the other major "window" is the visible and near ultra-violet region (cut off at 34,000 cm^{-1} by the ozone in the stratosphere). Photographic observations extend well beyond a distance of 10^9 light-years (about 10^{22} km) and one may wonder why visible light is so penetrating. With an average density of 10^{-29} g/cm^3 (these values are connected in a very sensitive way with the future expansion or eventual collapse of the Universe) a column 10^{27} cm long contains 10 mg/cm^2. In other words, the average molar extinction coefficient ϵ seems to be below 10. It is very difficult to measure the refractive index of the almost empty space but it is possible to give a higher limit to the dispersion, that is the difference between n for blue and for red light. This difference can hardly be less than 2 percent of $(n-1)$. The dispersion can be evaluated from binary variables showing occultation of the bright component by a dark (or much less luminous companion) such as Algol. If the distance is 300 light-years and it can be shown that the red image does not disappear more than at most 0.1 second before the blue image, the dispersion as inferred from the velocity of light c/n dependent on wave-length is below 10^{-11}. Because of the very small average density of the Universe, the average n is certainly much closer to 1, however, one might readily have large lenses of highly dilute gas floating around, and the interstellar gas in certain types of galaxies might very well refract the light perceptibly. Around several kinds of stars, considerable amounts of dust are suspended and it seems that iron(III) in mixed oxides and silicates produce characteristic absorption bands [34], making it possible to evaluate the nephelauxetic effect from the sharp transition having the wave-number of $^6S - ^4G$ in gaseous Fe^{+3} multiplied by a nephelauxetic ratio between 0.7 and 0.8.

If we communicate with our own space-ships (or with previously recognized receivers) it is possible to select sophisticated devices which use two or more sharp laser lines, or perhaps a definite time evolution, and chopping along the time axis in a highly characteristic fashion. The situation is far less favourable if the receivers (we or somebody else) do not know the codes and conventions of the time dependence. If we just scan large areas of the sky for signals (as has already been done by radio-astronomers) it is not necessarily an advantage to have a tremendously high resolution. We saw in Section 5D. that one has to look 16 million times on the sky with an opening with the diameter 1 milliradian. It is difficult to tell how effective such a scanning would be, since we do not know how much monochromatic light escapes from a typical civilization. The Earth irradiates large amounts of radio waves of several m wave-length transporting radio and television signals, but almost all the monochromatic spectral lines originate in sodium, mercury and neon lamps without any intention of communication (excepting the neon advertising for entirely local purpose). A considerable number of stars at comparatively short distance (some 10 light-years or 10^{14} km) have at least one planet. However, these systems are somewhat atypical, for our purpose in the sense that the planet has a mass of several percent (and not as Jupiter 0.1 percent) of the star. It may be noted that roughly half of all the stars are members of binary, ternary, . . . systems, but we have very little evidence for or against the existence of planets (such as Mercury, Venus, Earth and Mars) around 10^{-6} solar masses, which seem more likely carriers of life. Even if one could attain the science-fiction resolution 1 microradian, the light beam would still have the diameter 10^8 km for $R = 10^{14}$ km, and the power 1 watt corresponds to 320 photons/km^{2}sec. However, a greater difficulty may be that any such scale of resolution is insufficient to discriminate against the much stronger radiation density of the adjacent star (such as our Sun). It is clear that such unfavourable conditions can only be circumvented by a combination of lasers in the megawatt range and very subtle coding of messages. It may be noted that the R^{-2} dependence of isotropic light sources also occurs in the description of a laser beam as a very weakly divergent cone, but in our hypothetical microradian angle the intensity is $1.6 \cdot 10^{13}$ times the spherically symmetric average for a given R. If we consider $R = 1.5 \cdot 10^9$ km for Saturn, the spot has a diameter 1,500 km and each watt of visible light produces 140 photons/cm^{2}sec. In such a case, one would also need coding built into the time sequences and perhaps also the simple mode of repeating the message 100 times. Though there is not a clear-cut geometrical reason it may be preferable to build a square grid of 100 photon detectors with a mutual closest distance 100 km, though the budget committee of NASA may consternate at the thought of constructing a network of 100 Mount Palomar telescopes on one of the moons of Saturn. On the other hand, the Earth is seen as far as 6 degrees away from the Sun. If the laser beam goes through our atmosphere it is subject to refractive fluctuations of the same kind as by producing the scintillation of fixed stars, and this may pose a lower limit of 10^{-5} radian.

It is quite possible that laser interplanetary communication has a bright future, though it must be realized once more that radio waves may be a successful competitor. The major reason is the large antennae used as detectors in the latter case which allow interferometric effects, though the coherent character of a laser beam may be exploited in networks of photon detectors. As a matter of fact, the photographs taken by the Viking missions of the surface of Mars (*cf.* the 1. October 1976 issue of Science, volume *194*) and by the Pioneer fly-by at the close surroundings of Jupiter (Science *183*, 25. January 1974) were transmitted by radio, coded point for point. On the other hand, the content of information may be kept much higher in a coherent beam of visible light. However, interstellar communication over several light-years (our closest neighbours) needs imperatively pre-established coding of highly sophisticated complication. The mere isotropic search for signals would involve an overwhelming expenditure of energy either by the sender or by the receiver.

F. Antiferromagnetic Coupling with Adjacent Transition Group Ions

It is of great interest to investigate all the mechanisms influencing the line-width of a potential laser. One of the conceivable effects is the small energy differences between the states formed by a given J-level of a 4f group ion surrounded by other ions with positive S. The situation is particularly simple in systems having two (generally speaking, paramagnetic), constituents with positive S_1 and S_2. In such a case, the vector coupling described in Eq. (1.4)

$$S_1 \oplus S_2 = (S_1 + S_2) \text{ or } (S_1 + S_2 - 1) \text{ or } (S_1 + S_2 - 2) \ldots \text{ or } |S_1 - S_2| \qquad (5.5)$$

produces different $S = S_1 \oplus S_2$ with the energy relative to the baricentre of all the states

$$J_S \; \{S_1(S_1 + 1) + S_2(S_2 + 1) - S(S + 1)\} \qquad (5.6)$$

with a formula first proposed by Heisenberg. The situation where $S = |S_1 - S_2|$ (which is zero in the case $S_1 = S_2$) has the lowest energy corresponds to *antiferromagnetic coupling* with the Heisenberg exchange parameter J_S negative, whereas the situation where $S = S_1 + S_2$ constitutes the groundstate is called *ferromagnetic coupling*. It may be noted that many authors use the opposite sign of J_S in Eq. (5.6), and sometimes twice its absolute value. Ferromagnetism occurs in metallic iron, cobalt, nickel and gadolinium below their Curie temperature, and in many alloys, but it was thought for many years that isolating compounds cannot be ferromagnetic. However, it is now known that EuO, EuS, EuSe, $CrBr_3$ and several other compounds are ferromagnetic at low temperatures. By far the most common behaviour of paramagnetic compounds is to become antiferromagnetic by sufficient cooling. The magnetic properties of solids are discussed by Carlin and van Duyneveldt [35] in another book of the present series "Inorganic Chemistry Concepts". The coupling of N paramagnetic entities each with S_0 according to Eq. (5.5) to S values up to the higher limit NS_0 has been discussed previously [36] and it was shown that if all the states have the same energy (or negligible energy differences compared with kT), the average value of $\langle S(S + 1)\rangle$ is $NS_0(S_0 + 1)$ explaining the addivity of paramagnetic susceptibility, which is proportional to $S(S + 1)$. However, it must be noted with Griffith [37] that systems with N higher than 2 do not necessarily obey Eq. (5.6) but may show energy differences between states having the same S of the same size as between different S. The origin of the parameter J_S has been much discussed recently. The most clear-cut case is the energy difference between the singlet groundstate and the slightly higher triplet ($S = 1$)

of a hydrogen molecule [38] stretched to a much larger internuclear distance than the equilibrium value 0.75 Å. In the case of two Cr(III) coupled with an oxide bridge, Glerup [39] argues, that Eq. (5.6) is valid, with the highest septet ($S = 3$) state not modified, and the other states ($S = 2$, 1 and 0) stabilized, with $-J_S$ equal to 225 cm^{-1} in the blue "basic rhodo ion" $(NH_3)_5CrOCr(NH_3)_5^{+4}$ containing a linear Cr–O–Cr bridge. According to Glerup, $-J_S$ (which he calls 1/2 J) represents the summation of squared non-diagonal elements of the two-electron operator divided the distance to excited configurations involving Cr(II) and Cr(IV) with four and two d-like electrons. It is not easy to evaluate this rather large distance expressing the difference [40] between the ionization energy and the electron affinity of the partly filled 3d shell. However, it is interesting that Eq. (5.6) is obtained in this model with plausible assumptions. Kahn [41] maintains that a more important effect is a direct consequence of the overlap integral between the two distant orbitals carrying the density of an uncompensated spin. Anyhow, it is beyond doubt that most antiferromagnetic effects [35] are determined by "super-exchange" transmitted via anion bridges (such as oxide and the halides).

In the 3d group compounds the most spectacular spectroscopic effects of antiferromagnetic coupling are *simultaneous excitations* in two different systems with positive S_1 and S_2 in their groundstates. The quartet ($S = 3/2$) to doublet ($S = 1/2$) transition with lowest wave-number in Cr(III) octahedral chromophores occurs [10] between 13,000 and 15,000 cm^{-1}, and produces the two sharp lines at 14,419 and 14,448 cm^{-1} in ruby $Al_{2-x}Cr_xO_3$ providing some of the most extensively used lasers. Linz and Newnham [42] found transitions in the near ultra-violet (close to 29,000 cm^{-1}) of ruby, not obeying Beer's law Eq. (1.35) but having an intensity proportional to x^2 (the square of the chromium concentration). This is convincing evidence that these absorption bands (like sharp and rather intense bands of the basic rhodo ion [39]) originate in CrOCr groupings and correspond to transitions to the states with $S = 1$ and 0 formed by two doublet states according to Eq. (5.5) from the same S values constituting the two lowest of the four S alternatives available to the two quartet groundstates. There is a more general tendency for spin-forbidden transitions in 3d group compounds to become slightly spin-allowed by antiferromagnetic coupling, the same S occurring for the upper and the lower state in Eq. (5.5), whereas the usual mechanism for intensities of spin-forbidden transitions are the monatomic effects of spin-orbit coupling mixing states with different S in Russell-Saunders coupling, and a somewhat less frequent mechanism is an influence by large Landé parameters in adjacent atoms, such as bromine and iodine in the case of delocalization of the partly filled shell constituting anti-bonding MO. From recent studies of simultaneous excitations in solid 3d group compounds one should mention pairs [43] of manganese(II) in the colourless perovskites $KMgF_3$ and $KZnF_3$ and mixed pairs [44] of one manganese(II) and one nickel(II) in the same crystals, and that the individual values of $S = S_1 \oplus S_2$ have been identified.

Simultaneous J-level excitations in lanthanide compounds were reported by Varsanyi and Dieke [45, 46] in undiluted $PrCl_3$ and in mixed crystals $Pr_xLa_{1-x}Cl_3$. It is perhaps more important for our purposes, that Murphy and Ohlmann [47] found a striking difference between the luminescence of the lowest doublet state of Cr(III) in the diamagnetic perovskite $LaAl_{1-x}Cr_xO_3$ and in the corresponding gadolinium compound $GdAl_{0.998}Cr_{0.002}O_3$. The point is that each octahedral $Cr(III)O_6$ is surrounded by eight equivalent Gd(III) each having $S = 7/2$ and presenting each 8 mutually orthogonal wave-functions with almost identical energy. In such a case, $8^8 =$ 16,777,216 states are formed from the eight Gd(III) with S having all integral values from 0 to $8.7/2 = 28$, but the probabilities of each of these 29 alternatives are highly different, and peak at $S = 6$, as far goes the times of repeated S, and at $S = 9$ as far goes states. As mentioned above [36], the average value of $\langle S(S+1)\rangle$ is 126. The emission lines of Cr(III) are indeed accompanied by a kind of Maxwellian distribution, suggesting ferromagnetic coupling between S of the eight Gd(III) with the Cr(III) groundstate described by the parameter $J_S = +\ 2.1\ cm^{-1}$. The energetic effects of interaction between the gadolinium(III) were neglected.

It is possible [48, 49] to show perceptible effects on the sub-levels of individual J-levels of Dy(III) or Er(III) of antiferromagnetic crystals such as $DyFeO_3$ or $ErCrO_3$ cooled below their Néel temperature. In such circumstances, it is useful to remember the group-theoretical result that a linear magnetic field corresponds to the point-group C_∞ (which cannot be represented by a one-valued function [36] of the three Cartesian coordinates, in contrast to the two representable linear point-groups $C_{\infty v}$ and $D_{\infty h}$) where a reversal of the arrow of time corresponds to a multiplication of the magnetic field by -1. Such a point-group is able to split Kramers doublets (for an odd number of electrons) in their two component states, and may need the introduction of complex wave-functions consisting of the sum of a real and of an imaginary part.

In solid lanthanide compounds the oxide bridges are particularly apt to induce antiferromagnetic coupling also. Thus, Schugar *et al.* [50] demonstrated sharp lines between 20,490 and 21,858 cm^{-1} in undiluted Yb_2O_3. This is convincing evidence for simultaneous transitions (having twice the wave-numbers of $^2F_{7/2}$ to $^2F_{5/2}$) showing intensities about 0.1 percent of the transitions in the near infra-red, since no other absorption bands occur before the broad electron transfer bands above 40,000 cm^{-1}. Weaker simultaneous transitions can also be detected [50] in YbOF, but not in YbF_3, $Yb(OH)_3$ nor $LaYbO_3$. There seems to be a certain connection between this phenomenon and the energy transfer from one Yb(III) to another discussed in Chapter 4.

The reason why the present section on antiferromagnetic effects has been placed in this Chapter (including suggestions) is that one may readily imagine that the broadening or splitting of definite sub-levels of a given J-level due to a paramagnetic matrix with weak antiferromagnetic coupling, or due to interactions between neighbour lanthanides, may influence the properties as lasers.

G. Final Comments on Spectroscopy in Condensed Matter

The isolated atom or monatomic ion M^{+z} has discrete excited levels showing a first point of convergence (of an infinite number of nl values as a function of increasing n) at the lowest energy of ionization. This does not prevent that apparently very sharp excited levels can occur above this first ionization energy. Such auto-ionizing levels exist in several types, such as the configuration [Ar]3d4d of the calcium atom which has energy higher than the groundstate [Ar]4s of Ca^+; or levels [40] of the caesium atom belonging to the configuration terminating $5p^5 6s^2$ at much higher energy than the groundstate of Cs^+ terminating $5p^6$. More extreme kinds are the states lacking an electron in an inner shell studied by photo-electron spectra. These examples show that one should not be surprised to encounter sharp energy levels also in liquid samples and in vitreous and in crystalline solids.

In a gaseous molecule, the Born-Oppenheimer approximation consists of the resolution of the total wave-function in the product of a translational, a rotational, a vibrational and an electronic factor (a gaseous atom presents only a translational and an electronic factor). Whereas, the discrete levels of the rotational and the vibrational structure are discussed at length in many text-books, it is generally not mentioned that the translational spectrum of a species not confined in a small volume is immediately continuous (like the classical kinetic energy $1/2\ mv^2$), starting at $v = 0$ relative to the observer. The exceedingly polyatomic character of a solid modifies this picture, in sofar as the rotational frequences are lost in the lower end of the vibrational spectrum corresponding to librations. The excellent validity of the Born-Oppenheimer approximation corresponds to any coupling between the electronic states and *e. g.*, the translational spectrum being exceedingly weak.

When we concentrate our attention on the specific behaviour of partly filled shells, we observe once more that the excited states occur in a recognizable bunch. The essence of the "ligand field" theory is that the number of states for one partly filled *nl*-shell obtained by distributing q indiscernible particles on $(4l + 2)$ sites (in an abstract space that has nothing to do with our three-dimensional space) is conserved in chromophores of differing symmetry. In the specific case of a trivalent lanthanide, the states of $4f^q$ even remain bunched together, $(2J + 1)$ in each manifold, corresponding to the *J*-levels in spherical symmetry of a monatomic entity. The bunch of states is recognizable even when N identical lanthanide ions are present, producing $(2J + 1)^N$ mutually orthogonal states, and the weak antiferromagnetic coupling discussed in the previous section, only separates the energy levels within this bunch to a very small extent. This is a major difficulty [36] for the energy band theory of crystalline solids, because the number of states obtained by permutation treatment of Nq electrons on $N(4l + 2)$ sites is *far* larger than the N'th power of the number of states of q electrons distributed on $(4l + 2)$ sites, in spite of the latter number assuming astronomical proportions for N above 10, and beyond the number of electrons $\sim 10^{80}$ in the observable part of the Universe for slightly higher N.

In text-books it is frequently said that the difficulty for the energy band model in this case is related to it being a molecular orbital treatment, distributing the electrons on delocalized one-electron wave-functions in definite MO configurations, whereas the appropriate model for large internuclear distances (*e.g.*, H_2 for large R being *better* described as two hydrogen atoms than as the MO configuration $(\sigma_g)^2$ characterizing H_2 close to the equilibrium R) is the valence-bond (VB) treatment. It is indeed a part of the truth if one says that paramagnetic and weakly antiferromagnetic solids are better described in the VB model than by the energy band model, but a closer analysis shows that the decisive property is the huge difference between the ionization energy and the electron affinity of a partly filled shell with a small average radius and concomitant large parameters of interelectronic repulsion. Thus, in a solid chromium(III) compound the configuration $3d^3$ remains well separated below configurations (such as electron transfer states) containing $3d^4$, or the other way round $3d^2$, in spite of partly covalent bonding modifying the electrostatic model (assuming fully charged ions Na^+, K^+, Cr^{+3} and F^- in crystalline K_2NaCrF_6). This situation is fairly comparable to the configurations $4f^{q-1}5d$ and $4f^q$ remaining distinct in divalent lanthanides, even when there is not a centre of inversion present, preventing the mixture of the two configurations in gaseous M^{+2}.

One of the lasting conclusions of the "ligand field" theory is that interactions between the electronic structures of distant atoms vanish very rapidly as a function of increasing internuclear distance R. This is a necessary compliment to the indiscernibility of all electrons considered in quantum mechanics, and also the basis for atomic spectroscopy, in sofar arcs, sparks and electric discharge lamps do not contain vapours at infinite dilution. There is general agreement that the long-range interactions depend on the overlap integrals, though there is the difference [36] that covalent bonding between identical atoms tends to produce energetic effects proportional to the overlap integral, whereas atoms of highly different electronegativities have bonding and anti-bonding effects proportional to the square of the overlap integral. There has been considerable discussion as to whether the diagonal elements of the effective one-electron operator of M and X are identical if the ionization energies, say, of lanthanide M_2O_3 or MF_3 are found to be equal by photo-electron spectra. In earlier MO treatments of the LCAO type first proposed by Hückel and later elaborated by Wolfsberg and Helmholz, the identity of diagonal elements of the two atoms M and X produces a singularity, bringing the dependence on the overlap integral back to first-order. The concomitant strong covalency is in disagreement with the moderate nephelauxetic effect found from photo-electron spectra (ionization of the $4f^q$ groundstate to various *J*-levels of $4f^{q-1}$) of MSb and other solid compounds. It has been argued [40] that the Hückel-type models should then involve a denominator in the expression for the anti-bonding energy of the seven 4f orbitals according to the angular overlap model, being not only the difference between the two one-electron diagonal elements (which may vanish) but also being added to the A_* parameter of interelectronic repulsion (expressing the difference between the ionization energy and the electron affinity of the 4f shell). In this case, the extent of covalent bonding is expected to be much smaller. Recently, the isoelectronic series introduced by Kossel in 1916 were reviewed [54].

We saw above that the hydrogen molecule at large internuclear distance R can be used [38] as a prototype of antiferromagnetic coupling. The overlap integral $S(\mu)$ between the two hydrogen 1s orbitals with the Slater exponent μ (= 1 for isolated H atoms, and for large R, whereas the optimized value for R = 0.75 Å is μ = 1.17) has the explicit expression

$$S(\mu) = \left(1 + \mu R + \frac{\mu^2 R^2}{3}\right) e^{-\mu R} \tag{5.7}$$

with R measured in bohr units. The exponential decrease with large R has an important effect on the time-dependent evolution. It has been illustrated [36] how various spectroscopic techniques with a characteristic time-scale can legitimately disagree with crystallographic (X-ray or neutron diffraction) determinations of the time average content of the unit cell. Eq. (5.7) can be used to discuss a simple time-dependent problem of a proton suddenly brought to the distance R (here $\mu = 1$) from a hydrogen atom. This system is not in the eigen-state $(\sigma_g)^1$ of the Schrödinger equation, but can be described formally as (about) equal squared amplitudes of the two eigenstates $(\sigma_g)^1$ and $(\sigma_u)^1$. The average time τ in seconds for the evolution in direction of one of the eigen-states is the atomic unit of time $\tau_0 = 2.42 \cdot 10^{-17}$ sec divided by half the overlap integral (it may be noted that the equilibrium value of R for H_2^+ is exactly 2 bohr)

μR	= 5	10	15	20	30	60
$S(\mu)$	= 0.0966	0.0020	$2.78 \cdot 10^{-5}$	$3.18 \cdot 10^{-7}$	$3.10 \cdot 10^{-11}$	$1.10 \cdot 10^{-23}$
τ	= $5.0 \cdot 10^{-16}$	$2.4 \cdot 10^{-14}$	$1.74 \cdot 10^{-12}$	$1.52 \cdot 10^{-10}$	$1.56 \cdot 10^{-6}$	$4.4 \cdot 10^{6}$

(5.8)

Obviously, there is not a sharp distinction between "immediately" and "never" but it is also clear that a variation of R by a factor of 2 or 3 has dramatic effects on the characteristic time-scale. Such dynamics of a system evolving towards an eigen-state is well-known from the two enantiomers of an optically active molecule, where it was realized [52] in 1930 that the two eigen-states of the Schrödinger equation are the two orthogonal linear combinations (not possessing optical activity) of the wave-functions corresponding to the active enantiomers. By the way, one does not need to specify the point-group of a given nuclear skeleton, if one indicates all the internuclear distances (this set is actually overdetermined for more than four nuclei) but this information does not discriminate between the two enantiomers of an optically active molecule.

For the spectroscopist a much more important aspect of the time-dependent Schrödinger equation is the finite life-time (either by radiative transitions determined by Einstein's 1917 formula, or by non-radiative processes) giving reality to excited states. Recently a book [53] has appeared about these interesting problems.

In addition to the problems occurring already in isolated atoms, and at the next level, in gaseous oligo-atomic molecules, the overwhelming number of atoms in solids produce specific problems worthwhile being considered. The continuous spectra entering the Born-Oppenheimer approximation, and the hyper-astronomical (but finite) numbers of electronic states in paramagnetic and weakly antiferromagnetic compounds, leave a number of indefinite quantities floating around, of type zero divided by zero, and it is particularly interesting that the description of excitons in the case of mobile excitations in solids utilizes a tiny sub-set of the degrees of freedom in the solid. Though it is perfectly clear that non-radiative migration of excitations in lanthanides, such as in ytterbium(III), sometimes occur, it is also evident that mechanisms related to the time τ in Eq. (5.8) prevent the full mutual interactions between partly filled 4f shells at moderate distance. Actually, the long τ also prevents ionization of atoms and molecules exposed to electric potential differences above 25 V. Here we do not discuss the borderline cases between covalent bonding in d group compounds [10] and in conjugated aromatic systems, but we reiterate that a superposed, unexpected feature of the lanthanides in condensed matter is that also the individual *J*-levels remain recognizable.

Acknowledgment. We would very much like to thank Mrs. Esther Greenberg and Dr. Leah Boehm for their valuable technical assistance in several parts of this book.

References

1. McWeeny, R.: Nature *243,* 196 (1973)
2. Suess, H. E., Jensen, J. H. D.: Landolt-Börnstein, 6. Auflage, Band III: Astronomie und Geophysik. Berlin–Göttingen–Heidelberg: Springer 1952
3. Trimble, V.: Rev. Mod. Phys. *47,* 877 (1975)
4. Unsöld, A.: Naturwiss. *63,* 443 (1976)
5. Post, R. F., Ribe, F. L.: Science *186,* 397 (1974)
6. Metz, W. D.: Science *192,* 1320; *193,* 38 and 307 (1976)
7. Milam, D., Weber, M. J.: IEEE J. Quantum Electr. QE *12,* 512 (1976)
8. anonymous: SHIVA- the 10 kilojoule Nd: glass laser for the high energy laser facility. Livermore: University of California Lawrence Livermore Laboratory 1975
9. Dewhurst, R. J., Jacoby, D., Pert, G. J., Ramsden, S. A.: Phys. Rev. Letters *37,* 1265 (1976)
10. Jørgensen, C. K.: Oxidation Numbers and Oxidation States. Berlin–Heidelberg–New York: Springer 1969
11. Urbain, G.: Ann. chim. phys. (Paris) *18,* 222 and 289 (1909)
12. Brecher, C., Samelson, H., Lempicki, A., Riley, R., Peters, T.: Phys. Rev. *155,* 178 (1967)
13. Sovers, O. J., Yoshioka, T.: J. Chem. Phys. *51,* 5330 (1969)
14. Powell, C. J.: Rev. Mod. Phys. *48,* 33 (1976)
15. Siegbahn, K., Nordling, C., Johannson, G., Hedman, J., Hedén, P. F., Hamrin, K., Gelius, U., Bergmark, T., Werme, L. O., Manne, R., Baer, Y.: ESCA Applied to Free Molecules. Amsterdam: North-Holland Publishing Co. 1969
16. Jørgensen, C. K., Berthou, H.: Chem. Phys. Letters *25,* 21 (1974)
17. Wagner, C. D.: Analyt. Chem. *44,* 967 (1972)
18. Berthou, H., Jørgensen, C. K.: J. Electron Spectr. *5,* 935 (1974)
19. Wagner, C. D.: Discuss. Faraday Soc. *60,* 291 (1976)
20. Reisfeld, R., Jørgensen, C. K., Bornstein, A., Berthou, H.: Chimia (Zürich) *30,* 451 (1976)
21. Blasse, G.: Structure and Bonding *26,* 43 (1976)
22. Gmelin's Handbuch der Anorganischen Chemie, Ergänz. *8,* Teil A2: Transurane, die Elemente. Weinheim: Verlag Chemie 1973
23. McArthur, D. A., Tollefsrud, P. B.: Appl. Phys. Letters *26,* 187 (1975)
24. Helmick, H. H., Fuller, J. L. Schneider, R. T.: Appl. Phys. Letters *26,* 327 (1975)
25. Harry, J. E.: Industrial Lasers and their Applications. London: McGraw Hill 1974
26. Rosenberger, D.: Technische Anwendungen des Lasers. Berlin–Heidelberg–New York: Springer 1975
27. Shapiro, I. I., Counselman, C. C., King, R. W.: Phys. Rev. Letters *36,* 555 (1976)
28. Haugan, M. P., Will, C. M.: Phys. Rev. Letters *37,* 1 (1976)
29. Moesta, H.: Naturwiss. *63,* 491 (1976)
30. Jørgensen, C. K.: Revue chim. min. (Paris) *14,* 127 (1977)
31. Strömgren, B.: Handbuch der Experimentalphysik *26,* p. 361. Leipzig: Akademische Verlagsgesellschaft 1937
32. Jørgensen, C. K.: Chem. Phys. Letters *11,* 387 (1971)
33. Harrison, E. R.: Nature *204,* 271 (1964)
34. Manning, P. G.: Chem. Soc. Rev. (London) *5,* 233 (1976)
35. Carlin, R. L., Van Duyneveldt, A. J.: Magnetic Properties of Transition Metal Compounds. Berlin–Heidelberg–New York: Springer 1977
36. Jørgensen, C. K.: Modern Aspects of Ligand Field Theory. Amsterdam: North-Holland Publishing Co. 1971
37. Griffith, J. S.: Structure and Bonding *10,* 87 (1972)
38. Ammeter, J. H.: Chimia (Aarau) *29,* 504 (1975)
39. Glerup, J.: Acta Chem. Scand. *26,* 3775 (1972)
40. Jørgensen, C. K.: Structure and Bonding *22,* 49 (1975)
41. Charlot-Koenig, M. F., Kahn, O.: Chem. Phys. Letters *41,* 177 (1976)
42. Linz, A., Newnham, R. E.: Phys. Rev. *123,* 500 (1961)
43. Güdel, H. U.: Chem. Phys. Letters *36,* 328 (1975)

44. Ferguson, J., Güdel, H. U., Krausz, E. R.: Mol. Phys. *30,* 1139 (1975)
45. Varsanyi, F., Dieke, G. H.: Phys. Rev. Letters *7,* 442 (1961)
46. Dieke, G. H., Dorman, E.: Phys. Rev. Letters *11,* 17 (1963)
47. Murphy, J., Ohlmann, R. C.: Optical Properties of Ions in Crystals (eds. H. M. Crosswhite and H. W. Moss) p. 239. New York: Interscience (John Wiley) 1968
48. Faulhaber, R., Hüfner, S., Orlich, E., Schmidt, H., Schuchert, H.: *ibid.,* p. 329
49. Hasson, A., Hornreich, R. M., Komet, Y., Wanklyn, B. M., Yaeger, I.: Phys. Rev. *B12,* 5051 (1975)
50. Schugar, H. J., Solomon, E. I., Cleveland, W. L., Goodman, L.: J. Am. Chem. Soc. *97,* 6442 (1975)
51. Sargent, M., Scully, M. O., Lamb, W. E.: Laser Physics. Reading (Mass.): Addison-Wesley 1974
52. Ewald, P. P.: Trans. Faraday Soc. *36,* 313 (1930)
53. Macomber, J. D.: The Dynamics of Spectroscopic Transitions. New York: John Wiley 1976
54. Jørgensen, C. K.: Adv. Quantum Chem. *11,* in press

6. Subject Index

7. Author Index

This index only mentions authors explicitly named in the text, and does not include implicit citation as one of the 726 reference numbers

Structure and Bonding

Volume 22: **Rare Earths**
36 figures, IV, 175 pages. 1975
E. Nieboer: The Lanthanide Ions as Structural Probes in Biological and Model Systems
C. K. Jϕrgensen: Partly Filled Shells Constituting Anti-bonding Orbitals with Higher Ionization Energy than their Bonding Counterparts
R. D. Peacock: The Intensities of Lanthanide $f \longleftrightarrow f$ Transitions
R. Reisfeld: Radiative and Non-Radiative Transitions of Rare-Earth Ions in Glasses

Volume 24: **Photoelectron Spectrometry**
40 figures, 21 tables. IV, 169 pages. 1975
C. K. Jϕrgensen: Photo-electron Spectra of Non-metallic Solids and Consequences for Quantum Chemistry
P. A. Cox: Fractional Parentage Methods for Ionisation of Open Shells of *d* and *f* Electrons

Volume 26: **Spectra and Chemical Interactions**
15 figures, 52 tables. IV, 144 pages. 1976
P. S. Braterman: Spectra and Bonding in Metal Carbonyls. Part B: Spectra and their Interpretation
G. Blasse: The Influence of Charge-Transfer and Rydberg States on the Luminescence Properties of Lanthanides and Actinides
A. Müller, E. J. Baran, R. O. Carter: Vibrational Spectra of Oxo-, Thio-, and Selenometallates of Transition Elements in the Solid State

Volume 31: **Bonding and Compounds of Less Abundant Metals**
40 figures, 6 tables. IV, 112 pages. 1976
R. Ferreira: Paradoxical Violations of Koopmans' Theorem with Special Reference to the 3d Transition Elements and the Lanthanides
C. Bonnelle: Band and Localized States in Metallic Thorium, Uranium, and Plutonium, and in some Compounds, Studied by X-Ray Spectroscopy
V. Gutmann, H. Mayer: The Application of the Functional Approach to Bond Variations under Pressure
J. K. Burdett: The Shapes of Main-Group Molecules. A Simple Semi-Quantitative Molecular Orbital Approach

Springer-Verlag
Berlin Heidelberg New York